Statistics and Big Data

Foundations and Applications

Volume 1

Editor-in-Chief

Liping Zhu, Institute of Statistics and Big Data, Renmin University of China, Beijing, China

Series Editors

Wenlin Dai, Institute of Statistics and Big Data, Renmin University of China, Beijing, China

Runze Li, Department of Statistics, Pennsylvania State University, State College, PA, USA

Wei Ma, Institute of Statistics and Big Data, Renmin University of China, Beijing, China

Zheng Zhang, Institute of Statistics and Big Data, Renmin University of China, Beijing, China

The book series "Statistics and Big Data: Foundations and Applications" delves into the fundamental concepts, methodologies, and applications of statistics in the context of big data analytics. The series provides a comprehensive and accessible resource for graduate students, researchers, and professionals interested in understanding the principles and practices that underpin the analysis and interpretation of large-scale data. From classical statistical theories to modern computational techniques, the series primarily accepts monographs, lecture notes and proceedings that covering a broad spectrum of topics including statistical inference, regression analysis, experimental design, Bayesian statistics, data visualization, machine learning, and data mining. The series equips readers with the necessary tools and methodologies to tackle real-world challenges encountered in various domains, such as healthcare, finance, management, etc. "Statistics and Big Data: Foundations and Applications" fosters a holistic understanding of statistics in the era of big data. Whether you are a student, or researchers in academia and industry, this series serves as a guide to navigate the intersection of statistics and big data, advancing knowledge and promoting excellence in the field.

Zheng Zhang · Kun Zhang · Xing Yan ·
Songshan Yang · Yuqian Zhang

Big Data in Economics and Management

Springer

Zheng Zhang
Institute of Statistics and Big Data
Renmin University of China
Beijing, China

Kun Zhang
Institute of Statistics and Big Data
Renmin University of China
Beijing, China

Xing Yan
Institute of Statistics and Big Data
Renmin University of China
Beijing, China

Songshan Yang
Institute of Statistics and Big Data
Renmin University of China
Beijing, China

Yuqian Zhang
Institute of Statistics and Big Data
Renmin University of China
Beijing, China

ISSN 3091-4256 ISSN 3091-4264 (electronic)
Statistics and Big Data
ISBN 978-981-95-3124-0 ISBN 978-981-95-3125-7 (eBook)
https://doi.org/10.1007/978-981-95-3125-7

Supported by fund for building world-class universities (disciplines) of Renmin University of China.

Mathematics Subject Classification: 62Gxx, 62D20, 91B05, 91Gxx

This Springer imprint is published by the registered company Springer Nature Singapore Pte Ltd.
The registered company address is: 152 Beach Road, #21-01/04 Gateway East, Singapore 189721, Singapore

Acknowledgements This monograph is supported by the fund for building world-class universities (disciplines) of the Renmin University of China.

Competing Interests The authors have no competing interests to declare that are relevant to the content of this manuscript.

Contents

Part I
Causal Inference in Economics

The first part of this monograph focuses on the causal inference in observational studies. Chapter 1 starts the causal inference for a discrete treatment variable. Within the potential outcomes framework, we first introduce the average treatment effects (ATE) of a binary treatment and its identification methods under the unconfoundedness condition. To estimate the ATE, we introduce the calibration weighting method by exploiting the covariate balancing property of the propensity score and show that the calibration estimators are consistent and attain the semiparametric efficiency bound. Further, we introduce the semi-supervised machine learning method for estimating ATE. Finally, we discuss the estimation of the general causal effects of a multi-valued treatment variable based on the artificial neural networks, and show that the "curse of dimensionality" can be alleviated if nuisance functions belong to a mixed smoothness class.

Chapter 2 extends the causal inference for a continuous treatment variable. We first introduce a general continuous treatment model that encompasses a wide variety of causal parameters of interest including average, quantile, and asymmetric least squares treatment effects. For this general continuous treatment model, we propose the identification method and derive the semiparametric efficiency bounds under the unconfoundedness condition. We then propose a covariate balancing estimation method that attains the general semiparametric efficiency bounds. Finally, we study various advanced causal testing problems for a continuous treatment.

Chapter 3 concerns the causal inference for a continuous treatment variable in the presence of measurement errors. We first identify the average dose-response function (ADRF) for a continuously valued error contaminated treatment by a weighted conditional expectation. We then estimate the weights nonparametrically by maximizing a local generalized empirical likelihood subject to an expanding set of conditional moment equations incorporated into the deconvolution kernels. Thereafter, we construct a deconvolution kernel estimator of ADRF. We derive the asymptotic bias and variance of the proposed estimator and provide its asymptotic linear expansion, which helps conduct statistical inference. To select our smoothing parameters, we adopt the simulation-extrapolation method and propose a new extrapolation procedure to stabilize the computation.

Chapter 1
Causal Inference for a Discrete Treatment

1.1 Basic Framework

Evaluating the causal effect of a treatment or intervention on an outcome of interest is a fundamental statistical problem. This problem is pervasive across various disciplines, such as economics, political science, psychology, biomedical science, and numerous others. For instance, economists hope to examine the effect of enrolling in a job training program on the future earning. Likewise, in other fields, researchers may seek to evaluate the causal effect of psychological interventions or medical trials on individuals' mental or physical health status.

To formulate the causal inference problem, we introduce the potential outcomes framework, initially proposed by [1] in the context of completely randomized experiments and later extended to observational studies by [2]. Let us begin by considering a binary treatment with $T_i \in \{0, 1\}$, where $T_i = 1$ indicates that the ith individual has been treated, and $T_i = 0$ otherwise. Denote $Y_i(1)$ as the outcome variable that would have been observed if the individual had been treated and $Y_i(0)$ as the outcome variable that would have been observed if the individual had not been treated. The random variables $Y_i(1)$ and $Y_i(0)$ are known as *potential* (or *counterfactual*) *outcomes*. For each individual, we can only observe one of the potential outcomes—the one corresponding to the treatment status that the individual has received. The observed outcome Y_i can be expressed as a function of the treatment assignment and potential outcomes:

$$Y_i = Y_i(T_i) = T_i Y_i(1) + (1 - T_i) Y_i(0). \tag{1.1}$$

The condition (1.1) is referred to as *consistency*. Note that the definitions above implicitly assume that there is no interference between different individuals—the treatment received by an individual does not affect the potential outcomes of other individuals. Such an assumption is included in the *stable-unit-treatment-value assumption* (SUTVA), formally introduced by [3].

Z. Zhang et al., *Big Data in Economics and Management*, Statistics and Big Data 1,
https://doi.org/10.1007/978-981-95-3125-7_1

Within the potential outcome framework, we define the treatment effect of the treatment on the ith individual as

$$\Delta_i := Y_i(1) - Y_i(0).$$

However, as we can only observe one of the potential outcomes for each individual, the difference Δ_i is always unobservable. As an alternative, we can consider the *average treatment effect* (ATE) in a population of individuals:

$$\tau := \mathbb{E}[\Delta_i] = \mathbb{E}[Y_i(1) - Y_i(0)].$$

Randomized controlled trials (RCTs), also known as randomized experiments, are typically considered the "gold standard" to evaluate causal effects. The RCTs require researchers to have full control of the treatment assignments so that the treatment and control groups are *exchangeable*:

$$T_i \perp \{Y_i(1), Y_i(0)\}. \tag{1.2}$$

In the presence of exchangeability, we have

$$\mathbb{E}[Y_i(t)] = \mathbb{E}[Y_i(t) \mid T_i = t] = \mathbb{E}[Y_i \mid T_i = t] \quad \text{for each} \quad t \in \{0, 1\}. \tag{1.3}$$

Therefore, the ATE can be represented as

$$\tau = \mathbb{E}[Y_i(1)] - \mathbb{E}[Y_i(0)] = \mathbb{E}[Y_i \mid T_i = 1] - \mathbb{E}[Y_i \mid T_i = 0].$$

Based on the above representation, we can easily estimate the ATE through the difference of sample means within the treatment and control groups—this is typically known as the *difference-in-means estimator*:

$$\hat{\tau}_{\mathrm{DM}} := \frac{\sum_{i=1}^{N} T_i Y_i}{\sum_{i=1}^{N} T_i} - \frac{\sum_{i=1}^{N}(1 - T_i) Y_i}{\sum_{i=1}^{N}(1 - T_i)}.$$

The difference-in-means estimator is root-n consistent and asymptotically normal under regular conditions. If we can collect additional covariates $\boldsymbol{X}_i \in \mathbb{R}^d$ that are related to the potential outcomes, one can leverage regression adjustments to further enhance estimation efficiency; refer to, for instance, [4–6].

Despite the widespread recognition of RCTs, their usage is hindered by various practical limitations. Random treatment assignments are sometimes viewed as ethically problematic in certain situations, and participants in RCTs are typically required to meet specific enrollment conditions. This restricts the sample's representativeness and limits the generalizability of findings to broader populations. Additionally, conducting RCTs can be both costly and time-consuming. These limitations underscore the importance of studying and drawing causal conclusions through observational studies. Particularly in today's big data era, where large-scale observational data with rich information is easily accessible, understanding how to obtain effective information and draw accurate, robust conclusions becomes a key challenge.

In observational studies, researchers do not control treatment assignments as they do in RCTs. Instead, treatment assignments are influenced by factors outside researchers' control. In many applications, treatments are affected by a set of random variables $\boldsymbol{X}_i \in \mathbb{R}^d$ that also impact potential outcomes—such variables are known as *confounders*.

Due to the occurrence of confounding, T_i is dependent with $Y_i(t)$ through $\boldsymbol{X}_i$. Therefore, the equality (1.3) no longer holds, and the difference-in-means estimator

$$\hat{\tau}_{\mathrm{DM}} \xrightarrow{p} \mathbb{E}[Y_i(1) \mid T_i = 1] - \mathbb{E}[Y_i(0) \mid T_i = 0] \neq \mathbb{E}[Y_i(1)] - E[Y_i(0)] = \tau$$

is in general an inconsistent estimate of the ATE.

To identify the causal effects in observational studies, the following *unconfoundedness* condition is often imposed.

Assumption 1 (*Unconfoundedness*)

$$T_i \perp \{Y_i(1), Y_i(0)\} \mid \boldsymbol{X}_i.$$

The condition (1) is also known as *conditional exchangeability* or *no unmeasured confounding*. It essentially requires that we have measured all the related covariates that affect the treatment assignments and potential outcomes simultaneously.

Moreover, we impose the *positivity* condition (also referred to as *overlap*).

Assumption 2 (*Positivity*) For every $x \in \mathbb{R}^d$, $\pi(x) = P[T_i = 1 \mid \boldsymbol{X}_i = x]$ is called the *propensity score* function, we assume

$$P[0 < \pi(\boldsymbol{X}_i) < 1] = 1.$$

Under the aforementioned identification conditions, the ATE can be identified by using any of the following representations:

1. The regression representation:

$$\tau = \mathbb{E}[m_1(\boldsymbol{X}_i) - m_0(\boldsymbol{X}_i)], \tag{1.4}$$

   where $m_1(x) = \mathbb{E}[Y_i \mid \boldsymbol{X}_i = x, T_i = 1]$ and $m_0(x) = E[Y_i \mid \boldsymbol{X}_i = x, T_i = 0]$ are the *outcome regression* functions.

2. The inverse probability weighting (IPW) representation:

$$\tau = \mathbb{E}\left[\frac{T_i Y_i}{\pi(\boldsymbol{X}_i)} - \frac{(1 - T_i)Y_i}{1 - \pi(\boldsymbol{X}_i)}\right]. \tag{1.5}$$

3. The doubly robust (DR)/augmented inverse probability weighting (AIPW) representation:

$$\tau = \mathbb{E}\left[m_1(\boldsymbol{X}_i) + \frac{T_i(Y_i - m_1(\boldsymbol{X}_i))}{\pi(\boldsymbol{X}_i)} - m_0(\boldsymbol{X}_i) - \frac{(1-T_i)(Y_i - m_0(\boldsymbol{X}_i))}{1-\pi(\boldsymbol{X}_i)}\right]. \tag{1.6}$$

To construct an ATE estimator using any of the representations mentioned above, it is sufficient to estimate the *nuisance functions* $m_1(x)$, $m_0(x)$, and/or $\pi(x)$ and then substitute the expectations in (1.4)–(1.6) with empirical averages after plugging in the nuisance estimates.

Under Assumptions 1 and 2, the semiparametric efficiency bound of ATE is developed in [7], which provides a guide to estimation methods and gives an asymptotic efficiency standard.

Proposition 1.1 *Under Assumptions 1 and 2, an estimator of ATE, denoted by $\widehat{\tau}$, is called a semiparametric efficient estimator if it admits the following asymptotic representation:*

$$\begin{aligned}\sqrt{N}\{\widehat{\tau} - \tau\} = \frac{1}{\sqrt{N}}\sum_{i=1}^{N}\Bigg[&\frac{T_iY_i}{\pi(\boldsymbol{X}_i)} - \frac{T_i - \pi(\boldsymbol{X}_i)}{\pi(\boldsymbol{X}_i)}m_1(\boldsymbol{X}_i)\\ &- \frac{(1-T_i)Y_i}{1-\pi(\boldsymbol{X}_i)} - \frac{T_i - \pi(\boldsymbol{X}_i)}{1-\pi(\boldsymbol{X}_i)}m_0(\boldsymbol{X}_i) - \tau\Bigg] + o_p(1).\end{aligned}$$

1.2 Causal Inference Based on Covariate Balancing Calibration

Based on Assumption 1, an important feature for the propensity score $\pi(\mathbf{X})$ is the following covariate balancing property:

$$\mathbb{E}\left[\frac{Tu(\mathbf{X})}{\pi(\mathbf{X})}\right] = \mathbb{E}\left[\frac{(1-T)u(\mathbf{X})}{1-\pi(\mathbf{X})}\right] = \mathbb{E}[u(\mathbf{X})] \tag{1.7}$$

holds for any integrable function $u(\boldsymbol{X})$. This implies that the propensity score balances the covariates in the treated, control and combined groups.

Many authors have proposed estimators by combining the (1.5) and (1.7) in a variety of manners under the propensity score framework, see [8–15]. Note that (1.7) gives an overidentifying set of moment restrictions, estimation is generally done within the generalized method of moments or the empirical likelihood framework. This section introduces the calibration weighting method developed by [16] to estimate ATE.

1.2.1 Calibration Estimation

Let $D(v, v_0)$ denote a general metric, where $v_0 \in \mathbb{R}$ is fixed. This metric is continuously differentiable with respect to $v \in \mathbb{R}$, nonnegative, strictly convex in v, and satisfies $D(v_0, v_0) = 0$. The fundamental concept of calibration, as described in [17], involves minimizing the distance between the estimated weights $w = (w_1, \ldots, w_N)$ and a predefined set of design weights $d = (d_1, \ldots, d_N)$ while maintaining moment constraints. We adopt a set of uniform design weights $d = (1, 1, \ldots, 1)$ and the calibration weights w are defined by solving the following constrained minimization problem:

$$\min \sum_{i=1}^{N} D(w_i, 1),$$

subject to the empirical version of (1.7):

$$\frac{1}{N}\sum_{i=1}^{N} T_i w_i u(\mathbf{X}_i) = \frac{1}{N}\sum_{i=1}^{N} u(\mathbf{X}_i) \text{ and } \frac{1}{N}\sum_{i=1}^{N}(1 - T_i) w_i u(\mathbf{X}_i) = \frac{1}{N}\sum_{i=1}^{N} u(\mathbf{X}_i).$$

To ensure the consistent estimation, we consider that K increases to infinity as N grows. The constrained optimization problem stated above is equivalent to two separate constrained optimization problems:

$$\min \sum_{i=1}^{N} T_i D(Np_i, 1) \text{ s.t. } \sum_{i=1}^{N} T_i p_i u_K(\mathbf{X}_i) = \frac{1}{N}\sum_{i=1}^{N} u_K(\mathbf{X}_i), \tag{1.8}$$

and

$$\min \sum_{i=1}^{N}(1 - T_i) D(Nq_i, 1) \text{ s.t. } \sum_{i=1}^{N}(1 - T_i) q_i u_K(\mathbf{X}_i) = \frac{1}{N}\sum_{i=1}^{N} u_K(\mathbf{X}_i). \tag{1.9}$$

For implementation of the method, we derive the dual solution of (1.8) and (1.9) which are unconstrained convex maximization problems and can be efficiently computed. Let $D(v) = D(v, 1)$, $f(v) = D(1 - v)$, and its derivative be $f'(v)$, $\forall v \in \mathbb{R}$. The dual solutions are given as follows. For i such that $T_i = 1$:

$$\hat{p}_K(\mathbf{X}_i) \triangleq \frac{1}{N}\rho'(\hat{\lambda}_K^T u_K(\mathbf{X}_i)), \tag{1.10}$$

where ρ' is the first derivative of a strictly concave function:

$$\rho(v) = f((f')^{-1}(v)) + v - v \cdot (f')^{-1}(v). \tag{1.11}$$

and $\hat{\lambda}_K \in \mathbb{R}^K$ maximizes the following objective function:

$$\hat{G}_K(\lambda) \triangleq \frac{1}{N}\sum_{i=1}^{N}\left[T_i\rho(\lambda^T u_K(\mathbf{X}_i)) - \lambda^T u_K(\mathbf{X}_i)\right]. \tag{1.12}$$

Similarly, for i such that $T_i = 0$:

$$\hat{q}_K(\mathbf{X}_i) \triangleq \frac{1}{N}\rho'(\hat{\beta}_K^T u_K(\mathbf{X}_i)), \tag{1.13}$$

and $\hat{\beta}_K \in \mathbb{R}^K$ maximizes the following objective function:

$$\hat{H}_K(\beta) \triangleq \frac{1}{N}\sum_{i=1}^{N}\left[(1-T_i)\rho(\beta^T u_K(\mathbf{X}_i)) - \beta^T u_K(\mathbf{X}_i)\right]. \tag{1.14}$$

The derivation of the dual solutions (1.10) and (1.13) is given in the Sect. 1.2.2. We then define the calibration estimator for τ as follows:

$$\hat{\tau}_K \triangleq \sum_{i=1}^{N}\{T_i\hat{p}_K(\mathbf{X}_i)Y_i - (1-T_i)\hat{q}_K(\mathbf{X}_i)Y_i\}.$$

The relationship (1.11) between $\rho(v)$ and $f(v) = D(1-v)$ will be elucidated in the subsequent section. We will also demonstrate that the strict convexity of D is equivalent to the strict concavity of ρ. When $\rho(v) = -\exp(-v)$, the weights align with the implied weights of exponential tilting [18, 19]. For $\rho(v) = \log(1+v)$, the weights correspond to empirical likelihood [20, 21]. If $\rho(v) = -(1-v)^2/2$, the weights represent the implied weights of the continuous updating estimator of the generalized method of moments [22, 23], and also minimize the squared distance function. Lastly, when $\rho(v) = v - \exp(-v)$, the weights are equivalent to the inverse of a logistic function.

Despite its close ties to generalized empirical likelihood, the calibration estimator exhibits several significant differences from the generalized empirical likelihood literature. In econometrics, generalized empirical likelihood is frequently employed to estimate a p-dimensional parameter by specifying a q-dimensional estimating equation, where $q > p \geq 1$. However, we do not estimate the target parameter τ by directly solving an overidentified estimating equation. The calibration conditions in (1.8) and (1.9) can be viewed as a K-dimensional moment restriction with a degenerate parameter of interest, and (1.12) and (1.14) essentially represent degenerate cases of generalized empirical likelihood, with only the auxiliary parameters λ and β appearing, but not the target parameter τ. Even though the generalized empirical likelihood estimation problem is undefined because the moment restrictions are not functions of target parameters, implied weights can still be constructed. In econometrics, generalized empirical likelihood estimators typically solve saddle-point problems and can be computationally challenging. In our case, $\hat{\lambda}$ and $\hat{\beta}$ are solutions to unconstrained convex maximization problems rather than a saddle-point problem and can be computed using fast and stable Newton-type algorithms. Moreover, while the generalized empirical likelihood literature primarily deals with a fixed number of moment restrictions, the dimension K of moment restrictions increases with N in our context. Furthermore, the moment restrictions are misspecified for finite K in our case, whereas most theoretical results for generalized empirical likelihood are derived under correct model specifications.

1.2.2 Duality of Calibration Estimation

This section proceeds to derive the dual of the constrained optimization problem (1.8) using the methodology outlined in [24]. The dual of (1.9) can be obtained through a similar argument. Define $E_{K\times N} \triangleq (u_K(\mathbf{X}_1), \ldots, u_K(\mathbf{X}_N))$, $s_i \triangleq 1 - T_i N p_i, i = 1, \ldots, N, \mathbf{s} \triangleq (s_1, \ldots, s_N)^T$ and $f(v) \triangleq D(1-v)$, then we can rewrite the problem (2.7) as

$$\min_{\mathbf{s}} \sum_{i=1}^{N} T_i f(s_i) \quad \text{subject to} \quad E_{K\times N} \cdot \mathbf{s} = 0.$$

For every $j \in \{1, \ldots, N\}$, we define the conjugate convex function [24] of $T_j f(\cdot)$ to be

$$\begin{aligned} g_j(z_j) &= \sup_{s_j} \left\{ z_j s_j - T_j f(s_j) \right\} = \sup_{p_j} \left\{ -T_j N p_j z_j + z_j - T_j f(1 - T_j N p_j) \right\} \\ &= \sup_{p_j} \left\{ -T_j N p_j z_j + z_j - T_j f(1 - N p_j) \right\} \\ &= -T_j N p_j^* z_j + z_j - T_j f(1 - N p_j^*), \end{aligned}$$

where the third equality follows by noting that $Tf(1 - TNp_j) = Tf(1 - Np_j)$, and p_j^* satisfies the first-order condition:

$$-T_j z_j = -T_j f'(1 - N p_j^*) \Rightarrow p_j^* = \frac{1}{N} \left\{ 1 - (f')^{-1}(z_j) \right\}.$$

By defining $\rho(z) \triangleq f\left((f')^{-1}(z) \right) + z - z \cdot (f')^{-1}(z)$, then

$$g_j(z_j) = -T_j \rho(z_j) + z_j.$$

By [24], the dual problem of (1.8) is

$$\min_{\lambda} \sum_{j=1}^{N} g_j(\lambda^T E_j) = -\max_{\lambda} \sum_{j=1}^{N} \left\{ T_j \rho \left(\lambda^T u_K(\mathbf{X}_j) \right) - \lambda^T u_K(\mathbf{X}_j) \right\}$$
$$= -\max_{\lambda} \hat{G}_K(\lambda),$$

where E_j is the j-th column of $E_{K\times N}$, i.e., $E_j = u_K(\mathbf{X}_j)$.

Since D is strictly convex, $f''(v) = D''(1-v)$, and hence f is also strictly convex and f' is strictly increasing. Note that

$$\rho(v) = f((f')^{-1}(v)) + v - v(f')^{-1}(v) \Leftrightarrow \rho\left(f'(v)\right) = f(v) + f'(v) - vf'(v)\ ;$$

differentiating with respect to v both sides of the latter equation yields:

$$\rho'\left(f'(v)\right) f''(v) = f'(v) + f''(v) - f'(v) - vf''(v) = (1-v) f''(v),$$

which also implies

$$\rho'\left(f'(v)\right) = 1 - v,$$

since $f'' > 0$. Further differentiating with respect to v of the above equation, we get $\rho''\left(f'(v)\right) f''(v) = -1$, which implies

$$\rho''(v) = -\frac{1}{f''\left((f')^{-1}(v)\right)} < 0.$$

Thus, the convexity of D is equivalent to the concavity of ρ.

1.2.3 Large Sample Properties of Calibration Estimators

To establish the large sample properties, we require the following technical assumptions.

Assumption 3 $\mathbb{E}[Y(1)^2] < \infty$ and $\mathbb{E}[Y(0)^2] < \infty$.

Assumption 4 The support $\mathcal{X}$ of the r-dimensional covariate $\mathbf{X}$ is a Cartesian product of r compact intervals.

Assumption 5 $\pi(x)$ is uniformly bounded away from 0 and 1, i.e., there exist some constants $\frac{1}{\eta_1} \triangleq \inf_{x\in\mathcal{X}} \pi(x)$, $\frac{1}{\eta_2} \triangleq \sup_{x\in\mathcal{X}} \pi(x)$ such that

$$0 < \frac{1}{\eta_1} \le \pi(x) \le \frac{1}{\eta_2} < 1 \ \ \forall x \in \mathcal{X}$$

where $\mathcal{X}$ is the support of $\mathbf{X}$.

Assumption 6 $\pi(x)$ is s-times continuously differentiable, where $s > 13r$.

Assumption 7 $m_0(x)$ and $m_1(x)$ are t-times continuously differentiable, where $t > \frac{3r}{2}$.

Assumption 8 $K = O(N^{\nu})$ and $\frac{1}{\frac{s}{r} - 2} < \nu < \frac{1}{11}$.

Assumption 9 ρ is a strictly concave function defined on $\mathbb{R}$, i.e., $\rho''(\gamma) < 0,\ \forall \gamma \in \mathbb{R}$, and the range of ρ' contains $[\eta_2, \eta_1]$, which is a subset of the positive real line.

Assumption 3 is crucial for ensuring the finiteness of the asymptotic variance. Assumptions 4 and 5 are necessary to ensure the uniform boundedness of approximations. Assumption 5 constitutes a stringent overlap condition essential for the nonparametric identification of average treatment effects in the population. In cases where a region of $\mathbf{X}$ exists such that the probability of treatment assignment is 0 or 1, treatment effects cannot be identified unless certain extrapolatory modeling assumptions are imposed. In such instances, defining a subpopulation with sufficient overlap enables the nonparametric estimation of average treatment effects within this subpopulation.

Assumptions 6 and 7 are indispensable for controlling the remainder of approximations with a given basis function. These are standard assumptions for multivariate smoothing, where the required order of smoothness increases with the dimension of $\mathbf{X}$. There is typically no a priori reason to assume that $\pi(x)$, $m_1(x)$, and $m_0(x)$ are not smooth in x. Moreover, the dimension of $\mathbf{X}$ is not restricted by these assumptions, unlike in the kernel estimation of $\pi(x)$ discussed in [25], where their assumptions imply that the dimension of $\mathbf{X}$ cannot exceed 4.

Assumption 8 is essential for controlling the stochastic order of the residual terms, which is advantageous in practice because K increases slowly with N, making a relatively small number of moment conditions sufficient for the proposed method to perform effectively.

Assumption 9 constitutes a mild requirement on ρ selected by statisticians and encompasses all the significant special cases considered in the literature. In contrast, the theoretical results in [7, 26–28] were primarily developed for linear or logistic models for the propensity score.

Define $\mu_0 \triangleq \mathbb{E}[Y(0)]$, $\mu_1 \triangleq \mathbb{E}[Y(1)]$, $\sigma_1^2(\mathbf{X}) = Var(Y(1)|\mathbf{X})$, and $\sigma_0^2(\mathbf{X}) = Var(Y(0)|\mathbf{X})$, which are finite by Assumption 3. We have the following theorem whose proof can be found in the supplemental material of [16].

Theorem 1.1 *Under Assumptions 1–9, we have*

1. $\sum_{i=1}^{N}\{T_i\hat{p}_K(\boldsymbol{X}_i)Y_i - (1-T_i)\hat{q}_K(\boldsymbol{X}_i)Y_i\} \xrightarrow{p} \tau$;
2. $\sqrt{N}\left(\sum_{i=1}^{N}\{T_i\hat{p}_K(\boldsymbol{X}_i)Y_i - (1-T_i)\hat{q}_K(\boldsymbol{X}_i)Y_i\} - \tau\right) \xrightarrow{d} \mathcal{N}(0, V_{semi})$, *where*

$$V_{semi} \triangleq \mathbb{E}\left[(m_1(\boldsymbol{X}) - m_0(\boldsymbol{X}) - \tau)^2 + \frac{\sigma_1^2(\boldsymbol{X})}{\pi(\boldsymbol{X})} + \frac{\sigma_0^2(\boldsymbol{X})}{1-\pi(\boldsymbol{X})}\right]$$

attains the semiparametric efficiency bound as shown in [7, 29].

1.2.4 A Nonparametric Variance Estimator

By Theorem 1.1, the calibration estimator achieves the semiparametric efficiency bound, with the following asymptotic variance:

$$V_{\text{semi}} \triangleq \mathbb{E}\left[(m_1(\mathbf{X}) - m_0(\mathbf{X}) - \tau)^2 + \frac{\sigma_1^2(\mathbf{X})}{\pi(\mathbf{X})} + \frac{\sigma_0^2(\mathbf{X})}{1-\pi(\mathbf{X})}\right].$$

As the variance contains unknown functions $\pi(x)$, $m_1(x)$, and $m_0(x)$, plug-in estimators typically necessitate nonparametric estimation of functions other than $\pi(x)$, as discussed in [26]. One advantage of our proposed point estimator is that it circumvents the need to directly estimate those functions. It would be advantageous to have a variance estimator that also avoids the involvement of additional estimates of nonparametric functions.

Define

$$\begin{aligned}
g_{K1}(T,\mathbf{X};\lambda) &\triangleq T\rho'(\lambda^T u_K(\mathbf{X}))u_K(\mathbf{X}) - u_K(\mathbf{X}) ,\\
g_{K2}(T,\mathbf{X};\beta) &\triangleq (1-T)\rho'(\beta^T u_K(\mathbf{X}))u_K(\mathbf{X}) - u_K(\mathbf{X}) ,\\
g_{K3}(T,\mathbf{X},Y;\theta) &\triangleq T\rho'(\lambda^T u_K(\mathbf{X}))Y - (1-T)\rho'(\beta^T u_K(\mathbf{X}))Y - \tau\\
g_K(T,\mathbf{X},Y;\theta) &\triangleq \begin{pmatrix} g_{K1}(T,\mathbf{X};\lambda)\\ g_{K2}(T,\mathbf{X};\beta)\\ g_{K3}(T,\mathbf{X},Y;\theta)\end{pmatrix} ,
\end{aligned}$$

where $\theta \triangleq (\lambda,\beta,\tau)^T$. Also define $\hat{\theta}_K \triangleq (\hat{\lambda}_K,\hat{\beta}_K,\hat{\tau}_K)^T$, $\theta_K^* \triangleq (\lambda_K^*,\beta_K^*,\tau_K^*)^T$ and $\tau_K^* \triangleq \mathbb{E}[TNp_K^*(\mathbf{X})Y - (1-T)Nq_K^*(\mathbf{X})Y]$.

Note that $\hat{\theta}_K$ satisfies

$$\frac{1}{N}\sum_{i=1}^N g_K(T_i,\mathbf{X}_i,Y_i;\hat{\theta}_K) = 0.$$

Taylor series expansion on the left-hand side at θ_K^* yields

$$0 = \frac{1}{N}\sum_{i=1}^N g_K(T_i,\mathbf{X}_i,Y_i;\theta_K^*) + \frac{1}{N}\sum_{i=1}^N \frac{\partial g_K(T_i,\mathbf{X}_i,Y_i;\tilde{\theta}_K)}{\partial\theta}(\hat{\theta}_K - \theta_K^*) , \quad (1.15)$$

where $\tilde{\theta}_K = (\tilde{\lambda}_K,\tilde{\beta}_K,\tilde{\tau}_K)^T$ lies on the line joining $\hat{\theta}_K$ and θ_K^*. We shall show in the supplementary material that

$$\frac{1}{N}\sum_{i=1}^N \frac{\partial g_K(T_i,\mathbf{X}_i,Y_i;\tilde{\theta}_K)}{\partial\theta} = \mathbb{E}\left[\frac{\partial g_K(T,\mathbf{X},Y;\theta_K^*)}{\partial\theta}\right] + o_p(1) , \quad (1.16)$$

where

$$\mathbb{E}\left[\frac{\partial g_K(T, \mathbf{X}, Y; \theta_K^*)}{\partial \theta}\right] = \begin{pmatrix} A_{2K\times 2K}, & B_{2K\times 1} \\ C_{1\times 2K}, & D_{1\times 1} \end{pmatrix},$$

and

$$\begin{aligned}
A_{2K\times 2K} &\triangleq \begin{pmatrix} \mathbb{E}[T\rho''((\lambda_K^*)^T u_K(\mathbf{X}))u_K(\mathbf{X})u_K^T(\mathbf{X})], & 0_{K\times K} \\ 0_{K\times K}, & \mathbb{E}[(1-T)\rho''((\beta_K^*)^T u_K(\mathbf{X}))u_K(\mathbf{X})u_K^T(\mathbf{X})] \end{pmatrix}, \\
B_{2K\times 1} &\triangleq 0_{2K\times 1}, \\
C_{1\times 2K} &\triangleq (\mathbb{E}[T\rho''((\lambda_K^*)^T u_K(\mathbf{X}))Yu_K^T(\mathbf{X})], -\mathbb{E}[(1-T)\rho''((\beta_K^*)^T u_K(\mathbf{X}))Yu_K^T(\mathbf{X})]), \\
D_{1\times 1} &\triangleq -1.
\end{aligned}$$

Since we are only interested in the limiting of $Var(\sqrt{N}(\hat{\tau}_K - \tau_K^*))$, which is the last element of $Var(\sqrt{N}(\hat{\theta}_K - \theta_K^*))$ when $N \uparrow \infty$, this leads us to consider the last row of $\mathbb{E}\left[\frac{\partial g_K(T,\mathbf{X},Y;\theta_K^*)}{\partial\theta}\right]^{-1}$ which is

$$\left(C_{1\times 2K} \cdot A_{2K\times 2K}^{-1}, -1\right) = (L_K, R_K, -1),$$

where

$$\begin{aligned}
L_K &\triangleq \mathbb{E}[T\rho''((\lambda_K^*)^T u_K(\mathbf{X}))Yu_K^T(\mathbf{X})] \cdot \mathbb{E}[T\rho''((\lambda_K^*)^T u_K(\mathbf{X}))u_K(\mathbf{X})u_K^T(\mathbf{X})]^{-1}, \\
R_K &\triangleq -\mathbb{E}[(1-T)\rho''((\beta_K^*)^T u_K(\mathbf{X}))Yu_K^T(\mathbf{X})] \\
&\quad \cdot \mathbb{E}[(1-T)\rho''((\beta_K^*)^T u_K(\mathbf{X}))u_K(\mathbf{X})u_K^T(\mathbf{X})]^{-1}.
\end{aligned}$$

Since $\lim_{K\to\infty} \mathbb{E}[g_K(T, \mathbf{X}, Y; \theta_K^*)]^T] = 0$, we can have

$$\begin{aligned}
&\lim_{K\to\infty} Var(\sqrt{N}(\hat{\tau}_K - \tau_K^*)) \qquad (1.17) \\
&= \lim_{K\to\infty} (L_K, R_K, -1)\mathbb{E}[g_K(T, \mathbf{X}, Y; \theta_K^*)g_K^T(T, \mathbf{X}, Y; \theta_K^*)] \begin{pmatrix} L_K^T \\ R_K^T \\ -1 \end{pmatrix} \\
&= \lim_{K\to\infty} (L_K, R_K, -1)P_K \begin{pmatrix} L_K^T \\ R_K^T \\ -1 \end{pmatrix},
\end{aligned}$$

where $P_K \triangleq \mathbb{E}[g_K(T, \mathbf{X}, Y; \theta_K^*)g_K^T(T, \mathbf{X}, Y; \theta_K^*)]$.

This motivates us to define our estimator for the asymptotic variance by:

$$\hat{V}_K \triangleq (\hat{L}_K, \hat{R}_K, -1)\hat{P}_K \begin{pmatrix} \hat{L}_K^T \\ \hat{R}_K^T \\ -1 \end{pmatrix},$$

where

$$\hat{L}_K \triangleq \left[\sum_{i=1}^{N} T_i\rho''(\hat{\lambda}_K^T u_K(\mathbf{X}_i))u_K^T(\mathbf{X}_i)Y_i\right] \times \left[\sum_{i=1}^{N} T_i\rho''(\hat{\lambda}_K^T u_K(\mathbf{X}_i))u_K(\mathbf{X}_i)u_K^T(\mathbf{X}_i)\right]^{-1},$$

$$\hat{R}_K \triangleq -\left[\sum_{i=1}^{N}(1-T_i)\rho''(\hat{\beta}_K^T u_K(\mathbf{X}_i))u_K^T(\mathbf{X}_i)Y_i\right] \times \left[\sum_{i=1}^{N}(1-T_i)\rho''(\hat{\beta}_K^T u_K(\mathbf{X}_i))u_K^T(\mathbf{X}_i)u_K(\mathbf{X}_i)\right]^{-1},$$

$$\hat{P}_K \triangleq \frac{1}{N}\sum_{i=1}^{N} g_K(T_i, \mathbf{X}_i, Y_i; \hat{\theta}_K)g_K^T(T_i, \mathbf{X}_i, Y_i; \hat{\theta}_K).$$

Although the construction of the proposed variance estimator did not start with a direct approximation of the influence function, the variance estimator can be written as $\hat{V}_K = N^{-1} \times \sum_{i=1}^{N} \hat{\varphi}_{CAL}^2(T_i, \mathbf{X}_i, Y_i; \hat{\theta})$ where

$$\begin{aligned}\hat{\varphi}_{CAL}(T, \mathbf{X}, Y) &= -(L_K, -R_K, -1)g_K(T, \mathbf{X}, Y; \hat{\theta}) \\ &= g_{K3}(T, \mathbf{X}, Y; \hat{\theta}) - L_K g_{K1}(T, \mathbf{X}; \hat{\lambda}) + R_K g_{K2}(T, \mathbf{X}; \hat{\beta}),\end{aligned}$$

which is an estimator of the efficient influence function:

$$\varphi_{eff}(T, \mathbf{X}, Y) = \frac{TY}{\pi(\mathbf{X})} - \frac{(1-T)Y}{1-\pi(\mathbf{X})} - \tau + (T - \pi(\mathbf{X}))\beta(\mathbf{X}), \tag{1.18}$$

$$\beta(\mathbf{X}) = -\left[\frac{m_1(\mathbf{X})}{\pi(\mathbf{X})} + \frac{m_0(\mathbf{X})}{1-\pi(\mathbf{X})}\right].$$

The following theorem ensures the consistency of the proposed variance estimator $\hat{V}_K$ whose proof can be found in the supplemental material of [16].

Theorem 1.2 *Under Assumptions 1–9 with Assumption 3 being strengthened to* $\mathbb{E}(Y^4(1)) < \infty$ *and* $\mathbb{E}(Y^4(0)) < \infty$, $\hat{V}_K$ *is a consistent estimator for the asymptotic variance* V_{semi}.

1.2.5 Numerical Results

In this section we present results from simulation studies to investigate the finite sample performance of various weighting estimators and standard error estimators.

The first set of simulation scenarios was similar to those in [30] for the estimation of a population mean. Sample size for each simulated data set was 200 or 1000, and 10000 Monte Carlo datasets were generated for each scenario. For each observation, a random vector $Z = (Z_1, Z_2, Z_3, Z_4)$ was generated from the standard multivariate normal distribution. The potential outcome $Y(1)$ was generated from a normal distribution with mean $210 + b(Z)$ and unit variance, where $b(Z) = 27.4Z_1 + 13.7Z_2 + 13.7Z_3 + 13.7Z_4$; $Y(0)$ was generated from a normal distribution with mean $200 - 0.5b(Z)$ and unit variance. An individual was assigned to treatment $T = 1$ with probability $\exp(\eta_0(Z))/(1 + \exp(\eta_0(Z)))$, where $\eta_0(Z) = -Z_1 + 0.5Z_2 - 0.25Z_3 - 0.1Z_4$. The observed outcome was $Y = TY(1) + (1 - T)Y(0)$. Instead of observing covariates Z, we were only able to observe nonlinear transformations $X_1 = \exp(Z_1/2)$, $X_2 = Z_2/(1 + \exp(Z_1))$, $X_3 = (Z_1 Z_3/25 + 0.6)^3$ and $X_4 = (Z_2 + Z_4 + 20)^2$. Denote $\mathbf{X} \triangleq (X_1, X_2, X_3, X_4)$. We compared the Horvitz-Thompson estimator (HT)

$$\hat{\tau}_{HT} = \frac{1}{N}\sum_{i=1}^{N} \frac{T_i Y_i}{\hat{\pi}(\mathbf{X}_i)} - \frac{1}{N}\sum_{i=1}^{N} \frac{(1 - T_i)Y_i}{1 - \hat{\pi}(\mathbf{X}_i)},$$

the ratio-type inverse probability weighting estimator (IPW)

$$\hat{\tau}_{IPW} = \frac{\sum_{i=1}^{N} T_i(\hat{\pi}(\mathbf{X}_i))^{-1} Y_i}{\sum_{i=1}^{N} T_i(\hat{\pi}(\mathbf{X}_i))^{-1}} - \frac{\sum_{i=1}^{N}(1 - T_i)(1 - \hat{\pi}(\mathbf{X}_i))^{-1} Y_i}{\sum_{i=1}^{N}(1 - T_i)(1 - \hat{\pi}(\mathbf{X}_i))^{-1}},$$

and the calibration estimators (CAL) with $\rho(v) = -e^{-v}$ (exponential tilting; ET), $\rho(v) = \log(1 + v)$ (empirical likelihood; EL), $\rho(v) = -(1 - v)^2/2$ (quadratic; Q) and $\rho(v) = v - e^{-v}$ (inverse logistic; IL). For HT and IPW estimators, we used a working logistic regression for propensity score model, and we considered six different ways to estimate the propensity score parameters: i. The maximum likelihood estimator ($\hat{\gamma}_{MLE}$); ii. The moment estimator solving the empirical counterpart of (11) that balances the covariates of the treated and the full data ($\hat{\gamma}_{1F}$); iii. The moment estimator solving the empirical counterpart of (12) that balances the covariates of the controls and the full data ($\hat{\gamma}_{0F}$); iv. The moment estimator solving the empirical counterpart of (13) that balances the covariates of the treated and the controls ($\hat{\gamma}_{10}$); v. The generalized method of moment estimator for the overidentified system (11) and (12) that balances the covariates of the treated, the controls, and the full data ($\hat{\gamma}_{10F}$); vi. The covariate balancing propensity score estimator of [14] for an overidentified system defined by (13) and the likelihood score equation ($\hat{\gamma}_{CBPS}$). We presented the average bias and root mean squared error of the estimators over 10000 simulations, and the following standardized imbalance measures [14, 31]:

$$Imb_{10} = \left\{ \left(\frac{1}{N}\sum_{i=1}^{N} [(T_i w_{1i} - (1 - T_i) w_{0i}) \mathbf{X}_i]^T \right) \left(\frac{1}{N}\sum_{i=1}^{N} \mathbf{X}_i \mathbf{X}_i^T \right)^{-1} \left(\frac{1}{N}\sum_{i=1}^{N} (T_i w_{1i} - (1 - T_i) w_{0i}) \mathbf{X}_i \right) \right\}^{1/2},$$

$$Imb_{1F} = \left\{ \left(\frac{1}{N} \sum_{i=1}^{N} [(T_i w_{1i} - 1)\mathbf{X}_i]^T \right) \left(\frac{1}{N} \sum_{i=1}^{N} \mathbf{X}_i \mathbf{X}_i^T \right)^{-1} \left(\frac{1}{N} \sum_{i=1}^{N} (T_i w_{1i} - 1)\mathbf{X}_i \right) \right\}^{1/2},$$

and

$$Imb_{0F} = \left\{ \left(\frac{1}{N} \sum_{i=1}^{N} [1 - (1 - T_i) w_{0i})\mathbf{X}_i]^T \right) \left(\frac{1}{N} \sum_{i=1}^{N} \mathbf{X}_i \mathbf{X}_i^T \right)^{-1} \left(\frac{1}{N} \sum_{i=1}^{N} [1 - (1 - T_i) w_{0i}]\mathbf{X}_i \right) \right\}^{1/2},$$

where w_{1i} are the weights for the treated, which is $\hat{\pi}_i^{-1}$ for HT; $(\hat{\pi}_i \times \sum_{j=1}^{N} T_j \hat{\pi}_j^{-1})^{-1}$ for IPW; and $\hat{p}_i$ for calibration. And w_{0i} are the weights for the controls, which is $(1 - \hat{\pi}_i)^{-1}$ for HT; $((1 - \hat{\pi}_i) \times \sum_{j=1}^{N} (1 - T_j)(1 - \hat{\pi}_j)^{-1})^{-1}$ for IPW and $\hat{q}_i$ for calibration. Three imbalance measures were considered which measure three-way imbalance between the treated, the controls, and the full data. We examined all three measures because there is no guarantee that minimizing one imbalance measure could lead to a reduction in the others, as discussed in Sect. 2.4. A total imbalance measure is defined by $(Imb_{10}^2 + Imb_{1F}^2 + Imb_{0F}^2)^{1/2}$. Table 1.1 and 1.2 show the simulation results for $N = 200$ and $N = 1000$ respectively, using a linear specification of covariates $u_5 = (1, X_1, X_2, X_3, X_4)$ for both propensity score modeling and for calibration estimation.

The results showed that the Horvitz-Thompson estimators could worsen the problem of imbalance when the propensity score was estimated by maximum likelihood, or by method of moment that only balances a particular group to the full data. Balancing the treated to the full data only led to a noticeable improvement in estimating $\mathbb{E}(Y(1))$ but the estimator of $\mathbb{E}(Y(0))$ performed very poorly, as the imbalance between the controls and the full data actually increased. Therefore, the performance of the average treatment effects estimator was also very poor. Directly balancing the treated and the controls performed much better, and reduced the imbalance between the particular groups and the full data since the covariate distribution of the full data is a convex combination of the covariate distributions of the treated and controls. Generalized method of moment estimators $(\hat{\gamma}_{10F}, \hat{\gamma}_{CBPS})$ had more imbalance than balancing solely for the treated and controls, because the generalized method of moment estimates did not exactly solve the corresponding overidentified systems of equations. The calibration estimators achieved the exact three-way balance by design. In general, the mean squared errors of the estimates of average treatment effects were positively correlated with the overall imbalance. Finally, the exponential tilting estimator performed the best among the calibration estimators.

Next, we focused on the performance of weighting estimators under three sets of covariate specifications:

1. $u_5(\mathbf{X}) = (1, X_1, X_2, X_3, X_4)$,
2. $u_9(\mathbf{X}) = (1, X_1, X_2, X_3, X_4, X_1^2, X_2^2, X_3^2, X_4^2)$,
3. $u_{15}(\mathbf{X}) = (1, X_1, X_2, X_3, X_4, X_1^2, X_2^2, X_3^2, X_4^2, X_1X_2, X_1X_3, X_1X_4, X_2X_3, X_2 X_4, X_3X_4)$.

Table 1.1 Comparison of weighting estimators for the Kang and Schafer scenario, $n = 200$

	E(Y(1))		E(Y(0))		E(Y(1)–Y(0))		Imbalance * 100			
Estimator	Bias	RMSE	Bias	RMSE	Bias	RMSE	(1, 0)	(1, F)	(0, F)	Total
Unweighted	–10.09	10.68	–5.06	5.34	–5.03	6.38	85	43	42	104
HT($\hat{\gamma}_{MLE}$)	25.52	264.87	–4.93	8.36	30.45	266.19	165	145	31	228
HT($\hat{\gamma}_{1F}$)	–2.2	3.87	78.98	>999	–81.17	>999	454	0	454	642
HT($\hat{\gamma}_{0F}$)	>999	>999	–1.25	1.92	>999	>999	>999	>999	1	>999
HT($\hat{\gamma}_{10}$)	1.3	7.38	0.65	7.79	0.65	4.9	0	21	21	30
HT($\hat{\gamma}_{10F}$)	–3.38	7.36	–6.68	8.31	3.3	10.97	41	25	29	59
HT($\hat{\gamma}_{CBPS}$)	–5.33	10.57	–4.89	7.81	–0.44	13.59	54	35	31	74
IPW($\hat{\gamma}_{MLE}$)	1.7	9.65	–1.87	2.5	3.57	10.72	32	33	12	49
IPW($\hat{\gamma}_{1F}$)	–2.17	3.86	–1.38	3.65	–0.79	4.96	33	0	34	47
IPW($\hat{\gamma}_{0F}$)	10.53	21.49	–1.3	1.96	11.83	22.46	89	90	1	126
IPW($\hat{\gamma}_{10}$)	–1.33	3.41	–1.87	2.65	0.54	4.44	0	11	11	16
IPW($\hat{\gamma}_{10F}$)	–2.41	4.06	–2.14	2.73	–0.27	4.43	11	12	11	20
IPW($\hat{\gamma}_{CBPS}$)	–2.79	4.7	–2.24	2.85	–0.56	4.67	18	13	13	26
CAL(ET)	–1.45	3.56	–1.95	2.46	0.5	4.29	0	0	0	0
CAL(EL)	–1.72	3.76	–1.54	2.17	–0.19	4.36	0	0	0	0
CAL(Q)	–0.52	3.38	–2.49	2.9	1.97	4.77	0	0	0	0
CAL(IL)	–2.08	3.83	–1.18	1.87	–0.9	4.34	0	0	0	0

RMSE: root mean squared error; HT: Horvitz-Thompson estimators; IPW: ratio-type inverse probability weighting estimators; CAL: calibration estimators. For HT and IPW estimators, propensity score parameters were estimated in six ways: i. The maximum likelihood estimator ($\hat{\gamma}_{MLE}$); ii. The moment estimator that balances the covariates of the treated and the full data ($\hat{\gamma}_{1F}$); iii. The moment estimator that balances the covariates of the controls and the full data ($\hat{\gamma}_{0F}$); iv. The moment estimator that balances the covariates of the treated and the controls ($\hat{\gamma}_{10}$); v. The generalized method of moment estimator for an overidentified system that balances the covariates of the treated, the controls, and the full data ($\hat{\gamma}_{10F}$); vi. The covariate balancing propensity score estimator of [14] ($\hat{\gamma}_{CBPS}$). For calibration estimators, ET: exponential tilting; EL: empirical likelihood; Q: quadratic; IL: inverse logistic

In theory, global efficiency is attained when the number of moment conditions K increases with the sample size N, but intuition and theory both suggest that K cannot be too large. [26, 27] both suggest using $K = N^{\nu}$ where $\nu < 1/9$ for the inverse probability weighting estimators, therefore the theory is in favor of u_5 for $N = 200$ and 1000. Since vigorous theories for nonparametric inverse probability weighting estimators have only been developed for maximum likelihood estimation, we limit our discussion here to estimators where the propensity score parameters are estimated by maximum likelihood. The results are shown in Table 1.3. While the existing theory suggested that K should be small for the sample sizes considered, the performance of both HT and IPW estimators for u_5 was quite poor, while calibration estimations worked well, even for u_5. Propensity score weighting estimators performed better for u_9 and u_{15}, but the calibration estimators had uniformly smaller mean squared errors compared to that through the propensity score methods. We examined the sta-

Table 1.2 Comparison of weighting estimators for the Kang and Schafer scenario, $n = 1000$

	E(Y(1))		E(Y(0))		E(Y(1)–Y(0))		Imbalance * 100			
Estimator	Bias	RMSE	Bias	RMSE	Bias	RMSE	(1,0)	(1,F)	(0,F)	Total
Unweighted	−10.04	10.16	−5.01	5.07	−5.03	5.32	82	41	41	101
HT($\hat{\gamma}_{MLE}$)	47.35	414.45	−4.76	5.38	52.1	415.16	241	227	19	334
HT($\hat{\gamma}_{1F}$)	−2.79	3.19	−4.69	9.71	1.9	8.37	36	0	36	51
HT($\hat{\gamma}_{0F}$)	285.46	>999	−1.11	1.28	286.57	>999	>999	>999	0	>999
HT($\hat{\gamma}_{10}$)	−2.05	3.01	−2.81	3.37	0.77	2.09	0	10	10	15
HT($\hat{\gamma}_{10F}$)	2.03	5.05	−6.62	7.37	8.65	11.37	44	24	25	57
HT($\hat{\gamma}_{CBPS}$)	1.92	6.67	−4.74	5.48	6.66	10.37	39	27	19	52
IPW($\hat{\gamma}_{MLE}$)	5.08	12.22	−1.84	1.97	6.91	13.21	44	47	7	66
IPW($\hat{\gamma}_{1F}$)	−2.79	3.19	−2.51	2.69	−0.27	1.9	13	0	13	19
IPW($\hat{\gamma}_{0F}$)	13.25	21.36	−1.12	1.29	14.38	22.15	98	98	0	138
IPW($\hat{\gamma}_{10}$)	−1.44	2.05	−2.24	2.4	0.8	2.11	0	8	8	12
IPW($\hat{\gamma}_{10F}$)	−1.25	2.14	−2.32	2.45	1.07	2.5	7	11	9	16
IPW($\hat{\gamma}_{CBPS}$)	−0.92	2.33	−2.25	2.41	1.33	2.76	10	12	9	18
CAL(ET)	−1.97	2.49	−1.87	1.99	−0.09	1.96	0	0	0	0
CAL(EL)	−2.11	3	−1.39	1.56	−0.73	2.59	0	0	0	0
CAL(Q)	−0.81	1.7	−2.47	2.56	1.66	2.56	0	0	0	0
CAL(IL)	−2.78	3.18	−1.1	1.27	−1.68	2.59	0	0	0	0

RMSE: root mean squared error; HT: Horvitz-Thompson estimators; IPW: ratio-type inverse probability weighting estimators; CAL: calibration estimators. For HT and IPW estimators, propensity score parameters were estimated in six ways: i. The maximum likelihood estimator ($\hat{\gamma}_{MLE}$); ii. The moment estimator that balances the covariates of the treated and the full data ($\hat{\gamma}_{1F}$); iii. The moment estimator that balances the covariates of the controls and the full data ($\hat{\gamma}_{0F}$); iv. The moment estimator that balances the covariates of the treated and the controls ($\hat{\gamma}_{10}$); v. The generalized method of moment estimator for an overidentified system that balances the covariates of the treated, the controls, and the full data ($\hat{\gamma}_{10F}$); vi. The covariate balancing propensity score estimator of [14] ($\hat{\gamma}_{CBPS}$). For calibration estimators, ET: exponential tilting; EL: empirical likelihood; Q: quadratic; IL: inverse logistic

Table 1.3 Comparisons of weighting estimators for various covariate configurations

		u_5		u_9		u_{15}		(u_5, u_9)	(u_5, u_{15})	(u_9, u_{15})
		Bias	RMSE	Bias	RMSE	Bias	RMSE	Corr	Corr	Corr
HT	N = 200	30.45	266.19	−1.41	19.29	3.41	20.47	0.04	<0.01	0.37
	N = 1000	52.1	415.16	−0.8	7.35	4.63	10.17	0.01	<0.01	0.35
IPW	N = 200	3.57	10.72	−0.10	4.73	−0.16	5.04	0.45	0.32	0.79
	N = 1000	6.91	13.21	0.36	2.10	0.51	2.47	0.20	0.11	0.75
CAL	N = 200	0.5	4.29	−1.39	4.46	−0.66	4.50	0.82	0.70	0.79
	N = 1000	−0.09	1.96	−0.76	1.97	−0.24	1.88	0.84	0.76	0.87

HT: Horvitz-Thompson estimator with propensity score estimated from maximum likelihood; IPW: ratio-type inverse probability weighting estimator with propensity score estimated from maximum likelihood; CAL: calibration estimator with exponential tilting bility of the estimation procedure by estimating the sample correlation of the same estimators under different covariate specifications. The correlations of HT estimators between u_5 and each of u_9 and u_{15} were negligible, indicating that the inclusion of additional covariates could change the estimates arbitrarily. The correlations of the IPW estimators were also quite low, except for the correlation between u_9 and u_{15}. In contrast, the calibration estimators had high correlations under different covariate specifications. Therefore, adding higher order terms of covariates did not dramatically change the calibration estimate and its performance was minimally affected by the choice of K.

Next, we studied the performance of the proposed estimator for the efficient asymptotic variance V_{semi}, compared with a few other existing estimators. To describe the other estimators, we note that $V_{semi} = \mathbb{E}(\varphi_{eff}^2)$ where φ_{eff} is the efficient influence function given in (17). The variance estimator of [26] is based on plugging in a propensity score estimate $\hat{\pi}(\mathbf{X})$ and a polynomial series estimate $\hat{\beta}(\mathbf{X})$. To estimate $\beta(\mathbf{X})$, they note that $\beta(\mathbf{X}) = \mathbb{E}(Y^*|\mathbf{X})$ where

$$Y^* = -\frac{TY}{\pi^2(\mathbf{X})} - \frac{(1-T)Y}{(1-\pi(\mathbf{X}))^2}\,.$$

Therefore, they calculate Y^* by substituting $\pi(\mathbf{X})$ with a nonparametric estimate $\hat{\pi}(\mathbf{X})$, and $\beta(\mathbf{X})$ can be estimated by a linear regression of Y^* on $u(\mathbf{X})$, which is the same design matrix as in the logistic regression model for $\pi(\mathbf{X})$. The estimator of [26] is

$$\hat{V}_{HIR} = \frac{1}{N}\sum_{i=1}^{N}\left[\frac{T_iY_i}{\hat{\pi}(\mathbf{X}_i)} - \frac{(1-T_i)Y_i}{1-\hat{\pi}(\mathbf{X}_i)} - \hat{\tau}^2 + (T_i - \pi(\mathbf{X}_i))\hat{\beta}(\mathbf{X}_i)\right]^2.$$

An alternative plug-in estimator is to directly plug in $\hat{\pi}(\mathbf{X})$, $\hat{m}_1(\mathbf{X})$ and $\hat{m}_0(\mathbf{X})$ into the influence function, instead of estimating $\beta(\mathbf{X})$ which is a function of $(\pi(\mathbf{X}), m_1(\mathbf{X}), m_0(\mathbf{X}))$. The conditional expectations $m_1(\mathbf{X})$ and $m_0(\mathbf{X})$ can be estimated by polynomial series linear regression models of Y on $u(\mathbf{X})$ among the treated and controls respectively. We compared the plug-in estimators and the proposed estimator in Table 1.4. The Hirano et al. estimator performed poorly for all covariate specifications and sample sizes, the direct plug-in estimator performed poorly in covariate specification u_5, and the proposed estimator performed well in all covariate specifications and sample sizes. A hypothesis for explaining the failure of the Hirano et al. estimator is that Y^* depends on the squared inverse of propensity score, which is estimated very poorly and creates highly influential outlying values when the fitted propensity score is close to zero. To understand whether the problem is solely caused by poorly estimated propensity scores, we studied whether the performance of plug-in variance estimators can be improved when the propensity score was known. A known propensity score did not solve the entire problem for the Hirano et al. estimator because $\beta(\mathbf{X})$ is still highly nonlinear in $\mathbf{X}$ and the sieve estimator for $\beta(\mathbf{X})$ did not approximate $\beta(\mathbf{X})$ well enough in the given situations. To further understand

Table 1.4 Comparisons of variance estimators, where the true asymptotic variance was 54

		Covariate specifications					
		u_5		u_9		u_{15}	
Estimators	N	Estimate	Correlation	Estimate	Correlation	Estimate	Correlation
HIR(MLE)	200	>999	0.06	822	0.11	792	0.07
	1000	>999	–0.02	>999	0.19	>999	0.17
HIR(True)	200	511	0.12	888	–0.16	>999	–0.16
	1000	328	0.19	532	–0.16	732	–0.15
Plug-in(MLE)	200	133	0.72	59	0.87	61	0.91
	1000	494	0.47	59	0.88	59	0.92
Plug-in(True)	200	74	0.85	64	0.87	68	0.91
	1000	71	0.80	65	0.85	61	0.92
Proposed	200	58	0.94	55	0.92	61	0.91
	1000	59	0.92	56	0.95	57	0.96

HIR: the estimator of Hirano, Imbens and Ridder (2003), MLE: Maximum likelihood estimator for propensity score; True: substituting in true propensity score and average treatment effects

the performance of estimators, we studied the correlations between the true and the estimated influence functions. For estimators that showed a good performance, the correlations were above 0.8 in general. The correlations were very low for the Hirano et al. estimator because the sieve estimator for $\beta(\mathbf{X})$ performed poorly. The direct plug-in estimator did not perform well for covariate specification u_5, because the propensity score was poorly estimated in that case. The results showed that $\hat{\varphi}_{CAL}$ and the true (but practically unknown) efficient influence function were highly correlated with correlation coefficients being greater than 0.9 for all simulation scenarios, and the estimated standard deviations are consistently close to the true asymptotic standard deviation. Further simulations (not included in the manuscript) showed that averages of the proposed variance estimates were very close to the sampling variances, and the coverage of confidence intervals based on normal approximations were close to the nominal levels.

We further considered an additional simulation scenario as in Hainmueller (2012). Six covariates X_j, $j = 1, \ldots, 6$ were generated, where (X_1, X_2, X_3) were multivariate normal with means zero, variances of (2, 1, 1), and the covariances of X_1 and X_2, X_1 and X_3, X_2 and X_3, were 1, -1 and -0.5 respectively; X_4 was uniformly distributed on (-3,3), X_5 was $\chi^2(1)$-distributed and X_6 was Bernoulli random variable with mean 0.5. The treatment indicator followed $T = I(X_1 + 2X_2 - 2X_3 - X_4 - 0.5X_5 + X_6 + \epsilon > 0)$ where $\epsilon \sim N(0, 30)$. This corresponds to the case with the largest imbalance between the treated and control groups in Hainmueller's simulation setting. The outcome distribution was $Y = (X_1 + X_2 + X_5)^2 + \eta$ where η was the standard normal random variable. The outcome did not differ between the treated and controls, and the average treatment effects was zero. We compared the same set of estimators as in Tables 1.1 and 1.2. We report the results for $n = 300$ with a linear covariate specification $(X_1, \ldots, X_6)$ in Table 1.5. Similar to Tables 1.1 and 1.2 for

Table 1.5 Comparison of weighting estimators for the Hainmueller scenario, $n = 300$

	E(Y(1)–Y(0))		Imbalance * 100			
Estimator	Bias	RMSE	(1, 0)	(1, F)	(0, F)	Total
Unweighted	2.95	3.39	111	55	55	136
HT(γ_{MLE})	0.15	3.07	25	19	18	38
HT(γ_{1F})	<–999	>999	>999	0	>999	>999
HT(γ_{0F})	>999	>999	>999	>999	0	>999
HT(γ_{10})	0.1	1.9	0	14	14	20
HT(γ_{10F})	0.34	1.81	17	14	13	26
HT(γ_{CBPS})	0.38	1.84	21	16	16	31
IPW(γ_{MLE})	0.21	2.34	23	18	17	35
IPW(γ_{1F})	–1.96	8	59	0	59	84
IPW(γ_{0F})	0.5	5.76	61	61	0	86
IPW(γ_{10})	0.1	1.86	0	13	13	19
IPW(γ_{10F})	0.36	1.77	17	14	13	26
IPW(γ_{CBPS})	0.43	1.75	20	16	15	31
CAL(ET)	0.08	1.42	0	0	0	0
CAL(EL)	0.14	1.67	0	0	0	0
CAL(Q)	0.04	1.42	0	0	0	0
CAL(IL)	0.15	1.54	0	0	0	0

RMSE: root mean squared error; HT: Horvitz-Thompson estimators; IPW: ratio-type inverse probability weighting estimators; CAL: calibration estimators. For HT and IPW estimators, propensity score parameters were estimated in six ways: i. The maximum likelihood estimator ($\hat{\gamma}_{MLE}$); ii. The moment estimator that balances the covariates of the treated and the full data ($\hat{\gamma}_{1F}$); iii. The moment estimator that balances the covariates of the controls and the full data ($\hat{\gamma}_{0F}$); iv. The moment estimator that balances the covariates of the treated and the controls ($\hat{\gamma}_{10}$); v. The generalized method of moment estimator for an overidentified system that balances the covariates of the treated, controls and the full data ($\hat{\gamma}_{10F}$); vi. The covariate balancing propensity score estimator of [14] ($\hat{\gamma}_{CBPS}$). For calibration estimators, ET: exponential tilting; EL: empirical likelihood; Q: quadratic; IL: inverse logistic

the Kang and Schafer scenario, the calibration estimators performed the best in terms of mean squared error and created an exact three-way covariate balance.

1.3 Causal Inference Based on Semi-supervised Data

In this section, we address the problem of estimating the average treatment effect (ATE) with the assistance of additional unlabeled data.

In various economic, biomedical, and machine learning scenarios, in addition to a set of supervised samples $(\boldsymbol{X}_i, T_i, Y_i)_{i=1}^n$, we may also gather a substantial amount of unlabeled samples $(\boldsymbol{X}_i, T_i)_{i=n+1}^N$, where the labeled size is n and the unlabeled size is $m := N - n$. Here, $\boldsymbol{X}_i \in \mathbb{R}^d$ represents the covariates, $T_i \in \{0, 1\}$ is the binary treatment indicator, and $Y_i \in \mathbb{R}$ is the outcome variable. Let $\tau := \lim_{N,n\to\infty} n/N \in [0,1]$ be the labeling ratio. We are particularly interested in the scenario where $\tau = 0$, indicating a situation where $N \gg n$, and the dimensionality of the covariates d may be considerably larger than the labeled sample size n. Semi-supervised datasets typically arise when obtaining outcomes is expensive while the other variables are abundant and inexpensive to collect.

In semi-supervised learning, it is assumed that the outcomes are missing completely at random (MCAR). Our objective is to develop estimators that are more efficient than their supervised counterparts, leveraging the abundance of unlabeled data. Our goal is to estimate the average treatment effect (ATE):

$$\tau := \mathbb{E}[Y_i(1) - Y_i(0)],$$

where $Y_i(1), Y_i(0) \in \mathbb{R}$ represent the potential outcomes.

1.3.1 The Semi-supervised ATE Estimator

In the subsequent discussion, we present a semi-supervised ATE estimator built on the doubly robust representation (1.6).

Before formulating the final ATE estimator, we initially introduce estimates for the necessary nuisance functions: $\pi(x) := P[T_i = 1 \mid \boldsymbol{X}_i = x]$ and $m_t(x) := E[Y_i \mid \boldsymbol{X}_i = x, T_i = t]$ for each $t \in \{0, 1\}$.

Given that the propensity score function involves only the covariates and treatment variables, we can utilize the complete dataset (encompassing both labeled and unlabeled samples) to obtain an accurate estimate of $\pi(x)$. Given the potentially large size of the unlabeled sample, we allow for general nonparametric estimates denoted as $\hat{\pi}(x)$. Conversely, the identification of outcome regression functions $m_1(x)$ and $m_0(x)$ involves the outcome variable Y_i. In this case, we consider linear working models, particularly when the dimension d is considerably larger than the relatively small labeled size n. Define the population slopes as follows:

$$\beta_t^* := \arg\min_{\beta \in \mathbb{R}^d} E\left[(Y_i - \boldsymbol{X}_i^T \beta)^2 \mid T_i = t\right] \quad \text{for each} \quad t \in \{0, 1\}.$$

Let $\hat{\beta}_t$ be an estimate of β_t^*, for example, using the ℓ_1-regularized estimator Lasso [32]. Consequently, $\hat{m}_t(x) = x^T \hat{\beta}_t$ represents a linear estimate of the true outcome regression function $m_t(x)$.

In the following, we introduce a *semi-supervised doubly robust estimator* for the ATE, as originally proposed by [33]:

1. Split both labeled and unlabeled observations into K equal-sized sets, indexed by $(I_k)_{k=1}^K$ and $(I_k')_{k=1}^K$ respectively. Define $J_k := I_k \cup I_k'$, $I_{-k} := \{i \in \{1, \ldots, n\} \setminus I_k\}$, and $J_{-k} := \{1, \ldots, N\} \setminus J_k$.

2. For each $k \leq K$, obtain $\hat{\beta}_t^{(-k)}$ using samples $(\boldsymbol{X}_i, Y_i)_{i\in\{i\in I_{-k}:T_i=t\}}$ for each $t \in \{0,1\}$, and $\hat{\pi}^{(-k)}(\cdot)$ using samples $(\boldsymbol{X}_i, Y_i)_{i\in J_{-k}}$.
3. For each $k \leq K$ and $i \in I_k$, define

$$\nu_i := \frac{T_i(Y_i - \boldsymbol{X}_i^T\hat{\beta}_1^{(-k)})}{\hat{\pi}^{(-k)}(\boldsymbol{X}_i)} - \frac{(1-T_i)(Y_i - \boldsymbol{X}_i^T\hat{\beta}_0^{(-k)})}{1-\hat{\pi}^{(-k)}(\boldsymbol{X}_i)},$$
$$:= \boldsymbol{X}_i^T(\hat{\beta}_1^{(-k)} - \hat{\beta}_0^{(-k)}), \quad \text{and} \quad \bar{\xi}^{(k)} := |J_k|^{-1}\sum_{i\in J_k}\xi_i.$$

4. The semi-supervised doubly robust ATE estimator is

$$\hat{\tau}_{\text{SS}} := N^{-1}\sum_{k=1}^{K}\sum_{i\in J_k}\xi_i + n^{-1}\sum_{k=1}^{K}\sum_{i\in I_k}\nu_i.$$

Additionally, let $z_{1-\alpha/2}$ be the $(1-\alpha/2)$-quantile of a standard normal distribution, and $\hat{V}_{\text{SS}} := \sum_{k=1}^{K}\sum_{i\in I_k}((\nu_i - \hat{\tau}_{\text{SS}} + \bar{\xi}^{(k)})^2/n + (\xi_i - \bar{\xi}^{(k)})^2/N)$. Then, an asymptotic $(1-\alpha)$-level confidence interval for τ is defined as

$$(\hat{\tau}_{\text{SS}} - z_{1-\alpha/2}\sqrt{\hat{V}_{\text{SS}}/n},\ \hat{\tau}_{\text{SS}} + z_{1-\alpha/2}\sqrt{\hat{V}_{\text{SS}}/n}).$$

1.3.2 Theoretical Results

The following theorem characterizes the asymptotic behavior of the semi-supervised doubly robust ATE estimator $\hat{\tau}_{\text{SS}}$.

Theorem 1.3 (Theorem 6 of [33]) *Let conditions* (1.1) *and (1) hold. Denote* $C := \text{Var}[\boldsymbol{X}_i]$, $\mu := E[\boldsymbol{X}_i]$, $Z_i := C^{-1/2}(\boldsymbol{X}_i - \mu)$, $\varepsilon_i := Y_i - \boldsymbol{X}_i^T(D_i\beta_1^* + (1-D_i)\beta_0^*)$, *and* $\zeta_i := D_i - \pi(\boldsymbol{X}_i)$. *Let* $\lambda_{\min}(C) > 0$, $\lambda_{\max}(C) \leq c$, $\sup_{\|a\|_2=1} E|a^T Z|^{2+c} < c$, $E|Y|^{2+c} < c$, *and* $\pi(\boldsymbol{X}), \hat{\pi}^{(-k)}(\boldsymbol{X}) \in [c', 1-c']$ *almost surely for positive constants* $c, c' > 0$. *Suppose that* $a_{n,d}$, $b_{N,d}$, *and* c_d *are nonnegative sequences satisfying*

$$E_X\left[\boldsymbol{X}_i^T(\hat{\beta}_t^{(-k)} - \beta_t^*)\right]^2 = O_p(a_{n,p}^2),$$
$$E_X\left[\hat{\pi}^{(-k)}(\boldsymbol{X}_i) - \pi(\boldsymbol{X}_i)\right]^2 = O_p(b_{N,p}^2),$$
$$E_X\left[m_t(\boldsymbol{X}_i) - \boldsymbol{X}_i^T\beta_t^*\right]^2 = O_p(c_p^2),$$

for each $k \leq K$ *and* $t \in \{0,1\}$. *Then, whenever* $a_{n,p} = O(1)$, *as* $n, d \to \infty$,

$$\hat{\tau}_{\text{SS}} - \tau = O_p(n^{-1/2} + a_{n,p}b_{N,p} + b_{N,p}c_p).$$

Moreover, if $a_{n,p} + b_{N,p} = o(1)$ *and* $a_{n,p}b_{N,p} + b_{N,p}c_p = o(n^{-1/2})$, *then*

$$\sqrt{n}V_{\mathrm{SS}}^{-1/2}(\hat{\tau}_{\mathrm{SS}} - \tau) \xrightarrow{d} N(0, 1) \ \ \textit{and} \ \ \hat{V}_{\mathrm{SS}} = V_{\mathrm{SS}} + o_p(1),$$

provided that $V_{\mathrm{SS}} := \mathrm{Var}[\varepsilon_i\zeta_i/(\pi(\boldsymbol{X}_i)(1 - \pi(\boldsymbol{X}_i)))] + \tau(\beta_1^* - \beta_0^*)^T C(\beta_1^* - \beta_0^*) > c''$ *with some constant* $c'' > 0$.

As per Theorem 1.3, $\hat{\tau}_{\mathrm{SS}}$ is consistent as long as either $a_{n,p} + c_p = o(1)$ or $b_{N,p} = o(1)$, i.e., either the propensity score or outcome regression model is consistently estimated. Additionally, to ensure the root-n consistency and asymptotic normality, we require the product rate condition $b_{N,p}(a_{n,p} + c_p) = o(n^{-1/2})$. Since the convergence rate $b_{N,p}$ depends on the total sample size N and is potentially faster than $n^{-1/2}$ when m is large enough, valid inference can be established even if nonparametric methods are performed, and the outcome model is misspecified (i.e., $c_p \neq o(1)$). In comparison, the asymptotic normality of the supervised doubly robust (or double machine learning) estimator proposed by [34] requires all the nuisance models to be correctly specified. Additionally, even when asymptotic normality is achieved, the supervised estimator has an asymptotic variance $V_{\mathrm{sup}} := \mathrm{Var}[\varepsilon_i\zeta_i/(\pi(\boldsymbol{X}_i)(1 - \pi(\boldsymbol{X}_i)))] + (\beta_1^* - \beta_0^*)^T C(\beta_1^* - \beta_0^*) \geq V_{\mathrm{SS}}$, where the strict inequality holds as long as $\tau = \lim_{N,n\to\infty} n/N < 1$ and $\beta_1^* \neq \beta_0^*$ when $\lambda_{\min}(C)$ is bounded below. Under such scenarios, the semi-supervised estimator is more efficient than the supervised version.

Additionally, when the dimension d is relatively low compared with the supervised sample size n, nonparametric methods for the outcome regression models can be considered by replacing the linear projection $\boldsymbol{X}_i^T\hat{\beta}_t^{(-k)}$ with any general $\hat{m}_t^{(-k)}(\boldsymbol{X}_i)$:

$$\begin{aligned}\hat{\tau}_{\mathrm{SS,gen}} := {} & N^{-1}\sum_{k=1}^{K}\sum_{i\in J_k}(\hat{m}_1^{(-k)}(\boldsymbol{X}_i) - \hat{m}_0^{(-k)}(\boldsymbol{X}_i)) \\ & + n^{-1}\sum_{k=1}^{K}\sum_{i\in J_k}\left(\frac{T_i(Y_i - \hat{m}_1^{(-k)}(\boldsymbol{X}_i))}{\hat{\pi}^{(-k)}(\boldsymbol{X}_i)} - \frac{(1 - T_i)(Y_i - \hat{m}_0^{(-k)}(\boldsymbol{X}_i))}{1 - \hat{\pi}^{(-k)}(\boldsymbol{X}_i)}\right).\end{aligned}$$

1.4 Causal Inference Based on Neural Networks

This section considers a general treatment model with a multi-valued treatment and an increasing dimension of confounders. To distinguish the binary treatment in previous sections, we use D to denote a multi-valued treatment variable taking value in $\mathcal{D} = \{0, 1, ..., J\}$, where $J \geq 1$ is a positive integer. Let $Y(d)$ denote the potential outcome when the treatment status $D = d$ is assigned. The probability density of $Y(d)$ exists, denoted by $f_{Y(d)}$, is continuously differentiable. Let $\mathcal{L}(\cdot)$ denote a nonnegative and strictly convex loss function satisfying $\mathcal{L}(0) = 0$ and $\mathcal{L}(v) \geq 0$ for all $v \in \mathbb{R}$. The derivative of $\mathcal{L}(\cdot)$ exists almost everywhere and non-constant which is denoted by

$\mathcal{L}'(\cdot)$. Let $\boldsymbol{\beta}^* = (\beta_0^*, \beta_1^*, \ldots, \beta_J^*)^\top \in \mathbb{R}^{J+1}$ be the parameter of interest which is uniquely identified through the following optimization problem:

$$\boldsymbol{\beta}^* := \arg\min_{\boldsymbol{\beta}} \sum_{d=0}^{J} \mathbb{E}\left[\mathcal{L}\left(Y(d) - \beta_d\right)\right], \tag{1.19}$$

where $\boldsymbol{\beta} = (\beta_0, \beta_1, ..., \beta_J)^\top \in \mathbb{R}^{J+1}$ and $J \in \mathbb{N}$. The formulation (1.19) permits various definitions of treatment effect (TE) parameters, some of which have been considered in the literature. For example,

- $\mathcal{L}(v) = v^2$ and $J = 1$, then $\beta_0^* = \mathbb{E}[Y(0)]$ and $\beta_1^* = \mathbb{E}[Y(1)]$, and $\beta_1^* - \beta_0^*$ is the average treatment effects (ATE) studied by [7, 16, 26] and many others. When $J \geq 2$, then $\beta_d^* = \mathbb{E}[Y(d)]$ is the multi-valued ATE first studied by [35].
- $\mathcal{L}(v) = v \cdot \{\tau - \mathbb{1}(v \leq 0)\}$ for some $\tau \in (0, 1)$ and $J = 1$, then $\beta_0^* = F_{Y(0)}^{-1}(\tau)$ and $\beta_1^* = F_{Y(1)}^{-1}(\tau)$, and $\beta_1^* - \beta_0^*$ is the quantile treatment effects (QTE, [28, 36, 37]).
- $\mathcal{L}(v) = v^2 \cdot |\tau - \mathbb{1}(v \leq 0)|$ is the asymmetric least square treatment effects (ALSTE, [38]). ALSTE estimators have properties analogue to QTE estimators, but they are easier to compute. ALSTE has a variety of applications, such as the study of racial/ethnic disparities in health care, in which the data are often skewed.

The problem with (1.19) is that the potential outcomes $(Y(0), Y(1), ..., Y(J))$ cannot all be observed. The observed outcome is denoted by $Y := Y(D) = \sum_{d=0}^{J+1} \mathbb{1}(D = d)Y(d)$. One may attempt to solve the problem:

$$\min_{\boldsymbol{\beta}} \sum_{d=0}^{J} \mathbb{E}\left[\mathcal{L}\left(Y - \beta_d\right)\right].$$

However, due to the selection in treatment, the true value $\boldsymbol{\beta}^*$ is not the solution of the above problems. To address this problem, most literature imposes the following *unconfoundedness* condition [39]:

Assumption 10 For each $d \in \mathcal{D}$, $Y(d) \perp D|\boldsymbol{X}$.

This condition is also maintained in our work. Nevertheless, we depart from the classical semiparametric estimation and inference for various TEs by allowing the dimension p of the confounders $\boldsymbol{X}$ to grow with sample size N. Specifically, we work with triangular array data $\{((D_{i,N}, \boldsymbol{X}_{i,N}, Y_{i,N}), i = 1, ..., N), N = 1, 2, ...\}$ defined on some common probability space $(\Omega, \mathcal{A}, \mathbb{P})$. Each $\boldsymbol{X}_{i,N}$ is a vector whose dimension p_N may grow with N, the support of $\boldsymbol{X}_{i,N}$ is assumed to be $[0, 1]^{p_N}$. For each given N, these vectors are independent across i, but not necessarily identically distributed. The law $\mathbb{P}_N$ of $\{(D_{i,N}, \boldsymbol{X}_{i,N}, Y_{i,N}), i = 1, ..., N\}$ can change with N, though we do not make explicit use of $\mathbb{P}_N$. Thus, all parameters (including p_N) that characterize the distribution of $\{(D_{i,N}, \boldsymbol{X}_{i,N}, Y_{i,N}), i = 1, ..., N\}$ are implicitly indexed by the sample size N, but we omit the index n in what follows to simplify notation.

1.4.1 ANN-IPW Estimator for General TEs

Under Assumption 10, the causal parameters $\boldsymbol{\beta}^*$ can be identified by the minimizer of the following optimization problem:

$$\boldsymbol{\beta}^* = \arg\min_{\boldsymbol{\beta}} \sum_{d=0}^{J} \mathbb{E}\left[\frac{\mathbb{1}(D_i = d)}{\pi_d^*(\boldsymbol{X}_i)} \mathcal{L}(Y_i - \beta_d)\right], \tag{1.20}$$

where $\pi_d^*(\boldsymbol{X}_i) := \mathbb{P}(D_i = d|\boldsymbol{X}_i)$ is the propensity score (PS) function which is unknown in practice.

Based on (1.20), existing approaches rely on parametric or nonparametric estimation of the PS function $\pi_d^*(\cdot)$. Parametric methods suffer from model misspecification problems, while conventional nonparametric methods, such as linear sieve or kernel regression, fail to work if the dimension of covariates p is large which is known as the "curse of dimensionality." The goal of this article is to efficiently estimate $\boldsymbol{\beta}^*$ under this general framework when the dimension of covariates p is large, and it possibly increases as the sample size N grows. We propose to estimate the PS function $\pi_d^*(\cdot)$ using feedforward ANNs with one hidden layer described below.

All three ANNs can be applied to estimate the PS function $\pi_d^*(\cdot)$, and the resulting TE estimators have the same asymptotic properties based on the three ANNs. For convenience of presentation, we use the sigmoid type ANNs to present the theoretical results in this section. To facilitate our subsequent statistical applications, we allow $r = r_N$ and $B = B_N$ to depend on sample size N. We denote the resulting ANN sieve space as

$$\mathcal{G}_N := \mathcal{G}(\psi, B_N, r_N, p).$$

Denote $D_{di} := \mathbb{1}(D_i = d)$ for brevity. Let $L(a) := \exp(a)/(1 + \exp(a))$, for $a \in \mathbb{R}$, be the logistic function. The inverse logistic transform of the true PS is defined by

$$g_d^*(x) := L^{-1}\left(\pi_d^*(x)\right) = \log\left\{\pi_d^*(x)/(1 - \pi_d^*(x))\right\},$$

and it satisfies $\mathbb{E}[\ell_d(D_{di}, \boldsymbol{X}_i; g_d^*)] \geq \mathbb{E}[\ell_d(D_{di}, \boldsymbol{X}_i; g_d)]$ for all $g_d \in \mathcal{G}_N$, where

$$\begin{aligned}\ell_d(D_{di}, \boldsymbol{X}_i; g_d) :=& D_{di} \log L\left(g_d(\boldsymbol{X}_i)\right) + \{1 - D_{di}\} \log\left(1 - L\left(g_d(\boldsymbol{X}_i)\right)\right) \\ =& D_{di} g_d(\boldsymbol{X}_i) - \log\left[1 + \exp(g_d(\boldsymbol{X}_i))\right].\end{aligned}$$

Let $\widehat{g}_d$ be the ANN estimator of g_d^* based on the space $\mathcal{G}_N$, i.e.,

$$L_{d,N}(\widehat{g}_d) \geq \sup_{g_d \in \mathcal{G}_N} L_{d,N}(g_d) - O(\epsilon_N^2), \tag{1.21}$$

where $L_{d,N}(g_d) := N^{-1} \sum_{i=1}^{N} \ell_d(D_{di}, \boldsymbol{X}_i; g_d)$ is the empirical criterion, and $\epsilon_N = o(N^{-1/2})$.

The ANN estimator of g_d^* depends on the sample size N. For notational simplicity, we omit the index N. (1.21) states that the ANN estimator $\widehat{g}_d$ of g_d^* does not need to be the global maximizer of the objective function $L_{d,N}(g_d)$, which may not be obtained in practice. It can be any local solutions satisfying (1.21), i.e., the values of the objective function evaluated at the local solutions and at the global maximizer cannot be far away from each other, and their difference needs to satisfy the order $O(\epsilon_N^2)$. This assumption is also imposed for sieve extreme estimation; see [40–42]. The estimator of π_d^* is defined by $\widehat{\pi}_d := L(\widehat{g}_d)$, then we use the empirical version of (1.20) to construct the estimator of $\boldsymbol{\beta}^*$, denoted by $\widehat{\boldsymbol{\beta}} = (\widehat{\beta}_0, ..., \widehat{\beta}_J)^\top$ where

$$\widehat{\beta}_d := \arg\min_{\beta\in\Theta} \frac{1}{N}\sum_{i=1}^{N} \frac{D_{di}}{\widehat{\pi}_d(\boldsymbol{X}_i)}\mathcal{L}\left(Y_i - \beta\right), \tag{1.22}$$

for every $d \in \mathcal{D} = \{0, 1, ..., J\}$. $\widehat{\boldsymbol{\beta}}$ is called the artificial neural networks-based inverse probability weighting (ANN-IPW) estimator.

1.4.2 ANN-OR Estimator for ATE

In this subsection, we consider an alternative estimator for a particularly important parameter of interest, ATE, which corresponds to a loss function $\mathcal{L}(v) = v^2$. Using Assumption 10 and the property of conditional expectation, we can rewrite (1.19) as follows:

$$\boldsymbol{\beta}^* = \arg\min_{\boldsymbol{\beta}} \sum_{d=0}^{J} \mathbb{E}\left[\mathbb{E}[\mathcal{L}(Y_i - \beta_d)|\boldsymbol{X}_i, D_i = d]\right]. \tag{1.23}$$

Based on the above expression, an alternative estimation strategy for $\boldsymbol{\beta}^*$ is to first estimate the conditional expectation $\mathbb{E}[\mathcal{L}(Y_i - \beta_d)|\boldsymbol{X}_i, D_i = d]$ (with β_d being fixed), and then estimate $\boldsymbol{\beta}^*$ by minimizing the empirical version of (1.23) with $\mathbb{E}[\mathcal{L}(Y_i - \beta_d)|\boldsymbol{X}_i, D_i = d]$ replaced by its estimate.

In this case, $\beta_d^* = \mathbb{E}[Y(d)] = \mathbb{E}[z_d^*(\boldsymbol{X}_i)]$, where $z_d^*(\boldsymbol{X}_i) := \mathbb{E}[Y_i|\boldsymbol{X}_i, D_i = d]$ is the outcome regression (OR) function and satisfies $\mathbb{E}[\ell_d^{OR}(D_{di}, \boldsymbol{X}_i, Y_i; z_d^*)] \geq \mathbb{E}[\ell_d^{OR}(D_{di}, \boldsymbol{X}_i, Y_i; z_d)]$ for all $z_d \in \mathcal{G}_N$, where

$$\ell_d^{OR}(D_{di}, \boldsymbol{X}_i, Y_i; z_d) := -D_{di}\left\{Y_i - z_d(\boldsymbol{X}_i)\right\}^2.$$

Let $\widehat{z}_d$ be the ANN estimator of z_d^* based on the space $\mathcal{G}_N$, i.e.,

$$L_{d,N}^{OR}(\widehat{z}_d) \geq \sup_{z_d\in\mathcal{G}_N} L_{d,N}^{OR}(z_d) - O(\epsilon_N^2), \tag{1.24}$$

where $L_{d,N}^{OR}(z_d) := N^{-1}\sum_{i=1}^{n} \ell_d^{OR}(D_{di}, \boldsymbol{X}_i, Y_i; z_d)$ is the empirical criterion, and $\epsilon_N = o(N^{-1/2})$. Then the ANN-OR estimator of β_d^* is defined to be

$$\widehat{\beta}_d^{OR} = \frac{1}{N}\sum_{i=1}^{N} \widehat{z}_d(\boldsymbol{X}_i). \tag{1.25}$$

1.4.3 Large Sample Properties of Estimators

Properties of the ANN-IPW Estimator for General TEs

We first introduce sufficient conditions for the convergence rates of our ANN estimators $\{\widehat{\pi}_d\}_{d=0}^{J}$ for the unknown PS nuisance functions.

Assumption 11 For every $d \in \mathcal{D} = \{0, 1, ..., J\}$ and $m \geq 1$, we assume $g_d^*(\cdot) \in \mathcal{F}_p^m$.

Assumption 12 (i) The dimension of $\boldsymbol{X}_i$ is denoted by $p \in \mathbb{N}$ and the number of hidden units is denoted by $r_N \in \mathbb{N}$. They satisfy

$$\max\left\{ v_{g_d^*,m} \cdot r_N^{-\frac{1}{2}-\frac{1}{p}}, \sqrt{\frac{r_N \cdot p \cdot \log N}{N}} \right\} = o\left(N^{-\frac{1}{4}}\right).$$

(ii) The bound of the hidden-to-output weights, B_N, satisfies $B_N \leq 2v_{g_d^*,m}$.

Assumption 11 is a smoothness condition imposed on the transformed PS functions. Assumption 12 (i) allows the dimension of covariates going to infinity as the sample size grows, while it imposes restrictions on the growth rate of the dimension of covariates and that of the number of hidden units to ensure that the $L^2(dF_X)$-convergence rate of estimated PS attains $o_P(N^{-1/4})$, which is needed to establish the $\sqrt{N}$-asymptotic normality for the proposed TE estimator.

The following result establishes the convergence rates of g_d^* and π_d^*.

Theorem 1.4 *Suppose Assumptions 11 and 12 hold. Then*

$$\|\widehat{g}_d - g_d^*\|_{L^2(dF_X)} = O_P\left(\max\left\{ v_{g_d^*,m} \cdot r_N^{-\frac{1}{2}-\frac{1}{p}}, \sqrt{\frac{r_N \cdot p \cdot \log N}{N}} \right\}\right) = o_P\left(N^{-1/4}\right),$$

and

$$\|\widehat{\pi}_d - \pi_d^*\|_{L^2(dF_X)} = O_P\left(\max\left\{ v_{g_d^*,m} \cdot r_N^{-\frac{1}{2}-\frac{1}{p}}, \sqrt{\frac{r_N \cdot p \cdot \log N}{N}} \right\}\right) = o_P\left(N^{-1/4}\right),$$

where the constants hiding inside O_P and o_P do not depend on p and N.

The proof of Theorem 1.4 is given in the supplemental material of [43]. Theorem 1.4 shows that under a suitable smoothness condition, the M-estimates based on ANNs with a single hidden layer circumvent the curse of dimensionality and achieve a desirable rate for establishing the asymptotic normality of plug-in estimators [44]. Bauer and Kohler [45] showed that their least squares estimator based on multilayer neural networks with a smooth activation function can achieve the convergence rate of $N^{-2s/(2s+d^*)}$ (up to a log factor), if the regression function satisfies a s-smooth generalized hierarchical interaction model of order d^*, where d^* is fixed. Schmidt-Hieber [46] established a similar rate for the ReLU activation function. However, the target function class considered in [45, 46] is different from that used in our paper. The extension of our results for multilayer neural networks is beyond the scope of the current article. We refer to the Supplement for more discussion.

Let $\mathcal{E}_d(x, \beta_d^*) := \mathbb{E}[\mathcal{L}'(Y_i(d) - \beta_d^*)|\boldsymbol{X}_i = x]$, $u_d^*(x) := \mathcal{E}_d(x; \beta_d^*)/\pi_d^*(x)$, and $\overline{g}(g_d, \epsilon_N) := (1 - \epsilon_N) \cdot g_d + \epsilon_n \cdot \{u_d^* + g_d^*\}$ be a local alternative value around $g_d \in \mathcal{G}_N$. The directional derivative of $\ell_d(D_{di}, \boldsymbol{X}_i; g_d)$ is given by

$$\begin{aligned}\frac{\partial}{\partial g_d}\ell_d(D_{di}, \boldsymbol{X}_i; g_d)[u] &:= \lim_{t\to 0} \frac{\ell_d(D_{di}, \boldsymbol{X}_i; g_d + t \cdot u) - \ell_d(D_{di}, \boldsymbol{X}_i; g_d)}{t} \\ &= \{D_{di} - L(g_d(\boldsymbol{X}_i))\}\, u(\boldsymbol{X}_i), \quad \text{for } u \in L^2(dF_X).\end{aligned}$$

We now introduce sufficient conditions and additional notation for the asymptotic normality of our ANN-IPW estimators $\widehat{\boldsymbol{\beta}}$ for the general TE parameters.

Assumption 13 (i) Let Θ be a compact set of $\mathbb{R}^{J+1}$ containing the true parameters $\boldsymbol{\beta}^*$. (ii) The propensity scores are uniformly bounded away from zero, i.e., there exists a constant $\underline{c}$ such that $0 < \underline{c} \le \pi_d^*(x)$ for all $x \in \mathcal{X}$ and $d \in \{0, 1, ..., J\}$. (iii) For every $d \in \{0, 1, ..., J\}$ and $m \ge 1$, we assume the function $\mathcal{E}_d(\cdot, \beta_d^*)$ is uniformly bounded.

Assumption 14 (*Approximation error*) We assume the following conditions hold:

$$\sup_{\{g_d \in \mathcal{G}_N : \|g_d - g_d^*\|_{L^2(dF_X)} \le \delta_N\}} \|\text{Proj}_{\mathcal{G}_N} \overline{g}(g_d, \epsilon_N) - \overline{g}(g_d, \epsilon_N)\|_{L^2(dF_X)} = O\left(\frac{\epsilon_N^2}{\delta_N}\right),$$

and

$$\begin{aligned}&\sup_{\{g_d \in \mathcal{G}_N : \|g_d - g_d^*\|_{L^2(dF_X)} \le \delta_N\}} \frac{1}{N}\sum_{i=1}^N \left(\frac{\partial}{\partial g_d}\ell_d(D_{di}, \boldsymbol{X}_i; g_d^*)[\overline{g}(g_d, \epsilon_n) - \text{Proj}_{\mathcal{G}_N} \overline{g}(g_d, \epsilon_N)]\right) \\ &= o_P(\epsilon_N^2),\end{aligned}$$

where $\text{Proj}_{\mathcal{G}_N} \overline{g}(g_d, \epsilon_N)$ denotes the $L^2(dF_X)$-projection of $\overline{g}(g_d, \epsilon_n)$ on the ANN space $\mathcal{G}_N$ and δ_N is a sequence of positive real numbers satisfying $\|\widehat{g}_d - g_d^*\|_{L^2(dF_X)} = o_P(\delta_N)$.

Assumption 15

1. There exists a finite positive constant $\kappa \geq 1/2$ such that for any $\beta \in \Theta$ and any $\delta > 0$ in a neighborhood of zero,

$$\left\{\mathbb{E}\left[\sup_{\widetilde{\beta}:|\widetilde{\beta}-\beta|<\delta}\left\{\mathcal{L}'(Y-\widetilde{\beta})-\mathcal{L}'(Y-\beta)\right\}^2\right]\right\}^{1/2} \leq \text{const} \times \delta^{\kappa};$$

2. $\sup_{\beta\in\Theta}\mathbb{E}\left[|\mathcal{L}'(Y-\beta)|^2\right] < \infty$ and $\mathbb{E}\left[\sup_{\beta\in\Theta}|\mathcal{L}'(Y-\beta)|\right] < \infty$;
3. $\sup_{x\in\mathcal{X}}\mathbb{E}\left[|\mathcal{L}'(Y-\beta_d^*)||\boldsymbol{X}=x\right] < C < \infty$ for some finite constant $C > 0$;
4. $H_d := -\partial_{\beta_d}\mathbb{E}[\mathcal{L}'(Y(d)-\beta_d^*)] > 0$.

Assumption 13 (i) is a standard condition for the parameter space. Assumption 13 (ii) is a strict overlap condition ensuring the existence of participants at all treatment levels, which is commonly assumed in the literature. D'Amour et al. [47] discussed the applicability of the strict overlap condition with high-dimensional covariates, and provided a variety of circumstances under which this condition holds. They also argued that the strict overlap condition may not be necessary if other smoothness conditions are imposed on the potential outcomes, or it can be technically relaxed with some non-standard asymptotic analyses (e.g., [48, 49]) and the sacrifice of uniform inference on ATE. Assumption 13 (iii) is a smoothness condition for approximation. The functions $\{\pi_d^*(\cdot), \mathcal{E}_d(\cdot, \beta_d^*)\}_{d=0}^J$ generally depend on the sample size n. Assumption 14 specifies both approximation error and stochastic equicontinuity in neural network space. Such a condition is also imposed in [40, 41, 50]. Assumption 15 concerns L^2 continuity and envelope conditions, which are needed for the applicability of the uniform law of large numbers, establishing stochastic equicontinuity and weak convergence, see [28]. Again, they are satisfied by widely used loss functions such as $\mathcal{L}(v) = v^2$, $\mathcal{L}(v) = v\{\tau - \mathbb{1}(v \leq 0)\}$, and $\mathcal{L}(v) = v^2 \cdot |\tau - \mathbb{1}(v \leq 0)|$. Assumption 15 (3) implies $\sup_{x\in\mathcal{X}}|\mathcal{E}(x;\beta_d^*)| < C < \infty$ by Jensen's inequality.

The following theorem shows the asymptotic distribution of the proposed estimator $\widehat{\boldsymbol{\beta}}$, whose proof is given in the supplemental material of [43].

Theorem 1.5 *Under Assumptions 10–15, for any $d \in \{0, 1, .., J\}$, we have $\widehat{\beta}_d \xrightarrow{p} \beta_d^*$ and*

$$\sqrt{N}(\widehat{\beta}_d - \beta_d^*) = H_d^{-1} \cdot \frac{1}{\sqrt{N}}\sum_{i=1}^N S_d(Y_i, D_{di}, \boldsymbol{X}_i; \beta_d^*) + o_P(1), \tag{1.26}$$

where $H_d = -\partial_{\beta_d}\mathbb{E}[\mathcal{L}'(Y(d)-\beta_d^*)]$ and

$$S_d = S_d(Y_i, D_{di}, \boldsymbol{X}_i; \beta_d^*) := \frac{D_{di}}{\pi_d^*(\boldsymbol{X}_i)}\mathcal{L}'\{Y_i - \beta_d^*\} - \left\{\frac{D_{di} - \pi_d^*(\boldsymbol{X}_i)}{\pi_d^*(\boldsymbol{X}_i)}\right\}\mathcal{E}_d(\boldsymbol{X}_i, \beta_d^*).$$

Consequently,

$$V^{-1/2} \cdot \sqrt{N} \left\{ \widehat{\boldsymbol{\beta}} - \boldsymbol{\beta}^* \right\} \xrightarrow{d} \mathcal{N}\left(0, I_{(J+1)\times(J+1)}\right),$$

where $I_{(J+1)\times(J+1)}$ *is the* $(J+1)\times(J+1)$ *identity matrix,* $\boldsymbol{V} = \boldsymbol{H}^{-1}\mathbb{E}[\boldsymbol{S}\boldsymbol{S}^\top]\boldsymbol{H}^{-1}$, $\boldsymbol{H} = Diag\{H_0, ..., H_J\}$ *and* $\boldsymbol{S} = (S_0, ..., S_J)^\top$.

Based on the strict overlap condition and the integrability of the outcome, Assumptions 13 (ii) and 15 (ii), we have that the asymptotic variance is finite, which implies that the proposed estimator $\widehat{\boldsymbol{\beta}}$ is $\sqrt{N}$-consistent. In addition, when p is a fixed number, our estimator attains the semiparametric efficiency bound given in [51].

1.4.4 Property of the ANN-OR Estimator for ATE

The asymptotic normality of the ANN-IPW estimator requires the strict overlap condition, i.e., Assumption 13 (ii). In this section, we prove that such a condition can be possibly relaxed for the ANN-OR estimator of ATE defined in (1.25). From both the theoretical analysis and the numerical comparison in Sect. 7.3.5, we recommend the use of ANN-OR estimator for estimating ATE in practice and the use of ANN-IPW estimator for estimating other types of causal effects such as QTE.

Let $w_d^*(x) := f_X(x)/f_{X|D}(x|d)$, $z_d^*(x) = \mathbb{E}[Y|\boldsymbol{X} = x, D = d]$, and $\overline{z}(z_d, \epsilon_N) := (1-\epsilon_N)\cdot z_d + \epsilon_N \cdot \{w_d^* + z_d^*\}$ be a local alternative value around $z_d \in \mathcal{G}_N$.

Assumption 16 For every $d \in \mathcal{D} = \{0, 1, ..., J\}$ and $m \geq 1$, we assume $z_d^*(\cdot) \in \mathcal{F}_p^m$.

Assumption 17 (i) We assume

$$\max\left\{ v_{z_d^*,m} \cdot r_N^{-\frac{1}{2}-\frac{1}{p}}, \sqrt{\frac{r_N \cdot p \cdot \log N}{N}} \right\} = o\left(N^{-\frac{1}{4}}\right).$$

(ii) The bound of the hidden-to-output weights, B_N, satisfies $B_N \leq 2v_{z_d^*,m}$.

Assumption 18 (i) $\mathbb{P}(D_{di} = 1) \in (0, 1)$ and $\pi_d^*(\boldsymbol{X}) \in (0, 1)$; (ii) There exists a constant $\bar{c}$ such that

$$\mathbb{E}\left[\{w_d^*(\boldsymbol{X})\}^2\right] = \mathbb{E}\left[\left\{\frac{f_X(\boldsymbol{X})}{f_{X|D}(\boldsymbol{X}|d)}\right\}^2\right] < \bar{c} < \infty.$$

Assumption 19 (*Approximation error*) We assume the following conditions hold:

$$\sup_{\{z_d \in \mathcal{G}_N : \|z_d - z_d^*\|_{L^2(dF_X)} \leq \delta_n\}} \|\mathrm{Proj}_{\mathcal{G}_N}\overline{z}(z_d, \epsilon_N) - \overline{z}(z_d, \epsilon_N)\|_{L^2(dF_X)} = O\left(\frac{\epsilon_N^2}{\delta_N}\right), \text{ and}$$

$$\sup_{\{z_d \in \mathcal{G}_N : \|z_d - z_d^*\|_{L^2(dF_X)} \leq \delta_N\}} \frac{1}{N}\sum_{i=1}^N \left(\overline{z}(z_d, \epsilon_N)(\boldsymbol{X}_i) - \mathrm{Proj}_{\mathcal{G}_N}\overline{z}(z_d, \epsilon_N)(\boldsymbol{X}_i)\right) = O_P(\epsilon_N^2),$$

where $\mathrm{Proj}_{\mathcal{G}_N}\bar{z}(z_d, \epsilon_N)$ denotes the $L^2(dF_X)$-projection of $\bar{z}(z_d, \epsilon_N)$ on the ANN space $\mathcal{G}_N$.

Assumption 20 $\sup_{x\in\mathcal{X}} \mathbb{E}[\{Y(d)\}^2|\boldsymbol{X} = x] < C < \infty$ for some finite constant $C > 0$.

Assumptions 16–20 are comparable to Assumptions 11–15. It's worth noting that Assumption 18 does not restrict the propensity score $\pi_d^*(\boldsymbol{X})$ to be uniformly bounded below by a constant.

Theorem 1.6 *Under Assumptions 10, 16–20, for every $d \in \{0, 1, ..., J\}$, we have $\widehat{\beta}_d^{OR} \xrightarrow{p} \beta_d^*$ and*

$$\sqrt{N}(\widehat{\beta}_d^{OR} - \beta_d^*) = \frac{1}{\sqrt{N}} \sum_{i=1}^{N} S_d^{OR}(Y_i, D_{di}, \boldsymbol{X}_i; \beta_d^*) + o_P(1),$$

where

$$S_d^{OR} = S_d^{OR}(Y_i, D_{di}, \boldsymbol{X}_i; \beta_d^*) = \frac{D_{di}}{\pi_d^*(\boldsymbol{X}_i)} Y_i - \left\{ \frac{D_{di} - \pi_d^*(\boldsymbol{X}_i)}{\pi_d^*(\boldsymbol{X}_i)} \right\} \cdot z_d^*(\boldsymbol{X}_i) - \mathbb{E}[z_d^*(\boldsymbol{X}_i)].$$

Consequently,

$$\left\{\boldsymbol{V}^{OR}\right\}^{-1/2} \cdot \sqrt{N} \left\{\widehat{\boldsymbol{\beta}}^{OR} - \boldsymbol{\beta}^*\right\} \xrightarrow{d} \mathcal{N}\left(0, I_{(J+1)\times(J+1)}\right),$$

where $I_{(J+1)\times(J+1)}$ is the $(J+1) \times (J+1)$ identity matrix, $\boldsymbol{V}^{OR} = \mathbb{E}[\boldsymbol{S}^{OR} \cdot (\boldsymbol{S}^{OR})^\top]$, and $\boldsymbol{S}^{OR} = (S_0^{OR}, ..., S_J^{OR})^\top$.

The proof of Theorem 1.6 is given in the supplemental material of [43]. Note that $1/\pi_d^*(\boldsymbol{X}) = w_d^*(\boldsymbol{X})/\mathbb{P}(D_{di} = 1)$, with Assumptions 18 and 20, we have that the asymptotic variance is finite, which implies the proposed estimator $\widehat{\boldsymbol{\beta}}^{OR}$ is $\sqrt{N}$-consistent. Moreover, the ANN-OR estimator $\widehat{\boldsymbol{\beta}}^{OR}$ has the same asymptotic variance as the ANN-IPW estimator $\widehat{\boldsymbol{\beta}}$ when $\mathcal{L}(v) = v^2$ for ATE. We can take the same inferential strategies as given in Sect. 1.4.5 to conduct inference based on the ANN-OR estimator.

1.4.5 Statistical Inference

This section presents a weighted bootstrap procedure to conduct statistical inference for $\boldsymbol{\beta}^*$. Our TE estimator is obtained from directly optimizing an objective function, so a weighted bootstrap procedure can be performed to conduct inference without the need of estimating the asymptotic variance function. Estimation of the variance function can be challenging in the quantile TE setting. We discuss a possible method for the estimation of the asymptotic variance based on the asymptotic formula given in Theorem 1.5.

Let $\{\omega_{d1}, ..., \omega_{dN}\}$ be $i.i.d.$ positive random weights that are independent of data satisfying $\mathbb{E}[\omega_{di}] = 1$ and $Var(\omega_{di}) = 1$, where $d \in \{0, 1, ..., J\}$. The weighted bootstrap estimator of the inverse logistic PS g_d^* is defined by satisfying

$$L_{d,N}^B(\widehat{g}_d^B) \geq \sup_{g_d \in \mathcal{G}_N} L_{d,N}^B(g_d) - O(\epsilon_N^2),$$

where $L_{d,N}^B(g_d) := N^{-1}\sum_{i=1}^N \omega_{di}\ell_d(D_{di}, \boldsymbol{X}_i; g_d(\cdot))$ is the bootstrapped empirical criterion, and $\epsilon_N = o(N^{-1/2})$. Let $\widehat{\pi}_d^B := L(\widehat{g}_d^B)$. Then the weighted bootstrap IPW estimator of β_d^* is given by

$$\widehat{\beta}_d^B = \arg\min_{\beta\in\Theta} \frac{1}{N}\sum_{i=1}^N \frac{\omega_{di} D_{di}}{\widehat{\pi}_d^B(\boldsymbol{X}_i)} \mathcal{L}\left(Y_i - \beta\right), \ d \in \{0, 1, ..., J\}.$$

The weighted bootstrap OR estimator of ATE can be derived similarly. The weighted bootstrap estimator of the OR function z_d^* is defined by satisfying

$$L_{d,N}^{OR,B}(\widehat{z}_d^B) \geq \sup_{z_d \in \mathcal{G}_N} L_{d,N}^{OR,B}(z_d) - O(\epsilon_N^2),$$

where $L_{d,N}^{OR,B}(z_d) := N^{-1}\sum_{i=1}^N \omega_{di}\ell_d^{OR}(D_{di}, \boldsymbol{X}_i, Y_i; z_d(\cdot))$. Then the weighted bootstrap OR estimator of $\mathbb{E}[Y(d)]$ is given by

$$\widehat{\beta}_d^{OR,B} = \frac{1}{N}\sum_{i=1}^N \omega_{d,i}\widehat{z}_d^B(\boldsymbol{X}_i), \ d \in \{0, 1, ..., J\}.$$

Let $\widehat{\boldsymbol{\beta}}^B := (\widehat{\beta}_0^B, ..., \widehat{\beta}_{J+1}^B)^\top$ and $\widehat{\boldsymbol{\beta}}^{OR,B} := (\widehat{\beta}_0^{OR,B}, ..., \widehat{\beta}_{J+1}^{OR,B})^\top$. The following theorem justifies the validation of the proposed bootstrap inference whose proof is presented in the supplemental material of [43].

Theorem 1.7 *(i) Under Assumptions 10–15, for any $d \in \{0, 1, .., J\}$, then conditionally on the data we have*

$$\boldsymbol{V}^{-1/2} \cdot \sqrt{N}\left(\widehat{\boldsymbol{\beta}}^B - \widehat{\boldsymbol{\beta}}\right) \xrightarrow{d} \mathcal{N}(0, I_{(J+1)\times(J+1)}).$$

(ii) Under Assumptions 10, 12, 16–20, for any $d \in \{0, 1, .., J\}$, then conditionally on the data we have

$$\{\boldsymbol{V}^{OR}\}^{-1/2} \cdot \sqrt{N}\left(\widehat{\boldsymbol{\beta}}^{OR,B} - \widehat{\boldsymbol{\beta}}^{OR}\right) \xrightarrow{d} \mathcal{N}(0, I_{(J+1)\times(J+1)}).$$

Possible Challenge of Applying the EIF-Based Method to Quantile TE Estimation

The EIF can be applied to different loss functions. When EIF is given, the estimator of β_d^* can be obtained by solving the estimated efficient score function [52]. For example, when the loss function $\mathcal{L}(v) = v^2$ corresponding to ATE, the EIF of $\beta_d^* = \mathbb{E}[Y(d)]$ is

$$\frac{D_{di}}{\pi_d^*(\boldsymbol{X}_i)}Y_i - \left\{\frac{D_{di}}{\pi_d^*(\boldsymbol{X}_i)} - 1\right\}\mathbb{E}[Y_i|D_{di} = 1, \boldsymbol{X}_i] - \beta_d^*. \quad (1.27)$$

It involves the PS function $\pi_d^*(x)$ and the OR function $\mathbb{E}[Y_i|D_{di} = 1, \boldsymbol{X}_i = x]$ that can be estimated separately. As a result, the ATE of β_d^* can be obtained with the estimated PS and OR functions directly plug into the function given in (1.27).

When the loss function $\mathcal{L}(v) = v \cdot \{\tau - \mathbb{1}(v \leq 0)\}$ corresponding to the τth-quantile TE, the specific form of EIF for $\beta_d^* = F_{Y^*(d)}^{-1}(\tau)$ can also be derived from $H_d^{-1} S_d(Y_i, D_{di}, \boldsymbol{X}_i; \beta_d^*)$. As a result, its estimator can be obtained from solving the estimated efficient score equation

$$\sum_{i=1}^{N}\left[\frac{D_{di}}{\widehat{\pi}_d(\boldsymbol{X}_i)}\{\tau - \mathbb{1}(Y_i \leq \beta)\} - \left\{\frac{D_{di}}{\widehat{\pi}_d(\boldsymbol{X}_i)} - 1\right\}\{\tau - \widehat{\mathbb{E}}[\mathbb{1}(Y_i \leq \beta)|D_{di} = 1, \boldsymbol{X}_i]\}\right] = 0, \quad (1.28)$$

where $\widehat{\pi}_d(x)$ and $\widehat{\mathbb{E}}[\mathbb{1}(Y_i \leq \beta)|D_{di} = 1, \boldsymbol{X}_i = x]$ are estimates of $\pi_d^*(x)$ and $\mathbb{E}[\mathbb{1}(Y_i \leq \beta)|D_{di} = 1, \boldsymbol{X}_i = x]$, respectively. We can see that the estimation of quantile TEs from (1.28) is challenging when ANNs or other nonlinear machine learning methods are employed to obtain $\widehat{\mathbb{E}}[\mathbb{1}(Y_i \leq \beta)|D_{di} = 1, \boldsymbol{X}_i = x]$, as it intertwines with the unknown quantile TE parameter β nonlinearly.

Different from the aforementioned estimators constructed based on the estimated EIF, our TE estimators are directly obtained from optimizing an objective function that only involves the ANN-based estimated PS function. This approach greatly facilitates the computation of obtaining TE estimates and conducting causal inference without the need to estimate the EIF.

1.4.6 Extension to the General Treatment Effect on the Treated

The above results can be easily extended to other multi-valued causal parameters defined on the treated subgroup. Let

$$\boldsymbol{\beta}_{d'}^* := \arg\min_{\boldsymbol{\beta}} \sum_{d=0}^{J} \mathbb{E}\left[\mathcal{L}\left(Y(d) - \beta_d\right)|D = d'\right], \quad (1.29)$$

for some fixed $d' \in \{0, 1, ..., J\}$, where $\boldsymbol{\beta} = (\beta_0, \beta_1, ..., \beta_J)$ and $\boldsymbol{\beta}^*_{d'} = (\beta^*_{0,d'}, \beta^*_{1,d'}, ..., \beta^*_{J,d'})$. The formulation (1.29) includes the following important cases discussed in [53]:

- $\mathcal{L}(v) = v^2$, then $\beta^*_{d,d'} = \mathbb{E}[Y(d)|D = d']$ is the average treatment effects on the treated.
- $\mathcal{L}(v) = v\{\tau - \mathbb{1}(v \leq 0)\}$, then $\beta^*_{d,d'} = F^{-1}_{Y(d)|D}(\tau|d')$ is the τth quantile of $Y(d)$ conditioned on the treated group $\{D = d'\}$.

Under Assumption 10, using the property of conditional expectation, the parameter of interest $\boldsymbol{\beta}^*_{d'}$ is identified by

$$
\begin{aligned}
\boldsymbol{\beta}^*_{d'} :&= \arg\min_{\boldsymbol{\beta}} \sum_{d=0}^{J} \frac{1}{p_{d'}} \mathbb{E}\left[\mathbb{1}(D = d')\mathcal{L}\left(Y(d) - \beta_d\right)\right] \\
&= \arg\min_{\boldsymbol{\beta}} \sum_{d=0}^{J} \frac{1}{p_{d'}} \mathbb{E}\left[\pi^*_{d'}(\boldsymbol{X}) \cdot \mathbb{E}[\mathcal{L}\left(Y(d) - \beta_d\right)|\boldsymbol{X}]\right] \\
&= \arg\min_{\boldsymbol{\beta}} \sum_{d=0}^{J} \frac{1}{p_{d'}} \mathbb{E}\left[\pi^*_{d'}(\boldsymbol{X}) \cdot \mathbb{E}\left[\mathcal{L}\left(Y(d) - \beta_d\right)|\boldsymbol{X}\right] \cdot \mathbb{E}\left[\frac{\mathbb{1}(D = d)}{\pi^*_d(X)}\middle|\boldsymbol{X}\right]\right] \\
&= \arg\min_{\boldsymbol{\beta}} \sum_{d=0}^{J} \frac{1}{p_{d'}} \cdot \mathbb{E}\left[\mathbb{1}(D = d) \cdot \frac{\pi^*_{d'}(\boldsymbol{X})}{\pi^*_d(\boldsymbol{X})} \cdot \mathcal{L}\left(Y - \beta_d\right)\right],
\end{aligned}
$$

where $p_{d'} := \mathbb{P}(D = d')$. The estimator of $\boldsymbol{\beta}^*_{d'}$ is obtained by minimizing the empirical analogue of the above equation:

$$
\widehat{\boldsymbol{\beta}}_{d'} = \arg\min_{\boldsymbol{\beta}} \sum_{d=0}^{J} \frac{\sum_{i=1}^{N} D_{di}\widehat{\pi}_{d'}(\boldsymbol{X}_i)\mathcal{L}\left(Y_i - \beta_d\right)/\widehat{\pi}_d(\boldsymbol{X}_i)}{\sum_{i=1}^{N} D_{d'i}},
$$

where $\widehat{\pi}_d$ is the ANN estimator of π^*_d. The estimator of $\beta^*_{d,d'}$ for $d \in \{0, 1, ..., J\}$ can be defined as

$$
\widehat{\beta}_{d,d'} = \arg\min_{\beta\in\Theta} \frac{1}{N} \sum_{i=1}^{N} \frac{D_{di}}{\widehat{\pi}_d(\boldsymbol{X}_i)} \widehat{\pi}_{d'}(\boldsymbol{X}_i)\mathcal{L}\left(Y_i - \beta\right).
$$

Similar to the proof of Theorem 1.5, we obtain the following result for $\boldsymbol{\beta}^*_{d'}$. The detailed proof is presented in the supplemental material of [43].

Theorem 1.8 *Under Assumptions 10–15, for any $d, d' \in \{0, 1, .., J\}$, we have that*

$$
\sqrt{N}(\widehat{\beta}_{d,d'} - \beta^*_{d,d'}) = H^{-1}_{d,d'} \cdot \frac{1}{\sqrt{N}} \sum_{i=1}^{N} S_{d,d'}(\boldsymbol{X}_i, D_{di}, Y_i; \beta^*_{d,d'}) + o_P(1),
$$

where $H_{d,d'} = -\partial_{\beta_d}\mathbb{E}[\pi^*_{d'}(\boldsymbol{X}_i)\mathcal{L}'(Y_i(d) - \beta^*_d)]$ *and*

$$S_{d,d'} := \frac{D_{di}}{\pi^*_d(\boldsymbol{X}_i)}\pi^*_{d'}(\boldsymbol{X}_i)\mathcal{L}'(Y_i - \beta^*_{d,d'}) - \left\{\frac{D_{di}}{\pi^*_d(\boldsymbol{X}_i)}\pi^*_{d'}(\boldsymbol{X}_i) - D_{d'i}\right\}\mathcal{E}_d(\boldsymbol{X}_i, \beta^*_{d,d'}).$$

1.4.7 Numerical Results

In this section, we illustrate the finite sample performance of our proposed methods via simulations in which we generate data from models in Sect. 1.4.7. Our proposed IPW estimator can be applied to various types of treatment effects. We use ATE, ATT (average treatment effects on the treated), QTE and QTT (quantile treatment effects on the treated) for illustration of the performance of the IPW estimator. For QTE and QTT, we consider the 25th (Q1), 50th (Q2), and 75th (Q3) percentiles. We also illustrate the performance of the OR estimator for ATE and ATT. To obtain the IPW and OR estimators, we estimate the PS and OR functions by using our proposed ANN method as well as five other popular methods, including the generalized linear models (GLM), the generalized additive models (GAM), the random forests (RF), the gradient boosted machines (GBM), and the deep neural networks with three hidden layers (DNN). We make a comparison of the performance of the resulting TE estimators with the nuisance functions estimated by the aforementioned six methods. Moreover, we compare our IPW and OR estimators with the doubly robust (DR) estimator [54] and the Oracle estimator for ATE. For the DR estimator, the IPW and OR functions are also approximated by ANNs. The Oracle estimator is constructed based on the efficient influence function with the true PS and OR functions plugged in, see [7]. The Oracle estimators are infeasible in practice, but they are expected to perform the best for the estimation of ATE, and serve as a benchmark to compare with. In the quantile TE settings, both DR (EIF-based) and OR estimators are difficult to obtain, so we only show the performance of the IPW estimator.

We use the Rectified Linear Unit (ReLU) as the activation function for both ANN and DNN. We use cubic regression spline basis functions for GAM. We apply grid search with fivefold cross-validation to select hyperparameters for all methods, including the number of neurons for ANN DNN, the number of trees and max depths of trees for RF and GBM, and the learning rate for GBM. All the simulation studies are implemented in Python 3.9. The DNN, GLM, GAM, RF, and GBM methods are implemented using the packages TensorFlow, statsmodels, pyGAM, and scikit-learn, respectively.

Data Generating Process

We generate the treatment and outcome variables from a nonlinear model and a linear model, respectively, given as follows.

Model 1 (nonlinear model):

$$\begin{aligned}
\text{logit}\{\mathbb{E}(D_i|\boldsymbol{X}_i)\} &= 0.5(X_{i1}^*X_{i2}^* - 0.7sin((X_{i3}^* + X_{i4}^*)(X_{i5}^* - 0.2)) - 0.1),\\
\mathbb{E}(Y_i(1)|\boldsymbol{X}_i) &= \mathbb{E}(Y_i|\boldsymbol{X}_i, D_i = 1) = 0.3(X_{i1}^* - 0.9)^2 + 0.1(X_{i2}^* - 0.5)^2\\
&\quad - 0.6X_{i2}^*X_{i3}^* + sin(-1.7(X_{i1}^* + X_{i3}^* - 1.1) + X_{i4}^*X_{i5}^*) + 1,\\
\mathbb{E}(Y_i(0)|\boldsymbol{X}_i) &= \mathbb{E}(Y_i|\boldsymbol{X}_i, D_i = 0) = 0.64(X_{i1}^* - 0.9)^2 + 0.16(X_{i2}^* + 0.2)^2\\
&\quad - 0.6X_{i2}^*X_{i3}^* + sin(-1.7(X_{i1}^* + X_{i3}^* - 1.1) + X_{i4}^*X_{i5}^*) - 1;
\end{aligned}$$

Model 2 (linear model) :

$$\begin{aligned}
\text{logit}\{\mathbb{E}(D_i|\boldsymbol{X}_i)\} &= 0.1(X_{i1}^* + X_{i2}^* - 2X_{i3}^* + 3X_{i4}^* - 3X_{i5}^*),\\
\mathbb{E}(Y_i(1)|\boldsymbol{X}_i) &= \mathbb{E}(Y_i|\boldsymbol{X}_i, D_i = 1) = 4X_{i1}^* + 3X_{i2}^* - X_{i3}^* - 5X_{i4}^* + 7X_{i5}^* + 1,\\
\mathbb{E}(Y_i(0)|\boldsymbol{X}_i) &= \mathbb{E}(Y_i|\boldsymbol{X}_i, D_i = 0) = 4X_{i1}^* + 3X_{i2}^* - X_{i3}^* - 5X_{i4}^* + 7X_{i5}^* - 1,
\end{aligned}$$

where $X_{ij'}^* = c_p \frac{5}{p} \sum_{j=p(j'-1)/5+1}^{pj'/5} X_{ij}$ for $1 \le j' \le 5, 1 \le i \le N$, and $Y_i(d) = \mathbb{E}(Y_i(d) \mid \boldsymbol{X}_i) + \epsilon_i, d = \{0, 1\}, \epsilon_i \overset{i.i.d.}{\sim} \mathcal{N}(0, 1)$ for $1 \le i \le N$.

We generate the confounders from $X_{ij} = 2(F(Z_{ij}) - 0.5)$, where $Z_i = (Z_{i1}, ..., Z_{ip})^\top \overset{i.i.d.}{\sim} \mathcal{N}(0, \boldsymbol{\Sigma})$, $\boldsymbol{\Sigma} = \{\sigma_{kk'}\}$, $\sigma_{kk'} = 0.2^{|k-k'|}$ for $1 \le k, k' \le p$, and $F(\cdot)$ is the cumulative distribution function of the standard normal for $1 \le i \le n, 1 \le j \le p$. Let c_p=1. We partition the confounders into 5 subgroups, and $X_{ij'}^*$ is the average of the $p/5$ confounders in the j'-th subgroup for $j' = 1, ..., 5$, so that every confounder is included in our models. We consider p = 5, 10 and N = 1000, 2000, 5000. All simulation results are based on 400 realizations.

We also use the nonlinear model (Model 1) to illustrate the performance of our proposed methods for $p = 100$ and $N = 2000$, with the confounders $X_{ij} = 2(F(Z_{ij}/\sigma_j) - 0.5)$, where σ_j is the standard deviation of Z_{ij}, and $Z_i = (Z_{i1}, ..., Z_{ip})^\top$ are generated from two designs.

- Design 1 (factor model): $Z_{ij} = F_i^\top L_j + \eta_{ij}$, where $F_i \overset{i.i.d.}{\sim} \mathcal{N}(0, \boldsymbol{\Sigma}^*)$, $\boldsymbol{\Sigma}^* = \{\sigma_{kk'}\}$, in which $\sigma_{kk'} = 0.2^{|k-k'|}$ for $1 \le k, k' \le 10$, L_j is a constant vector kept fixed for each realization and is generated from $L_j \overset{i.i.d.}{\sim} \mathcal{N}(0, \boldsymbol{\Sigma}^*)$, and $\eta_{ij} \overset{i.i.d.}{\sim} \mathcal{N}(0, 1)$. Let c_p=7.
- Design 2 (multivariate normal): $Z_i \overset{i.i.d.}{\sim} \mathcal{N}(0, \boldsymbol{\Sigma})$, $\boldsymbol{\Sigma} = \{\sigma_{kk'}\}$, $\sigma_{kk'} = 0.2^{m(|k-k'|)}$ for $1 \le k, k' \le p$, where $m(x) = \lceil x/10 \rceil$, and $\lceil a \rceil$ denotes the smallest integer no less than a. Let c_p=4.

Simulation Results

To compare the performance of different methods for estimating the TEs, we report the following statistics: the absolute value of bias (bias), the empirical standard deviation (emp_sd), the average value of the estimated standard deviations based on the asymptotic formula (est_sd) and obtained from the weighted bootstrapping (est_sd_boot), and the empirical coverage rates of the 95% confidence intervals based on the estimated asymptotic standard deviations (cover_rate) and the weighted bootstrapping method (cover_rate_boot). The 95% confidence intervals based on bootstrapping are obtained from the 2.5th percentile and 97.5th percentile of the weighted bootstrapping estimates. The bootstrap confidence intervals and the estimated standard deviations (est_sd_boot) are obtained based on 400 bootstrap replicates for each simulation sample. The bootstrap weights are randomly generated from the exponential distribution with mean 1 according to [55].

Tables 1.6 and 1.7 report the numerical results for different estimators of ATE for Model 1 with $p = 5$, 10, respectively. We see that as n increases, the empirical coverage rates (cover_rate and cover_rate_boot) based on our proposed ANN-based IPW and OR estimates become closer to the nominal level 95%. The biases are close to zero, and the values of emp_sd, est_sd, and est_sd_boot decrease as n increases. These results corroborate our asymptotic theories. We observe that our ANN-based IPW and OR estimators have comparable performance to the DR and the Oracle estimators when estimating ATE. The proposed ANN-based OR estimator slightly outperforms the ANN-based IPW and DR estimators in the sense that it has the smallest emp_sd value. It is possible that the estimated PS functions have a few values close to zero. This can affect the emp_sd value of the IPW estimate for ATE. The DR estimator which is constructed based on the estimates of both IPW and OR functions has larger emp_sd values than the OR estimator, but it yields smaller emp_sd values than the IPW estimator. Our numerical results suggest that the proposed ANN-based OR estimator is preferred for the estimation of ATE. However, in practice, it can be difficult to construct OR and DR estimators for other types of TEs, such as quantile TEs. Then the proposed ANN-based IPW estimator becomes a more appealing tool. Moreover, our numerical results given in Tables 1.8 and 1.9 show that the performance of the ANN-based IPW estimators for quantile TEs is less influenced by the small values of the estimated PS functions because of the robustness of the quantile objective functions. For our proposed ANN-based IPW and OR estimators, it is convenient to apply the proposed weighted bootstrap procedure for conducting inference. We find that the empirical coverage rates of 95% confidence intervals obtained from the weighted bootstrapping are closer to the nominal level than those obtained from the estimated asymptotic standard deviations.

Next, we compare the performance of different machine learning (ML) methods for the estimation of ATE. We see that the GLM and GAM methods yield large estimation biases due to the model misspecification problem. Our numerical results show that the proposed ANN method outperforms the other two ML methods, RF and GBM, for the estimation of TEs. The empirical coverage rates based on the ANN method are closer to the nominal level in all cases than the rates obtained from RF and GBM. It is worth noting that our ANN-based TE estimators enjoy the properties of root-n consistency and semiparametric efficiency. In general, our numerical results corroborate those theoretical properties. Moreover, for RF and GBM, the OR estimator also performs better than the IPW estimator for ATE estimation. The empirical coverage rates of the 95% confidence intervals obtained from the weighted bootstrapping are improved compared to the rates obtained from the estimated asymptotic standard deviation. The DNN method has comparable performance to the ANN method.

Table 1.6 The summary statistics of the estimated ATEs for Model 1 with p = 5

	IPW						OR						DR	Oracle
	ANN	GLM	GAM	RF	GBM	DNN	**ANN**	GLM	GAM	RF	GBM	DNN		
n = 1000														
Bias	**0.0008**	0.0584	0.0594	0.0523	0.0562	0.0015	**0.0010**	0.0586	0.0574	0.0151	0.0144	0.0012	0.0026	0.0006
emp_sd	**0.0758**	0.0907	0.0924	0.0830	0.0844	0.0789	**0.0713**	0.0906	0.0930	0.0769	0.0828	0.0726	0.0752	0.0668
est_sd	**0.0701**	0.0923	0.0903	0.0511	0.0483	0.0715	**0.0701**	0.0923	0.0903	0.0511	0.0483	0.0715	0.0701	0.0686
cover_rate	**0.9275**	0.9025	0.8750	0.6700	0.6400	0.9250	**0.9425**	0.9050	0.8900	0.8000	0.7450	0.9425	0.9325	0.9600
est_sd_boot	**0.0771**	0.0901	0.0922	0.0821	0.0853	0.0791	**0.0751**	0.0902	0.0924	0.0773	0.0841	0.0737		
cover_rate_boot	**0.9375**	0.8850	0.0888	0.9025	0.8950	0.9375	**0.9475**	0.8875	0.8875	0.9275	0.9350	0.9500		
n = 2000														
Bias	**0.0007**	0.0578	0.0585	0.0495	0.0425	0.0008	**0.0009**	0.0591	0.0571	0.0127	0.0131	0.0010	0.0007	0.0010
emp_sd	**0.0524**	0.0684	0.0675	0.0620	0.0634	0.0563	**0.0499**	0.0685	0.0677	0.0564	0.0575	0.0502	0.0523	0.0498
est_sd	**0.0488**	0.0652	0.0637	0.0371	0.0370	0.0508	**0.0488**	0.0652	0.0637	0.0371	0.0370	0.0508	0.0493	0.0485
cover_rate	**0.9375**	0.8175	0.8100	0.6250	0.6500	0.9175	**0.9500**	0.8175	0.8050	0.7875	0.7575	0.9475	0.9475	0.9475
est_sd_boot	**0.0573**	0.0679	0.0665	0.0612	0.0645	0.0599	**0.0493**	0.0685	0.0697	0.0563	0.0572	0.0505		
cover_rate_boot	**0.9475**	0.8275	0.8400	0.8400	0.8575	0.9400	**0.9500**	0.8275	0.8225	0.9200	0.9075	0.9500		
n = 5000														
Bias	**0.0001**	0.0558	0.0545	0.0407	0.0225	0.0002	**0.0003**	0.0558	0.0545	0.0043	0.0073	0.0003	0.0009	0.0010
emp_sd	**0.0334**	0.0397	0.0390	0.0362	0.0376	0.0335	**0.0312**	0.0397	0.0389	0.0324	0.0327	0.0305	0.0322	0.0309
est_sd	**0.0309**	0.0413	0.0403	0.0244	0.0258	0.0307	**0.0309**	0.0413	0.0403	0.0244	0.0258	0.0307	0.0306	0.0307
cover_rate	**0.9350**	0.7350	0.7175	0.5700	0.7400	0.9250	**0.9475**	0.7375	0.7150	0.8500	0.8450	0.9525	0.9400	0.9600
est_sd_boot	**0.0331**	0.0402	0.0401	0.0355	0.0365	0.0337	**0.0308**	0.0399	0.0402	0.0327	0.0331	0.0305		
cover_rate_boot	**0.9475**	0.7350	0.7225	0.7850	0.8725	0.9475	**0.9575**	0.7350	0.7250	0.9325	0.9250	0.9525		

Table 1.7 The summary statistics of the estimated ATEs for Model 1 with p = 10

	IPW						OR						DR	Oracle
	ANN	GLM	GAM	RF	GBM	DNN	**ANN**	GLM	GAM	RF	GBM	DNN		
n = 1000														
Bias	**0.0010**	0.0635	0.0655	0.0574	0.0536	0.0010	**0.0064**	0.0637	0.0640	0.0264	0.0279	0.0037	0. 0073	0.0077
emp_sd	**0.0855**	0.0900	0.1127	0.0887	0.0915	0.0859	**0.0793**	0.0901	0.1041	0.0825	0.0832	0.0785	0.0823	0.0666
est_sd	**0.0789**	0.0943	0.0982	0.0547	0.0512	0.0784	**0.0789**	0.0943	0.0982	0.0547	0.0512	0.0784	0.0789	0.0695
cover_rate	**0.9250**	0.9075	0.8675	0.6725	0.6500	0.9275	**0.9400**	0.9075	0.8800	0.7775	0.7375	0.9450	0.9325	0.9550
est_sd_boot	**0.0884**	0.0923	0.1089	0.0893	0.0921	0.0864	**0.0798**	0.0913	0.0965	0.0833	0.0856	0.0778		
cover_rate_boot	**0.9350**	0.8950	0.8825	0.8925	0.8950	0.9375	**0.9475**	0.8950	0.8650	0.9150	0.9125	0.9450		
n = 2000														
Bias	**0.0034**	0.0659	0.0661	0.0584	0.0462	0.0032	**0.0001**	0.0660	0.0675	0.0227	0.0266	0.0009	0.0012	0.0050
emp_sd	**0.0597**	0.0660	0.0692	0.0634	0.0629	0.0657	**0.0543**	0.0661	0.0683	0.0588	0.0590	0.0519	0.0558	0.0501
est_sd	**0.0532**	0.0668	0.0667	0.0397	0.0382	0.0539	**0.0532**	0.0668	0.0667	0.0397	0.0382	0.0539	0.0530	0.0492
cover_rate	**0.9250**	0.8275	0.8225	0.6050	0.6400	0.9250	**0.9400**	0.8250	0.8275	0.7825	0.7475	0.9650	0.9400	0.9500
est_sd_boot	**0.0604**	0.0668	0.0687	0.0624	0.0628	0.0649	**0.0560**	0.0664	0.0679	0.0590	0.0595	0.0530		
cover_rate_boot	**0.9475**	0.8250	0.8250	0.8125	0.8375	0.9425	**0.9500**	0.8275	0.8350	0.9025	0.8875	0.9525		
n = 5000														
Bias	**0.0015**	0.0679	0.0670	0.0565	0.0362	0.0012	**0.0001**	0.0679	0.0670	0.0158	0.0177	0.0009	0. 0017	0.0020
emp_sd	**0.0363**	0.0439	0.0439	0.0416	0.0419	0.0357	**0.0332**	0.0440	0.0439	0.0376	0.0376	0.0319	0.0345	0.0319
est_sd	**0.0322**	0.0422	0.0418	0.0258	0.0267	0.0319	**0.0322**	0.0422	0.0418	0.0258	0.0267	0.0319	0.0323	0.0311
cover_rate	**0.9275**	0.6100	0.5975	0.4225	0.6150	0.9250	**0.9425**	0.6125	0.6000	0.7875	0.7700	0.9500	0.9475	0.9550
est_sd_boot	**0.0368**	0.0441	0.0443	0.0420	0.0423	0.0349	**0.0333**	0.0440	0.0442	0.0376	0.0378	0.0323		
cover_rate_boot	**0.9475**	0.6550	0.6475	0.6575	0.7875	0.9450	**0.9525**	0.6250	0.6175	0.8275	0.8150	0.9525		

Table 1.8 The summary statistics of the estimated QTEs by the IPW method for Model 1 with p = 5

	Q1						Q2						Q3					
	ANN	GLM	GAM	RF	GBM	DNN	**ANN**	GLM	GAM	RF	GBM	DNN	**ANN**	GLM	GAM	RF	GBM	DNN
n = 1000																		
Bias	**0.0045**	0.0601	0.0589	0.0630	0.0684	0.0036	**0.0012**	0.0534	0.0574	0.0507	0.0539	0.0025	**0.0098**	0.0363	0.0407	0.0286	0.0302	0.0067
emp_sd	**0.1266**	0.1358	0.1418	0.1274	0.1301	0.1235	**0.1075**	0.1116	0.1166	0.1055	0.1074	0.1099	**0.1196**	0.1151	0.1201	0.1121	0.1130	0.1206
est_sd	**0.1414**	0.1403	0.1596	0.1398	0.1405	0.1404	**0.1238**	0.1240	0.1447	0.1241	0.1249	0.1222	**0.1251**	0.1232	0.1412	0.1237	0.1247	0.1243
cover_rate	**0.9500**	0.9275	0.9375	0.9350	0.9250	0.9525	**0.9675**	0.9550	0.9500	0.9675	0.9650	0.9575	**0.9600**	0.9650	0.9650	0.9650	0.9625	0.9625
est_sd_boot	**0.1259**	0.1378	0.1432	0.1275	0.1305	0.1278	**0.1091**	0.1154	0.1171	0.1087	0.1072	0.1093	**0.1231**	0.1131	0.1241	0.1172	0.1187	0.1236
cover_rate_boot	**0.9350**	0.9225	0.9300	0.9275	0.9175	0.9375	**0.9600**	0.9500	0.9475	0.9500	0.9475	0.9500	**0.9575**	0.9450	0.9500	0.9575	0.9550	0.9600
n = 2000																		
Bias	**0.0046**	0.0604	0.0560	0.0631	0.0646	0.0037	**0.0004**	0.0547	0.0523	0.0454	0.0467	0.0014	**0.0056**	0.0494	0.0506	0.0372	0.0369	0.0043
emp_sd	**0.0928**	0.1029	0.1012	0.0944	0.0944	0.0954	**0.0819**	0.0883	0.0884	0.0828	0.0844	0.0831	**0.0852**	0.0872	0.0888	0.0848	0.0860	0.0864
est_sd	**0.0900**	0.0986	0.0976	0.0984	0.0989	0.0933	**0.0791**	0.0870	0.0872	0.0871	0.0875	0.0804	**0.0814**	0.0860	0.0871	0.0862	0.0867	0.0824
cover_rate	**0.9325**	0.8650	0.8750	0.9000	0.8875	0.9425	**0.9350**	0.9050	0.9050	0.9300	0.9225	0.9325	**0.9550**	0.9100	0.9175	0.9425	0.9350	0.9375
est_sd_boot	**0.0911**	0.1014	0.1023	0.0954	0.0968	0.0961	**0.0811**	0.0889	0.0892	0.0857	0.0853	0.0842	**0.0841**	0.0872	0.0891	0.0857	0.0871	0.0858
cover_rate_boot	**0.9400**	0.8675	0.8925	0.8750	0.8650	0.9450	**0.9425**	0.9100	0.9025	0.9125	0.9200	0.9500	**0.9375**	0.9150	0.9200	0.9375	0.9350	0.9450
n = 5000																		
Bias	**0.0034**	0.0594	0.0560	0.0456	0.0255	0.0029	**0.0033**	0.0489	0.0478	0.0367	0.0188	0.0036	**0.0026**	0.0454	0.0460	0.0320	0.0176	0.0014
emp_sd	**0.0560**	0.0600	0.0591	0.0555	0.0559	0.0570	**0.0514**	0.0545	0.0539	0.0520	0.0522	0.0517	**0.0494**	0.0513	0.0508	0.0497	0.0503	0.0490
est_sd	**0.0558**	0.0624	0.0607	0.0624	0.0639	0.0531	**0.0493**	0.0546	0.0537	0.0546	0.0557	0.0510	**0.0508**	0.0542	0.0540	0.0542	0.0551	0.0513
cover_rate	**0.9575**	0.8625	0.8600	0.9125	0.9575	0.9575	**0.9400**	0.8675	0.8675	0.9200	0.9525	0.9475	**0.9625**	0.8900	0.8850	0.9300	0.9575	0.9675
est_sd_boot	**0.0560**	0.0614	0.0603	0.0576	0.0578	0.0560	**0.0509**	0.0553	0.5420	0.0534	0.0539	0.0509	**0.5010**	0.0517	0.0515	0.0508	0.0512	0.5010
cover_rate_boot	**0.9575**	0.8575	0.8550	0.8825	0.9325	0.9575	**0.9500**	0.8725	0.8725	0.9075	0.9500	0.9500	**0.9600**	0.8575	0.8550	0.9200	0.9325	0.9625

Table 1.9 The summary statistics of the estimated QTEs by the IPW method for Model 1 with p = 10

	Q1						Q2						Q3					
	ANN	GLM	GAM	RF	GBM	DNN	**ANN**	GLM	GAM	RF	GBM	DNN	**ANN**	GLM	GAM	RF	GBM	DNN
n = 1000																		
Bias	**0.0019**	0.0717	0.0750	0.0700	0.0696	0.0023	**0.0047**	0.0544	0.0541	0.0507	0.0472	0.0033	**0.0030**	0.0488	0.0476	0.0394	0.0367	0.0036
emp_sd	**0.1585**	0.1345	0.1634	0.1325	0.1364	0.1592	**0.1437**	0.1195	0.1485	0.1179	0.1206	0.1437	**0.1332**	0.1185	0.1529	0.1193	0.1208	0.1331
est_sd	**0.2530**	0.1444	0.4770	0.1442	0.1487	0.2630	**0.2639**	0.1269	0.4708	0.1269	0.1312	0.2684	**0.2034**	0.1253	0.4458	0.1255	0.1294	0.2109
cover_rate	**0.9675**	0.9200	0.9950	0.9350	0.9375	0.9675	**0.9425**	0.9450	1.0000	0.9475	0.9525	0.9450	**0.9575**	0.9325	1.0000	0.9350	0.9500	0.9575
est_sd_boot	**0.1623**	0.1345	0.1552	0.1376	0.1425	0.1687	**0.1523**	0.1231	0.1253	0.1198	0.1225	0.1523	**0.1376**	0.1203	0.1623	0.1198	0.1256	0.1392
cover_rate_boot	**0.9325**	0.9150	0.9125	0.9300	0.9250	0.9350	**0.9350**	0.9350	0.9375	0.9400	0.9475	0.9500	**0.9325**	0.9225	0.9375	0.9000	0.9225	0.9350
n = 2000																		
Bias	**0.0030**	0.0663	0.0678	0.0632	0.0535	0.0023	**0.0015**	0.0576	0.0580	0.0529	0.0421	0.0005	**0.0105**	0.0613	0.0620	0.0527	0.0426	0.0087
emp_sd	**0.1070**	0.1040	0.1083	0.1008	0.1016	0.1109	**0.0937**	0.0902	0.0943	0.0874	0.0872	0.0954	**0.0873**	0.0855	0.0890	0.0833	0.0830	0.0923
est_sd	**0.1057**	0.1021	0.1149	0.1019	0.1052	0.1134	**0.0919**	0.0889	0.1026	0.0889	0.0917	0.0939	**0.0913**	0.0878	0.1001	0.0878	0.0904	0.0919
cover_rate	**0.9425**	0.8775	0.8850	0.8900	0.9150	0.9475	**0.9200**	0.8900	0.9175	0.9100	0.9250	0.9350	**0.9425**	0.9125	0.9325	0.9275	0.9400	0.9525
est_sd_boot	**0.1072**	0.1043	0.1097	0.1011	0.1045	0.1122	**0.0932**	0.0901	0.0994	0.0883	0.0892	0.0952	**0.0885**	0.0861	0.0923	0.0869	0.0873	0.0925
cover_rate_boot	**0.9450**	0.8800	0.8000	0.8900	0.9125	0.9450	**0.9325**	0.8975	0.9150	0.9100	0.9175	0.9450	**0.9375**	0.9050	0.9125	0.9250	0.9275	0.9575
n = 5000																		
Bias	**0.0065**	0.0774	0.0764	0.0696	0.0472	0.0055	**0.0046**	0.0637	0.0633	0.0547	0.0356	0.0051	**0.0021**	0.0537	0.0535	0.0433	0.0273	0.0031
emp_sd	**0.0657**	0.0682	0.0680	0.0650	0.0656	0.0634	**0.0568**	0.0580	0.0584	0.0560	0.0567	0.0532	**0.0554**	0.0555	0.0562	0.0549	0.0563	0.0531
est_sd	**0.0629**	0.0644	0.0638	0.0644	0.0659	0.0615	**0.0537**	0.0556	0.0555	0.0556	0.0568	0.0531	**0.0542**	0.0550	0.0551	0.0549	0.0560	0.0522
cover_rate	**0.9275**	0.7675	0.7725	0.8025	0.8875	0.9275	**0.9350**	0.7800	0.7725	0.8225	0.9075	0.9450	**0.9375**	0.8375	0.8300	0.8650	0.9250	0.9375
est_sd_boot	**0.0655**	0.0686	0.0678	0.0648	0.0658	0.0635	**0.0565**	0.0577	0.0579	0.0552	0.0571	0.0545	**0.0548**	0.0561	0.0568	0.0552	0.0561	0.0528
cover_rate_boot	**0.9325**	0.7825	0.7875	0.8125	0.8875	0.9450	**0.9475**	0.7900	0.7850	0.8200	0.9100	0.9525	**0.9400**	0.8500	0.8525	0.8675	0.9275	0.9425

Tables 1.8 and 1.9 show the numerical results of different methods for the estimation of QTEs for Model 1 with $p = 5$, 10, respectively. It is difficult to construct OR and DR estimators for QTEs, so we only report the results for the IPW estimators, which are very convenient to be obtained in this context. The PS functions are estimated by different ML methods, and the numerical results of the resulting IPW estimates are summarized in Tables 1.8 and 1.9. In general, we observe similar patterns of numerical performance of different methods as shown in Table 1.6 and 1.7. It is worth noting that the proposed ANN-based IPW method has very stable performance for the estimation of QTEs. The resulting emp_sd values are not influenced by possibly small values of the estimated PS functions because of the robustness nature of the quantile objective function. Moreover, in the QTE settings, estimation of the asymptotic standard deviations can involve a complicated procedure, and several approximations are needed. As a result, the estimation is not guaranteed to perform well. Figure 1.1 shows the boxplots of the estimated asymptotic standard deviations of QTE (Q1) for Model 1 with $p = 5, 10, N = 1000$. We see that the estimated values are large for some simulation replicates. In contrast, the estimated standard deviations obtained from the weighted bootstrapping have more reliable performance. In complex TE settings such as QTEs, the proposed weighted bootstrap method that avoids the estimation of the asymptotic variance provides a robust way to conduct statistical inference, and thus it is recommended in practice. It is convenient to apply the weighted bootstrap method in our proposed TE estimation procedure, as the TE estimators are obtained from optimizing a general objective function. We apply different ML methods to estimate the PS function. The numerical results show that the ANN and DNN methods have comparable performance, and they still outperform other methods for the estimation and inference of QTEs.

At last, we evaluate the performance of our proposed TE estimators in the settings with $p = 100$ and $N = 2000$. In this scenario, the number of confounders is very large compared to the sample size, and it does not satisfy the order requirement given in Assumption 4. Note that when dealing with high-dimensional covariates, one often assumes a parametric structure on the regression model and imposes a

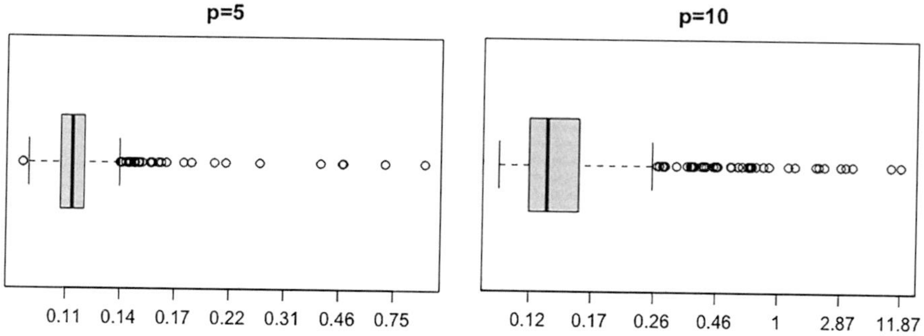

Fig. 1.1 Boxplots of the estimated asymptotic standard deviations of QTE(Q1) for Model 1, $n = 1000$.

sparsity condition such that a small number of covariates are useful for the prediction The sparsity assumption and the parametric structure are not required in our setting. For the purpose of dimensionality reduction, we apply Principal Component Analysis (PCA) to extract the first 20 leading principal components, and use them to estimate the PS and OR functions via ANNs. For comparison, we also use the original covariates matrix without PCA to fit the nuisance models via ANNs. The resulting TE estimators with and without the PCA procedure are called ANN-PCA and ANN, respectively. Tables 1.10 and 1.11 report the summary statistics of the ANN-based TE estimators for ATE, ATT, QTE and QTT for Model 1 with $p = 100$ and $N = 2000$, based on 400 simulation realizations, when the confounders are generated from Designs 1 & 2 given in Sect. 1.4.7. For QTE and QTT, we only report the estimated standard deviations and empirical coverage rates from the weighted bootstrapping, as it is difficult to estimate the asymptotic standard deviations in the quantile settings. The ATE and ATT are estimated by the IPW, OR and DR methods, respectively, while the QTE and QTT are only estimated by the IPW method.

From Table 1.10, for the estimation of ATE and ATT, we see that the empirical coverage rates obtained from all of the three methods, IPW, OR and DR, are smaller than the nominal level 0.95, and the values of bias and emp_sd are larger than those values given in Tables 1.6 and 1.7 for $p = 5, 10$. The ANN-PCA method yields larger biases but smaller emp_sd than the ANN method. The empirical coverage rates from the ANN-PCA method are closer to the nominal level than those from the ANN method for both designs, but they still cannot reach the nominal level. It is expected that these ANN-based methods have inferior performance for $p = 100$ compared to the $p = 5, 10$ settings, as the order assumption on the dimension p required for ANN approximations does not hold anymore when $p = 100$. As a result, the ANN-based estimators of the nuisance functions (OR and PS functions) are not guaranteed to be consistent estimators, yielding deteriorated performance, and those estimates further affect the estimation of ATE and ATT. The formula of est_sd involves the estimates of both OR and PS functions, so it is not surprising that its value is also affected. From Table 1.11 for the estimation of QTE and QTT, we can observe similar patterns as the results in Table 1.10, except that the ANN method has slightly larger empirical coverage rates than the ANN-PCA method. In sum, the TE estimation using ANNs in the context of ultra-high-dimensional covariates is a challenging task. A sparse model assumption may be needed for high-dimensional settings. The investigation of its methodology and theories is beyond the scope of this paper, and it can be an interesting topic to pursue in the future.

1.4.8 Application

In this section, we apply the proposed methods to the data from the National Health and Nutrition Examination Survey (NHANES) to investigate the causal effect of smoking on body mass index (BMI). The collected data consist of 6647 subjects, including 3359 smokers and 3288 nonsmokers. The confounding variables include

Table 1.10 The summary statistics of the estimated ATEs and ATTs for Model 1 with p = 100 and n = 2000

		Design 1				Design 2			
		ATE		ATT		ATE		ATT	
		ANN	ANN-PCA	ANN	ANN-PCA	ANN	ANN-PCA	ANN	ANN-PCA
IPW	bias	0.0104	0.0444	0.0205	0.0516	0.0163	0.0446	0.0197	0.0578
	emp_sd	0.0983	0.0719	0.1293	0.0790	0.1080	0.0776	0.1673	0.1005
	est_sd	0.0512	0.0651	0.0764	0.0789	0.0618	0.0733	0.0782	0.0865
	cover_rate	0.8225	0.8650	0.8025	0.8875	0.8100	0.8800	0.7650	0.8500
	est_sd_boot	0.0694	0.0668	0.0899	0.0796	0.0798	0.0737	0.1036	0.0934
	cover_rate_boot	0.8650	0.8675	0.8350	0.8900	0.8550	0.8925	0.8100	0.8825
OR	bias	0.0097	0.0379	0.0132	0.0356	0.0222	0.0388	0.0193	0.0511
	emp_sd	0.0892	0.0722	0.1163	0.0870	0.0988	0.0760	0.1356	0.0961
	est_sd	0.0512	0.0651	0.0764	0.0789	0.0618	0.0733	0.0782	0.0865
	cover_rate	0.8350	0.8900	0.8475	0.8950	0.7975	0.9150	0.8150	0.8400
	est_sd_boot	0.0625	0.0639	0.0871	0.0847	0.0714	0.0698	0.0939	0.0942
	cover_rate_boot	0.8775	0.8850	0.8575	0.9150	0.8450	0.8950	0.8575	0.8850
DR	bias	0.0101	0.0382	0.0158	0.0457	0.0195	0.0390	0.0204	0.0548
	emp_sd	0.0894	0.0723	0.1207	0.0862	0.0992	0.0765	0.1397	0.0964
	est_sd	0.0521	0.0669	0.0785	0.0807	0.0633	0.0724	0.0801	0.0849
	cover_rate	0.8375	0.8850	0.8500	0.9050	0.8025	0.8975	0.8275	0.8375

Table 1.11 The summary statistics of the estimated QTEs and QTTs by the IPW method for Model 1 with p = 100 and N = 2000

		Design 1				Design 2			
		QTE		QTT		QTE		QTT	
		ANN	ANN-PCA	ANN	ANN-PCA	ANN	ANN-PCA	ANN	ANN-PCA
Q1	bias	0.0157	0.0259	0.0139	0.0313	0.0132	0.0337	0.0111	0.0418
	emp_sd	0.1379	0.1101	0.1917	0.1257	0.1988	0.1210	0.2551	0.1542
	est_sd_boot	0.1305	0.1004	0.1812	0.1234	0.1753	0.1190	0.2236	0.1497
	cover_rate_boot	0.9050	0.8850	0.9125	0.8875	0.9125	0.8775	0.9050	0.8575
Q2	bias	0.0131	0.0283	0.0249	0.0302	0.0154	0.0326	0.0055	0.0386
	emp_sd	0.1256	0.1017	0.1687	0.1123	0.1680	0.1101	0.2056	0.1315
	est_sd_boot	0.1104	0.0977	0.1447	0.1026	0.1544	0.1041	0.1869	0.1278
	cover_rate_boot	0.9225	0.9150	0.9075	0.9125	0.9300	0.9025	0.9225	0.9050
Q3	bias	0.0265	0.0324	0.0446	0.0301	0.0248	0.0354	0.0272	0.0314
	emp_sd	0.1250	0.0966	0.1539	0.1041	0.1784	0.1077	0.2196	0.1192
	est_sd_boot	0.1213	0.0935	0.1495	0.1066	0.1647	0.1003	0.1967	0.1138
	cover_rate_boot	0.9175	0.9100	0.9025	0.8925	0.9200	0.9050	0.9175	0.8975

Table 1.12 Group comparisons

Covariates		Non-smoker (N_{ns}=3288)		Smoker (N_{ns}=3359)		Std. Dif.	p-value
Gender	1 = Male	1404	(41.8%)	2019	(61.41%)	−15.99	<0.001
	0 = Female	1955	(58.2%)	1269	(38.59%)		
Age	Mean(SD)	48.97	(19)	51.73	(17.57)	–6.14	<0.001
Marital	1 = Yes	1989	(59.21%)	1867	(56.78%)	2.01	0.0446
	0 = No	1370	(40.79%)	1421	(43.22%)		
Education	1 = College or above	1626	(48.41%)	1297	(39.45%)	7.36	<0.001
	0 = Less than college	1733	(51.59%)	1991	(60.55%)		
Family PIR	Mean(SD)	2.79	(1.63)	2.57	(1.6)	5.62	<0.001
Alcohol	1 = Yes	1897	(56.48%)	2708	(82.36%)	−22.87	<0.001
	0 = No	1462	(43.52%)	580	(17.64%)		
PHSVIG	1 = Yes	1102	(32.81%)	908	(27.62%)	4.61	<0.001
	0 = No	2257	(67.19%)	2380	(72.38%)		
PHSMOD	1 = Yes	1491	(44.39%)	1376	(41.85%)	2.09	0.0366
	0 = No	1868	(55.61%)	1912	(58.15%)		
SBP	Mean(SD)	126.42	(21.04)	126.63	(19.98)	–0.43	0.6684
DBP	Mean(SD)	72.1	(13.56)	71.61	(14.1)	1.44	0.15

four continuous variables: age, family poverty income ratio (Family PIR), systolic blood pressure (SBP), and diastolic blood pressure (DBP); six binary variables: gender, marital status, education, alcohol use, vigorous activity over past 30 days (PHSVIG), and moderate activity over past 30 days (PHSMOD). Table 1.12 presents the group comparisons of all confounding variables in the full dataset. Mean and standard deviation (SD) are presented for continuous variables, while the count and percentage (%) of observations for each group are presented for categorical variables. Standardized difference(Std. Dif.) is calculated as $(\bar{x}_{ns} - \bar{x}_s)/\sqrt{s_{ns}^2/n_{ns} + s_s^2/n_s}$ for continuous variables, and $(p_{ns} - p_s)/\sqrt{pq/n_{ns} + pq/n_s}$ for categorical variables, where $\bar{x}$, s^2, and p denote sample mean, sample variance, and sample proportion, and the subscripts ns and s refer to nonsmokers and smokers respectively, and p, q are the overall proportions. The last column shows the p-value of group comparison for each covariate. We notice that the smoking group and nonsmoking group differ greatly in their group characteristics. A naive comparison of the sample mean between smoking and nonsmoking groups will lead to a biased estimation of the smoking effects on BMI.

We apply our proposed ANN methods to estimate the PS and OR functions, respectively. We estimate ATE by the proposed IPW and OR methods, and estimate QTE by the IPW method only. The number of neurons is selected using grid search

Table 1.13 The estimates and standard errors of ATE and QTE

	ATE		QTE		
	IPW	OR	Q1	Q2	Q3
Estimate	–0.224	–0.241	–0.400	–0.269	-0.040
est_sd	0.154	0.154	0.157	0.184	0.247
z-value	–1.454	–1.564	–2.547	–1.467	–0.162
p-value	0.073	0.058	0.005	0.071	0.436
est_sd_boot	0.162	0.149	0.156	0.187	0.254
z-value_boot	–1.383	–1.617	–2.564	-1.443	–0.157
p-value_boot	0.083	0.053	0.005	0.074	0.437

with fivefold cross-validation. Table 1.13 reports the estimates of ATE and QTE, the estimated standard deviations based on the asymptotic formula (est_sd) and obtained from the weighted bootstrapping (est_sd_boot), and the corresponding z-values and p-values for testing ATE and QTE. The negative values of the estimates indicate that smoking has adverse effects on BMI. From the numerical results based on the estimated asymptotic standard deviations, we see that the p-values of testing ATE are 0.073 and 0.058 by the IPW and OR methods, respectively. We also notice that the p-value for testing QTE at the 25% quantile is very small, which is 0.005. However, the p-value increases to 0.071 at the 50% quantile (median), and further to 0.436 at the 75% quantile. This indicates that smoking has a more prominent effect on the population with smaller BMI, and its effect diminishes as BMI increases; i.e., the effect of smoking becomes less significant as the value of BMI becomes larger. This interesting pattern cannot be reflected in ATE. We can draw the same inferential conclusions as above when the weighted bootstrap method is applied.

We also examine the relationship between BMI and two continuous confounding variables, age and family poverty income ratio (Family PIR). Figure 1.2 depicts the estimated conditional mean functions (OR functions) $\tau_1(\cdot)$ and $\tau_0(\cdot)$ versus the two continuous variables for the smoking and nonsmoking groups, and for males and females, respectively. For each comparison, all the other confounding variables are fixed as constants: the continuous variables take the values of their means while the categorical variables are kept as married, college or above, drinks alcohol, no vigorous activity and no moderate activity. It is interesting to notice that for the same age or Family PIR, the estimated conditional mean in the smoking group is smaller than that in the nonsmoking group for both males and female, and the estimated conditional mean in the male group is also smaller than that in the female group for both smoker and nonsmoker. We can clearly see nonlinear relationships between age and BMI as well as between Family PIR and BMI. Age is positively associated with BMI when it is less than 50, and the association between age and BMI becomes more negative as people get older. We also see that the smoking effects on BMI are very different between the male group and the female group. Smoking has a more significant effect on BMI for males than for females at the same age. In the male group,

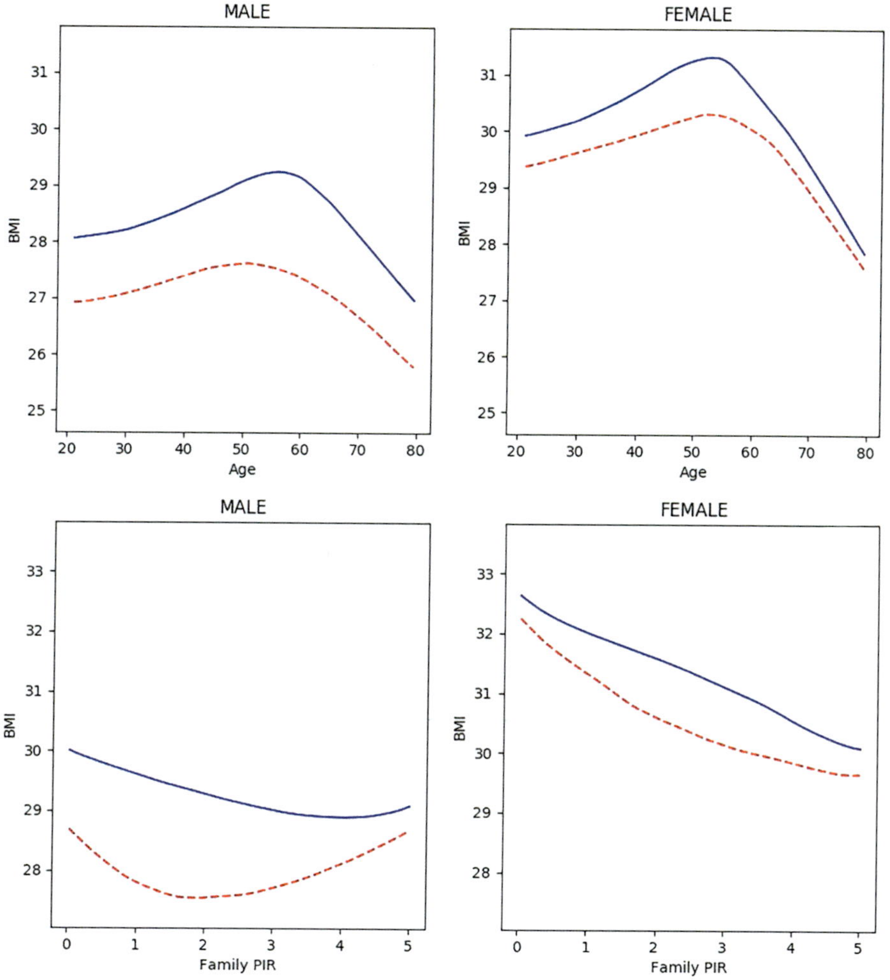

Fig. 1.2 The plots of $\tau_1(\cdot)$ and $\tau_0(\cdot)$ versus two continuous variables for the smoking and non-smoking groups, and for males and females, respectively, where the blue solid curves represent nonsmoking group and red dashed line represent smoking group.

the BMI decreases as family income increases until it reaches the poverty threshold, and then the BMI increases with family income for smokers. For nonsmokers, it shows a relatively flatter trend. In the female group, the BMI keeps decreasing as family income increases for both smokers and nonsmokers.

References

1. Neyman, J.: Sur les applications de la théorie des probabilités aux experiences agricoles: Essai des principes. Roczniki Nauk Rolniczych **10**(1), 1–51 (1923)
2. Rubin, D.B.: Estimating causal effects of treatments in randomized and nonrandomized studies. J. Educ. Psychol. **66**(5), 688 (1974)
3. Rubin, D.B.: Discussion of "randomization analysis of experimental data in the fisher randomization test" by D. Basu. J. Am. Stat. Assoc. **75**(371), 591–593 (1980)
4. Fisher, R.A.: The Design of Experiments, 1st edn. Oliver and Boyd, Edinburgh (1935)
5. Freedman, D.A.: On regression adjustments to experimental data. Adv. Appl. Math. **40**(2), 180–193 (2008)
6. Lin, W.: Agnostic notes on regression adjustments to experimental data: reexamining freedman's critique. Ann. Appl. Stat. **7**, 295–318 (2013)
7. Hahn, J.: On the role of the propensity score in efficient semiparametric estimation of average treatment effects. Econometrica **66**, 315–331 (1998)
8. Qin, J., Zhang, B.: Empirical-likelihood-based inference in missing response problems and its application in observational studies. J. R. Stat. Soc. Ser. B (Statistical Methodology) **69**(1), 101–122 (2007)
9. Tan, Z.: Bounded, efficient and doubly robust estimation with inverse weighting. Biometrika **97**(3), 661–682 (2010)
10. Chan, K.C.G.: Uniform improvement of empirical likelihood for missing response problem. Electron. J. Stat. **6**, 289–302 (2012)
11. Graham, B.S., Pinto, C.C.D.X., Egel, D.: Inverse probability tilting for moment condition models with missing data. Rev. Econ. Stud. **79**(3), 1053–1079 (2012)
12. Vansteelandt, S., Bekaert, M., Claeskens, G.: On model selection and model misspecification in causal inference. Stat. Methods Med. Res. **21**(1), 7–30 (2010)
13. Han, P., Wang, L.: Estimation with missing data: beyond double robustness. Biometrika **100**(2), 417–430 (2013)
14. Imai, K., Ratkovic, M.: Covariate balancing propensity score. J. R. Stat. Soc. Ser. B (Statistical Methodology) **76**(1), 243–263 (2014)
15. Chan, K.C.G., Yam, S.C.P.: Oracle, multiple robust and multipurpose calibration in a missing response problem. Stat. Sci. **29**(3), 380–396 (2014)
16. Chan, K.C.G., Yam, S.C.P., Zhang, Z.: Globally efficient non-parametric inference of average treatment effects by empirical balancing calibration weighting. J. R. Stat. Soc. Ser. B (Statistical Methodology) **78**(3), 673–700 (2016)
17. Deville, J., Särndal, C.: Calibration estimators in survey sampling. J. Am. Stat. Ass. **87**(418), 376–382 (1992)
18. Kitamura, Y., Stutzer, M.: An information-theoretic alternative to generalized method of moments estimation. Econometrica: J. Econ. Soc. 861–874 (1997)
19. Imbens, G., Johnson, P., Spady, R.H.: Information theoretic approaches to inference in moment condition models. Econometrica **66**(2), 333–357 (1998)
20. Owen, A.: Empirical likelihood ratio confidence intervals for a single functional. Biometrika **75**(2), 237–249 (1988)
21. Qin, J., Lawless, J.: Empirical likelihood and general estimating equations. Ann. Stat. **22**, 300–325 (1994)
22. Hansen, L.: Large sample properties of generalized method of moments estimators. Econometrica **50**, 1029–1054 (1982)
23. Hansen, L., Heaton, J., Yaron, A.: Finite-sample properties of some alternative GMM estimators. J. Bus. Econ. Stat. **14**(3), 262–280 (1996)
24. Tseng, P., Bertsekas, D.P.: Relaxation methods for problems with strictly convex separable costs and linear constraints. Math. Program. **38**(3), 303–321 (1987)
25. Chen, S.X., Qin, J., Tang, C.Y.: Mann-whitney test with adjustments to pretreatment variables for missing values and observational study. J. R. Stat. Soc. B (Statisticial Methodology) **75**(1), 81–102 (2013)

26. Hirano, K., Imbens, G.W., Ridder, G.: Efficient estimation of average treatment effects using the estimated propensity score. Econometrica **71**(4), 1161–1189 (2003)
27. Imbens, G., Newey, W., Ridder, G.: Mean-Squared-error Calculations for Average Treatment Effects. Unpublished manuscript, University of California Berkeley (2006)
28. Chen, X., Hong, H., Tarozzi, A.: Semiparametric efficiency in gmm models with auxiliary data. Ann. Stat. **36**(2), 808–843 (2008)
29. Robins, J.M., Rotnitzky, A., Zhao, L.P.: Estimation of regression coefficients when some regressors are not always observed. J. Am. Stat. Assoc. **89**(427), 846–866 (1994)
30. Kang, J.D., Schafer, J.L.: Demystifying double robustness: a comparison of alternative strategies for estimating a population mean from incomplete data. Stat. Sci. **22**(4), 523–539 (2007)
31. Rosenbaum, P.R., Rubin, D.B.: Constructing a control group using multivariate matched sampling methods that incorporate the propensity score. Am. Stat. **39**(1), 33–38 (1985)
32. Tibshirani, R.: The lasso method for variable selection in the cox model. Stat. Med. **16**(4), 385–395 (1997)
33. Zhang, Y., Bradic, J.: High-dimensional semi-supervised learning: in search of optimal inference of the mean. Biometrika **109**(2), 387–403 (2022)
34. Chernozhukov, V., Chetverikov, D., Demirer, M., Duflo, E., Hansen, C., Newey, W.: Double/debiased/neyman machine learning of treatment effects. Am. Econ. Rev. **107**(5), 261–265 (2017)
35. Cattaneo, M.D.: Efficient semiparametric estimation of multi-valued treatment effects under ignorability. J. Econ. **155**(2), 138–154 (2010)
36. Firpo, S.: Efficient semiparametric estimation of quantile treatment effects. Econometrica **75**(1), 259–276 (2007)
37. Han, P., Kong, L., Zhao, J.: A general framework for quantile estimation with incomplete data. J. R. Stat. Soc. B **81**, 305–333 (2019)
38. Newey, W.K., Powell, J.L.: Asymmetric least squares estimation and testing. Econometrica **55**(4), 819–847 (1987)
39. Rosenbaum, P.R., Rubin, D.B.: The central role of the propensity score in observational studies for causal effects. Biometrika **70**(1), 41–55 (1983)
40. Shen, X.: On methods of sieves and penalization. Ann. Stat. **25**(6), 2555–2591 (1997)
41. Chen, X., Shen, X.: Sieve extremum estimates for weakly dependent data. Econometrica, pp. 289–314 (1998)
42. Chen, X., White, H.: Improved rates and asymptotic normality for nonparametric neural network estimators. IEEE Trans. Inf. Theory **45**(2), 682–691 (1999)
43. Chen, X., Liu, Y., Ma, S., Zhang, Z.: Causal inference of general treatment effects using neural networks with a diverging number of confounders. J. Econometrics **238**(1), 105555 (2024)
44. Chen, X., Linton, O., Van Keilegom, I.: Estimation of semiparametric models when the criterion function is not smooth. Econometrica **71**(5), 1591–1608 (2003)
45. Bauer, B., Kohler, M.: On deep learning as a remedy for the curse of dimensionality in nonparametric regression. Ann. Stat. **47**(4), 2261–2285 (2019)
46. Schmidt-Hieber, J.: Nonparametric regression using deep neural networks with relu activation function. Ann. Stat. **48**, 1875–1897 (2020)
47. D'Amour, A., Ding, P., Feller, A., Lei, L., Sekhon, J.: Overlap in observational studies with high-dimensional covariates. J. Econometrics **221**(2), 644–654 (2021)
48. Hong, H., Leung, M.P., Li, J.: Inference on finite-population treatment effects under limited overlap. Econometrics J. **23**(1), 32–47 (2020)
49. Ma, X., Wang, J.: Robust inference using inverse probability weighting. J. Am. Stat. Assoc. **115**(532), 1851–1860 (2020)
50. Chen, X., Liao, Z.: Sieve semiparametric two-step gmm under weak dependence. J. Econometrics **189**(1), 163–186 (2015)
51. Ai, C., Linton, O., Motegi, K., Zhang, Z.: A unified framework for efficient estimation of general treatment models. Quant. Econ. **12**(3), 779–816 (2021)
52. Tsiatis, A.: Semiparametric Theory and Missing Data. Springer Science & Business Media (2007)

53. Lee, Y.Y.: Efficient propensity score regression estimators of multivalued treatment effects for the treated. J. Econometrics **204**(2), 207–222 (2018)
54. Farrell, M.H., Liang, T., Misra, S.: Deep neural networks for estimation and inference. Econometrica **89**(1), 181–213 (2021)
55. Ma, S., Kosorok, M.R.: Robust semiparametric m-estimation and the weighted bootstrap. J. Multivar. Anal. **96**(1), 190–217 (2005)

Chapter 2
Causal Inference for a Continuous Treatment

2.1 Basic Framework

Chapter 1 focuses on the discrete treatment where an individual, ignoring the treatment intensity. In many applications, however, the treatment intensity is a part of the treatment, and its causal effect is also of great interest to decision-makers. For example, in evaluating how financial incentives affect health care providers, the causal effect may depend on not only the introduction of incentive but also the level of incentive. Similarly, in studying how taxes affect addictive substance usages, the causal effect may depend not only on the imposition of tax but also on the tax rate. In finance, there are many plausible examples of interest. For example, in evaluating the effect of corporate bond purchase schemes on market quality, the causal effect may depend not just on whether the bond is selected into the scheme but on how much of it is purchased. In this chapter focuses on the causal inference for a continuous treatment variable.

Let T denote a continuous treatment status variable with support $\mathcal{T} \subset \mathbb{R}$, and T has a probability density function $f_T(t)$. Let $Y(t)$ denote the potential response when treatment $T = t$ is assigned. Let $L(\cdot)$ denote a known convex loss function whose derivative, denoted by $L'(\cdot)$, exists almost everywhere. For the leading part of the paper, we shall maintain that there exists a parametric causal effect function $g(t; \beta)$ with the unknown value $\boldsymbol{\beta}^* \in \mathbb{R}^p$ (with $p \in \mathbb{N}$) uniquely solving the minimization problem below, i.e.,

$$\boldsymbol{\beta}^* = \arg\min_{\boldsymbol{\beta}} \int_{\mathcal{T}} \mathbb{E}\left[L\left(Y(t) - g(t; \boldsymbol{\beta})\right)\right] f_T(t)dt. \tag{2.1}$$

The parameterization of the causal effect is restrictive, but quite common in applications. Some extensions to the unspecified causal effect function shall be discussed later in the paper (see Sect. 2.6).

Model (2.1) includes many prominent models as special cases. For example, it includes the average causal effect of continuous treatment studied in [1] (i.e.,

Z. Zhang et al., *Big Data in Economics and Management*, Statistics and Big Data 1,
https://doi.org/10.1007/978-981-95-3125-7_2

$L(v) = v^2$ and $\mathbb{E}[Y(t)] = g(t; \boldsymbol{\beta})$ is a parametric model indexed by $\boldsymbol{\beta}$ for the potential outcome means, which is also called a *marginal structural model* in [2]. Examples include the linear marginal structure model $\mathbb{E}[Y^*(t)] = \beta_0 + \beta_1 \cdot t$, and the nonlinear marginal structure model $\mathbb{E}[Y(t)] = \beta_0 \cdot t + 1/(t + \beta_1)^2$ studied in [1]). It also includes the quantile causal effect of multi-valued (i.e., $L(v) = v(\tau - I(v \leq 0))$ with $\tau \in (0, 1)$ and $g(t; \boldsymbol{\beta}) = \sum_{j=0}^{J} \beta_j I(t = j)$) and continuous treatment (i.e., $L(v) = v(\tau - I(v \leq 0))$ and $\inf\{q : \mathbb{P}(Y(t) \geq q) \leq \tau\} = g(t; \boldsymbol{\beta})$ is a parametric model indexed by $\boldsymbol{\beta}$ for the potential outcome quantiles. Examples include the linear model $\inf\{q : \mathbb{P}(Y(t) \geq q) \leq \tau\} = \beta_0 + \beta_1 \cdot t$ and the Box-Cox transformation model $\inf\{q : \mathbb{P}(Y(t) \geq q) \leq \tau\} = h_\lambda(\beta_0 + \beta_1 \cdot t)$ studied in [3], where $h_\lambda(z) = (\lambda z + 1)^{-1/\lambda}$).

The problem with (2.1) is that the potential outcome $Y(t)$ is not observed for all t. Let $Y := Y(T)$ denote the observed response. One may attempt to solve the following optimization problem:

$$\min_{\boldsymbol{\beta}} \mathbb{E}[L(Y - g(T; \boldsymbol{\beta}))].$$

However, if there exists a selection into treatment, the true value $\boldsymbol{\beta}_0$ does not solve the above minimization problem. Indeed, in this case, the observed response and treatment assignment data alone cannot identify $\boldsymbol{\beta}^*$. To address this identification issue, studies in the literature impose a selection on observable condition (e.g., [4–6]). Specifically, let $\boldsymbol{X}$ denote a vector of covariates. The following condition shall be maintained throughout the paper.

Assumption 1 (*Unconfounded Treatment Assignment*) T is independent of $Y(t)$ for all $t \in \mathcal{T}$ given $\boldsymbol{X}$, i.e., $Y(t) \perp T | \boldsymbol{X}$.

Let $f_{T|X}$ denote the conditional probability density function of T given the observed covariates $\boldsymbol{X}$. In the literature, $f_{T|X}$ is called the *generalized propensity score* [1, 5]. Suppose that $f_{T|X}(T|\boldsymbol{X})$ is positive everywhere and let

$$\pi_0(T, \boldsymbol{X}) := \frac{f_T(T)}{f_{T|X}(T|\boldsymbol{X})}.$$

The function $\pi_0(T, \boldsymbol{X})$ is called the *stabilized weight* in [2]. Under Assumption 1, we obtain

$$\mathbb{E}[\pi_0(T, \boldsymbol{X})L(Y - g(T; \boldsymbol{\beta}))] = \int \mathbb{E}\left[L(Y^*(t) - g(t; \boldsymbol{\beta}))\right] dF_T(t) \tag{2.2}$$

and hence the true value $\boldsymbol{\beta}^*$ solves the weighted optimization problem

$$\boldsymbol{\beta}^* = \arg\min_{\boldsymbol{\beta}} \mathbb{E}[\pi_0(T, \boldsymbol{X})L(Y - g(T; \boldsymbol{\beta}))]. \tag{2.3}$$

This result is very insightful. It tells us that the selection bias in the *unconfounded treatment assignment* can be corrected through covariate-balancing. More importantly, it says that the true value $\boldsymbol{\beta}^*$ can be identified from the observed data. The weighted optimization (2.3) provides a unified framework for estimating of a variety of continuous treatment effects under a general loss function. We will compute the semiparametric efficiency bound and to present an efficient estimator for $\boldsymbol{\beta}^*$ under this general framework.

2.2 Semiparametric Efficiency Bound

We begin by applying the approach of [7] to compute the semiparametric efficiency bound of the parameter $\boldsymbol{\beta}^*$ defined by (2.1) under Assumption 1. This gives the least possible variance achievable by a regular estimator in the semiparametric model. The result is presented in the following theorem whose proof is given in the supplemental material of [8].

Theorem 2.1 *Suppose that $g(T;\boldsymbol{\beta})$ is twice differentiable with respect to $\boldsymbol{\beta}$ in the parameter space $\Theta \subset \mathbb{R}^p$, with $m(T;\boldsymbol{\beta}^*) := \nabla_\beta g(T;\boldsymbol{\beta}^*)$, and $\mathbb{E}\left[L'(Y-g(T;\boldsymbol{\beta}))|Y,\boldsymbol{X}\right]$ is differentiable with respect to $\boldsymbol{\beta}\in\Theta$. Denote $\varepsilon(T,\boldsymbol{X};\boldsymbol{\beta}^*) := \mathbb{E}[L'(Y-g(T;\boldsymbol{\beta}^*))|T,\boldsymbol{X}]$, $H_0 := -\nabla_\beta \mathbb{E}\left[\pi_0(T,\boldsymbol{X})L'(Y-g(T;\boldsymbol{\beta}))m(T;\boldsymbol{\beta})\right]\big|_{\boldsymbol{\beta}=\boldsymbol{\beta}^*}$, and*

$$
\begin{aligned}
\psi(Y,T,\boldsymbol{X};\boldsymbol{\beta}^*) &:= \pi_0(T,\boldsymbol{X})m(T;\boldsymbol{\beta}^*)L'(Y-g(T;\boldsymbol{\beta}^*)) - \pi_0(T,\boldsymbol{X})m(T;\boldsymbol{\beta}^*)\varepsilon(T,\boldsymbol{X};\boldsymbol{\beta}^*) \\
&\quad + \mathbb{E}\left[\varepsilon(T,\boldsymbol{X};\boldsymbol{\beta}^*)\pi_0(T,\boldsymbol{X})m(T;\boldsymbol{\beta}^*)|T\right] + \mathbb{E}\left[\varepsilon(T,\boldsymbol{X};\boldsymbol{\beta}^*)\pi_0(T,\boldsymbol{X})m(T;\boldsymbol{\beta}^*)|\boldsymbol{X}\right].
\end{aligned}
$$

Suppose that H_0 is nonsingular and $\mathbb{E}\left[\psi(Y,T,\boldsymbol{X};\boldsymbol{\beta}^*)\psi(Y,T,\boldsymbol{X};\boldsymbol{\beta}^*)^\top\right]$ exists and is finite. Under Assumption 1 and model (2.1), the efficient influence function of $\boldsymbol{\beta}^*$ is given by

$$S_{eff}(Y,T,\boldsymbol{X};\boldsymbol{\beta}^*) = H_0^{-1}\psi(Y,T,\boldsymbol{X};\boldsymbol{\beta}^*).$$

Consequently, the efficient variance bound of $\boldsymbol{\beta}^*$ is

$$V_{eff} = \mathbb{E}\left[S_{eff}(Y,T,\boldsymbol{X};\boldsymbol{\beta}^*)S_{eff}(Y,T,\boldsymbol{X};\boldsymbol{\beta}^*)^\top\right]. \tag{2.4}$$

It is worth noting that our bound V_{eff} is equal to: the bound of [9] for the case of binary average treatment, the bound of [10] for the case of multi-valued average treatment, and the bound of [11] for the case of binary quantile treatment (see [12], Sects. 2.2–2.4). Moreover, our bound applies to a much wider class of models, including quantile causal effect of multi-valued, continuous, and mixture of discrete and continuous treatment as well as the asymmetric least squares estimation of the causal effect of all kinds of treatments.

Based on the expression of the efficient influence function, many papers construct an efficient estimator by solving the estimated efficient score equation [13, 14]. Such estimators typically have the double or multiple robustness property. However, in our case the efficient influence function $S_{eff}(T, \boldsymbol{X}, Y; \boldsymbol{\beta})$ involves five unknown functionals $f_T(T)$, $f_{T|X}(T|\boldsymbol{X})$, $\varepsilon(T, \boldsymbol{X}; \boldsymbol{\beta})$, $\mathbb{E}[\pi_0(T, \boldsymbol{X})\varepsilon(T, \boldsymbol{X}; \boldsymbol{\beta})m(T, \boldsymbol{\beta})|T]$, and $\mathbb{E}[\pi_0(T, \boldsymbol{X})\varepsilon(T, \boldsymbol{X}; \boldsymbol{\beta})m(T, \boldsymbol{\beta})|\boldsymbol{X}]$. Estimation of these functionals is difficult in practice, and we expect that the finite sample performance of the estimated $\boldsymbol{\beta}^*$ would be poor. Instead of explicitly estimating the efficient influence function S_{eff}, we propose a simple weighted optimization estimator based on (2.3) by estimating the stabilized weights $\pi_0(T, \boldsymbol{X})$. This procedure is remarkably stable numerically and performs well statistically in small samples as we demonstrate in the Monte Carlo section.

It is also worth noting that, if the stabilized weights are known and $g(t; \boldsymbol{\beta}^*)$ is correctly specified, one can estimate $\boldsymbol{\beta}^*$ by solving the sample analogue of the weighted optimization (2.3). The asymptotic variance of this estimator is

$$V_{ineff} = \mathbb{E}\left[S_{ineff}(Y, T, \boldsymbol{X}; \boldsymbol{\beta}^*)S_{ineff}(Y, T, \boldsymbol{X}; \boldsymbol{\beta}^*)^\top\right],$$

with

$$S_{ineff}(Y, T, \boldsymbol{X}; \boldsymbol{\beta}^*) = H_0^{-1} \cdot \pi_0(T, \boldsymbol{X})m(T; \boldsymbol{\beta}^*)L'\left\{Y - g(T; \boldsymbol{\beta}^*)\right\}.$$

It is easy to show that $V_{ineff} > V_{eff}$, implying that the weighted optimization estimator is not efficient. This follows because the weighted optimization does not account for the restriction on the stabilized weight $\pi_0(t, \boldsymbol{x})$ that

$$\mathbb{E}\left[\pi_0(T, \boldsymbol{X})u(T)v(\boldsymbol{X})\right] = \mathbb{E}[u(T)] \cdot \mathbb{E}[v(\boldsymbol{X})] \tag{2.5}$$

holds for any suitable functions $u(t)$ and $v(\boldsymbol{x})$. Incorporating restriction (2.5) into the estimation of the causal effect can improve efficiency. A similar observation was made by [4] in the binary treatment. Exactly how to incorporate restriction (2.5) into the estimation is the subject of the next section.

2.3 Maximum Entropy Weighting

One way to incorporate (2.5) into the estimation is to estimate the stabilized weights from (2.5) and then implement (2.3) with the estimated weights. But before doing so, we must verify that (2.5) uniquely identifies $\pi_0(T, \boldsymbol{X})$.

Theorem 2.2 *For any integrable functions $u(T)$ and $v(\boldsymbol{X})$, $\mathbb{E}\left[\pi(T, \boldsymbol{X})u(T)v(\boldsymbol{X})\right] = \mathbb{E}[u(T)] \cdot \mathbb{E}[v(\boldsymbol{X})]$ holds if and only if $\pi(T, \boldsymbol{X}) = \pi_0(T, \boldsymbol{X})$ a.s.*

The proof Theorem 2.2 is given in the appendix of [8]. Therefore, Condition (2.5) identifies the stabilized weights. The challenge now is that (2.5) implies an infinite number of moment conditions. With a finite sample of observations, it is impossible to solve an infinite number of equations. To overcome this difficulty, we approximate the (infinite-dimensional) function space with the (finite-dimensional) sieve space. Specifically, let $u_{K_1}(T) = (u_{K_1,1}(T), \ldots, u_{K_1,K_1}(T))^\top$ and $v_{K_2}(\boldsymbol{X}) = \left(v_{K_2,1}(\boldsymbol{X}), \ldots, v_{K_2,K_2}(\boldsymbol{X})\right)^\top$ denote the known basis functions with dimensions $K_1 \in \mathbb{N}$ and $K_2 \in \mathbb{N}$, respectively, and let $K := K_1 \cdot K_2$. The functions $u_{K_1}(t)$ and $v_{K_2}(\boldsymbol{x})$ are called the *approximation sieves* that can approximate any suitable functions $u(t)$ and $v(\boldsymbol{x})$ arbitrarily well (see [15, 16], for more discussion on sieve approximation). Since the sieve approximating space is also a subspace of the function space, $\pi_0(T, \boldsymbol{X})$ satisfies

$$\mathbb{E}\left[\pi_0(T, \boldsymbol{X}) u_{K_1}(T) v_{K_2}(\boldsymbol{X})^\top\right] = \mathbb{E}[u_{K_1}(T)] \cdot \mathbb{E}[v_{K_2}(\boldsymbol{X})]^\top. \tag{2.6}$$

Let $\{T_i, \boldsymbol{X}_i, Y_i\}_{i=1}^N$ denote an independently and identically distributed sample of observations drawn from the joint distribution of $(T, \mathbf{X}, Y)$. We propose to estimate the stabilized weights $\pi_i = \pi_0(T_i, \mathbf{X}_i)$ by solving the entropy maximization problem

$$\begin{cases} \max\left\{-\sum_{i=1}^N \pi_i \log \pi_i\right\} \\ \text{subject to } \frac{1}{N}\sum_{i=1}^N \pi_i u_{K_1}(T_i) v_{K_2}(\boldsymbol{X}_i)^\top = \left(\frac{1}{N}\sum_{i=1}^N u_{K_1}(T_i)\right)\left(\frac{1}{N}\sum_{j=1}^N v_{K_2}(\boldsymbol{X}_j)^\top\right). \end{cases} \tag{2.7}$$

Noting $\sum_{i=1}^N N^{-1}\pi_i = 1$ (since both $u_{K_1}(T)$ and $v_{K_2}(\boldsymbol{X})$ contain the constant 1) and

$$\max\left\{-\sum_{i=1}^N \pi_i \log \pi_i\right\} = -\min\left\{\sum_{i=1}^N \{N^{-1}\pi_i\} \cdot \log \frac{N^{-1}\pi_i}{N^{-1}}\right\},$$

the formulation (2.7) can be interpreted as the minimization of the Kullback-Leibler divergence between the estimated weights $\{N^{-1}\pi_i\}_{i=1}^N$ and the empirical frequencies $\{N^{-1}\}$ subject to the empirical moment constraints (2.6). This idea is similar to the exponential tilting (ET) idea developed in [17, 18]. The difference is that they consider a parametric problem and we consider a nonparametric problem.

The primal problem (2.7) is hard to solve numerically. We instead consider its dual problem, which can be solved by numerically efficient and stable algorithms. Specifically, let $\rho(v) := -e^{-v-1}$ for any $v \in \mathbb{R}$, by [19], we can show that the dual solution is given by

$$\hat{\pi}_K(T_i, \boldsymbol{X}_i) := \rho'\left(u_{K_1}(T_i)^\top \hat{\Lambda}_{K_1 \times K_2} v_{K_2}(\boldsymbol{X}_i)\right), \tag{2.8}$$

where $\hat{\Lambda}_{K_1 \times K_2}$ is the maximizer of the strictly concave function $\hat{G}_{K_1 \times K_2}$ defined by

$$\begin{aligned} \widehat{\Lambda}_{K_1 \times K_2} &= \arg\max_{\Lambda} \hat{G}_{K_1 \times K_2}(\Lambda) \\ &:= \frac{1}{N}\sum_{i=1}^N \rho\left(u_{K_1}(T_i)^\top \Lambda v_{K_2}(\boldsymbol{X}_i)\right) - \left(\frac{1}{N}\sum_{i=1}^N u_{K_1}(T_i)\right)^\top \Lambda \left(\frac{1}{N}\sum_{j=1}^N v_{K_2}(\boldsymbol{X}_j)\right). \end{aligned} \tag{2.9}$$

By the first-order condition, the constraints of (2.7) are automatically satisfied by $\{\hat{\pi}_K(T_i, \boldsymbol{X}_i)\}_{i=1}^N$. The duality between (2.7) and (2.9) can be derived similarly to that in Sect. 1.2.2 in Chap. 1. Having estimated the weights, we now estimate $\boldsymbol{\beta}^*$ by solving the generalized optimization problem, that is,

$$\widehat{\boldsymbol{\beta}} = \arg\min_{\boldsymbol{\beta}} \sum_{i=1}^N \hat{\pi}_K(T_i, \boldsymbol{X}_i) L\left(Y_i - g(T_i; \boldsymbol{\beta})\right). \tag{2.10}$$

2.4 Efficient Estimation Results

To establish the large sample properties of the generalized optimization estimator, we first show that the estimated weight function $\hat{\pi}_K(t, \boldsymbol{x})$ is consistent and compute its convergence rates under both the L_∞ norm and the L_2 norm. The following conditions shall be imposed.

Assumption 2 (i) The support $\mathcal{X}$ of $\boldsymbol{X}$ is a compact subset of $\mathbb{R}^r$. The support $\mathcal{T}$ of the treatment variable T is a compact subset of $\mathbb{R}$. (ii) There exist two positive constants η_1 and η_2 such that

$$0 < \eta_1 \leq \pi_0(t, \boldsymbol{x}) \leq \eta_2 < \infty, \ \forall (t, \boldsymbol{x}) \in \mathcal{T} \times \mathcal{X}.$$

Assumption 3 There exist $\Lambda_{K_1 \times K_2} \in \mathbb{R}^{K_1 \times K_2}$ and a positive constant $\alpha > 0$ such that

$$\sup_{(t,x)\in\mathcal{T}\times\mathcal{X}} \left|(\rho'^{-1}(\pi_0(t, \boldsymbol{x})) - u_{K_1}(t)^\top \Lambda_{K_1\times K_2} v_{K_2}(\boldsymbol{x})\right| = O(K^{-\alpha}),$$

where $\rho(v) = -\exp(-v - 1)$.

Assumption 4 (i) For every K_1 and K_2, the smallest eigenvalues of $\mathbb{E}\left[u_{K_1}(T)u_{K_1}(T)^\top\right]$ and $\mathbb{E}\left[v_{K_2}(\boldsymbol{X})v_{K_2}(\boldsymbol{X})^\top\right]$ are bounded away from zero uniformly in K_1 and K_2. (ii) There are two sequences of constants $\zeta_1(K_1)$ and $\zeta_2(K_2)$ satisfying $\sup_{t\in\mathcal{T}} \|u_{K_1}(t)\| \leq \zeta_1(K_1)$ and $\sup_{\boldsymbol{x}\in\mathcal{X}} \|v_{K_2}(\boldsymbol{x})\| \leq \zeta_2(K_2)$, $K = K_1(N)K_2(N)$ and $\zeta(K) := \zeta_1(K_1)\zeta_2(K_2)$, such that $\zeta(K)K^{-\alpha} \to 0$ and $\zeta(K)\sqrt{K/N} \to 0$ as $N \to \infty$.

Assumption 2 (i) restricts both the covariates and treatment level to be bounded. This condition is restrictive but convenient for computing the convergence rate under L_∞ norm. It is commonly imposed in the nonparametric regression literature. This condition can be relaxed, however, if we restrict the tail behavior of the joint distribution of $(\boldsymbol{X}, T)$. Assumption 2 (ii) restricts the weight function to be bounded and bounded away from zero. Given Assumption 2 (i), this condition is equivalent to $dF_{T|X}(T|\boldsymbol{X})$ being bounded away from zero, meaning that each type of individual (denoted by $\boldsymbol{X}$) always have a sufficient portion participating in each level of treatment. This restriction is important for our analysis since each individual participates only in one level of treatment and this condition allows us to construct her statistical counterparts from all other treatments. Although Assumption 2 (ii) is useful in causal analysis and establishing the convergence rates, it is not essential and could be relaxed by allowing η_1 (resp. η_2) to depend on N and to go to zero (resp. infinity) slowly, as $N \to \infty$. Notice that $u_{K_1}(t)^\top \Lambda v_{K_2}(\boldsymbol{x})$ is a linear sieve approximation to any suitable function of $(\boldsymbol{X}, T)$.

Assumption 3 requires the sieve approximation error of $\rho'^{-1}(\pi_0(t, \boldsymbol{x}))$ to shrink at a polynomial rate. This condition is satisfied for a variety of sieve basis functions. For power series and B-splines, $\alpha = -s/r$, where s is the smoothness of approximand and r is the dimension of $\boldsymbol{X}$. Hence, the proposed method still suffers from the curse of dimensionality that typically occurs in nonparametric estimation. We will show that the convergence rate of the estimated weight function (and consequently the rate of the generalized optimization estimator) is bounded by this polynomial rate.

Assumption 4 (i) essentially ensures the sieve approximation estimator is nondegenerate. Similar conditions are common in the sieve regression literature [15, 20]. If the approximation error is nonzero, Assumption 4 (ii) requires it to shrink to zero at an appropriate rate as the sample size increases. Newey [15] show that if $u_{K_1}(t)$ (resp. $u_{K_2}(\boldsymbol{x})$) is a power series then $\zeta_1(K_1) = O(K_1)$ (resp. $\zeta_2(K_2) = O(K_2)$), and if $u_{K_1}(t)$ (resp. $u_{K_2}(\boldsymbol{x})$) is a B-spline then $\zeta_1(K_1) = O(\sqrt{K_1})$ (resp. $\zeta_2(K_2) = O(\sqrt{K_2})$).

Under these conditions, we are able to establish the following theorem:

Theorem 2.3 *Suppose that Assumptions 2–4 hold. Then, we obtain the following:*

$$\int_{\mathcal{T}\times\mathcal{X}} |\hat{\pi}_K(t, \boldsymbol{x}) - \pi_0(t, \boldsymbol{x})|^2 dF_{T,X}(t, \boldsymbol{x}) = O_p\left(\max\left\{K^{-2\alpha}, \frac{K}{N}\right\}\right),$$

$$\frac{1}{N}\sum_{i=1}^{N} |\hat{\pi}_K(T_i, \boldsymbol{X}_i) - \pi_0(T_i, \boldsymbol{X}_i)|^2 = O_p\left(\max\left\{K^{-2\alpha}, \frac{K}{N}\right\}\right).$$

The proof of Theorem 2.3 immediately follows from Lemma 3.1 and Corollary 3.3 of the supplemental material [12].

The following additional condition is needed to establish the consistency of the proposed estimator $\hat{\boldsymbol{\beta}}$.

Assumption 5 (i) The parameter space $\Theta \subset \mathbb{R}^p$ is a compact set and the true parameter $\boldsymbol{\beta}^*$ is in the interior of Θ, where $p \in \mathbb{N}$. (ii) $L\left(Y - g(T; \boldsymbol{\beta})\right)$ is continuous in $\boldsymbol{\beta}$, $\sup_{\boldsymbol{\beta}\in\Theta} \mathbb{E}\left[|L\left(Y - g(T; \boldsymbol{\beta})\right)|^2\right] < \infty$ and $\mathbb{E}\left[\sup_{\boldsymbol{\beta}\in\Theta} |L\left(Y - g(T; \boldsymbol{\beta})\right)|\right] < \infty$.

Assumption 5 (i) is commonly imposed in the nonlinear regression literature, but can be relaxed if $g(t; \boldsymbol{\beta})$ is linear in $\boldsymbol{\beta}$. Assumption 5 (ii) is an envelope condition that is sufficient for the applicability of the uniform law of large numbers.

Under these and other conditions, we establish the consistency of the generalized optimization estimator. The proof of Theorem 2.4 is given in Sect. 4.1 of the supplemental material [12].

Theorem 2.4 *Suppose that Assumptions 1-5 hold. Then,* $\|\hat{\boldsymbol{\beta}} - \boldsymbol{\beta}^*\| \xrightarrow{p} 0$.

To establish the asymptotic distribution of the proposed estimator, we need some smoothness condition on the regression function and some under-smoothing condition on the sieve approximation (i.e., larger K than needed for consistency). We also have to address the possibility of a nonsmooth loss function. These conditions are presented below.

Assumption 6 (i) The loss function $L(v)$ is differentiable almost everywhere, $g(t;\boldsymbol{\beta})$ is twice continuously differentiable in $\boldsymbol{\beta} \in \Theta$ and we denote its first derivative by $m(t;\boldsymbol{\beta}) := \nabla_{\boldsymbol{\beta}} g(t;\boldsymbol{\beta})$;
(ii) $\mathbb{E}\left[\pi_0(T,\boldsymbol{X})L'(Y-g(T;\boldsymbol{\beta}))m(T;\boldsymbol{\beta})\right]$ is differentiable with respect to $\boldsymbol{\beta}$ and $H_0 := -\nabla_{\boldsymbol{\beta}}\mathbb{E}\left[\pi_0(T,\boldsymbol{X})L'(Y-g(T;\boldsymbol{\beta}))m(T;\ \boldsymbol{\beta})\right]\Big|_{\boldsymbol{\beta}=\boldsymbol{\beta}^*}$ is nonsingular;
(iii) $\varepsilon(t,\boldsymbol{x};\boldsymbol{\beta}^*) := \mathbb{E}[L'(Y-g(T;\boldsymbol{\beta}^*))|T=t,\boldsymbol{X}=\boldsymbol{x}]$ is continuously differentiable in $(t,\boldsymbol{x})$;
(iv) Suppose that $N^{-1}\sum_{i=1}^N \hat{\pi}_K(T_i,\boldsymbol{X}_i)L'(Y_i - g(T_i;\hat{\boldsymbol{\beta}}))m(T_i;\hat{\boldsymbol{\beta}}) = o_p(N^{-1/2})$ holds with probability approaching one.

Assumption 7 (i) $\mathbb{E}\left[\sup_{\boldsymbol{\beta}\in\Theta}|L'(Y-g(T;\boldsymbol{\beta}))|^{2+\delta}\right] < \infty$ for some $\delta > 0$; (ii) The function class $\{L'(y-g(t;\ \boldsymbol{\beta})) : \boldsymbol{\beta} \in \Theta\}$ satisfies:

$$\mathbb{E}\left[\sup_{\boldsymbol{\beta}_1:\|\boldsymbol{\beta}_1-\boldsymbol{\beta}\|<\delta}\left|L'(Y-g(T;\ \boldsymbol{\beta}_1)) - L'(Y-g(T;\boldsymbol{\beta}))\right|^2\right]^{1/2} \le a\cdot\delta^b$$

for any $\boldsymbol{\beta} \in \Theta$ and any small $\delta > 0$ and for some finite positive constants a and b.

Assumption 6 (i) imposes sufficient regularity conditions on both the regression function and the loss function. These conditions permit nonsmooth loss functions and are satisfied by the examples mentioned in previous sections. Assumption 6 (ii) ensures that the efficient variance to be finite. Assumption 6 (iv) is essentially saying that the almost sure first-order condition is approximately satisfied. Assumption 7 is a stochastic equicontinuity condition, which is needed for establishing weak convergence, see [21]. Again, it is satisfied by widely used loss functions such as $L(v) = v^2$, $L(v) = v\{\tau - I(v \le 0)\}$, and $L(v) = v^2 \cdot |\tau - I(v \le 0)|$ discussed in Sect. 2.1.

Under the above sufficient conditions, we have the following theorem.

Theorem 2.5 *Suppose that Assumptions 1–7 hold, and strengthen Assumption 4 (ii) to*

$$\textit{Assumption 24 (ii)}'\ \ \zeta(K)\sqrt{K^2/N} \to 0 \textit{ and } \sqrt{N}K^{-\alpha} \to 0.$$

Then, $\sqrt{N}(\widehat{\boldsymbol{\beta}} - \boldsymbol{\beta}^*) \xrightarrow{d} N(0, V_{eff})$, *where* $V_{eff} = \mathbb{E}\left[S_{eff}(T,\boldsymbol{X},Y;\boldsymbol{\beta}^*)S_{eff}(T,\boldsymbol{X},Y;\boldsymbol{\beta}^*)^\top\right]$. *Therefore,* $\widehat{\boldsymbol{\beta}}$ *attains the semiparametric efficiency bound of Theorem 2.1.*

Assumption 4 (ii)$'$ imposes further restrictions on the smoothing parameter (K) so that the sieve approximation is under-smoothed. This condition is stronger than Assumption 4 (ii) but it is commonly imposed in the semiparametric regression literature. The proof of Theorem 2.5 is given in the supplemental material of [8].

The asymptotic normality of $\widehat{\boldsymbol{\beta}} = (\widehat{\beta}_0, \widehat{\beta}_1, ..., \widehat{\beta}_{p-1})^\top$ established in Theorem 2.5 has a direct implication for constructing the confidence interval of $\boldsymbol{\beta}^* = (\beta_0^*, \beta_1^*, ..., \beta_{p-1}^*)^\top$. The 95% symmetric confidence interval for β_j^* is given by

$$\left[\hat{\beta}_j - 1.96 \cdot \widehat{SE}_j, \quad \hat{\beta}_j + 1.96 \cdot \widehat{SE}_j\right], \tag{2.11}$$

where $\widehat{SE}_j = \widehat{V}_{jj}^{1/2}/\sqrt{N}$ is the standard error of $\hat{\beta}_j$, and $\widehat{V}_{jj}$ is a consistent estimator for V_{jj}. Here, V_{ij} denotes the (i, j)-element of V_{eff}, the asymptotic covariance matrix of the estimator (recall (2.4)). We describe the Jackknife and bootstrap methods for computing the standard error $\widehat{SE}_j$.

First, the Jackknife method proceeds as follows. The ith Jackknife sample is constructed by deleting the ith observation from the dataset:

$$\boldsymbol{J}^{[-i]} := \left\{T_j, \boldsymbol{X}_j, Y_j : j \in \{1, 2, ..., i-1, i+1, ..., N\}\right\}.$$

The ith Jackknife replicate, denoted as $\widehat{\boldsymbol{\beta}}^{[-i]} = (\widehat{\beta}_0^{[-i]}, \widehat{\beta}_1^{[-i]}, ..., \widehat{\beta}_{p-1}^{[-i]})^\top$, is defined as the point estimator for $\boldsymbol{\beta}^*$ computed on the ith Jackknife sample $\boldsymbol{J}_{[-i]}$. The Jackknife-based standard error of estimated β_j^* is given by

$$\widehat{SE}_j^{jack} = \left\{\frac{N-1}{N}\sum_{i=1}^{N}\left(\widehat{\beta}_j^{[-i]} - \widehat{\beta}_j^{[\cdot]}\right)^2\right\}^{\frac{1}{2}}, \tag{2.12}$$

where $\widehat{\beta}_j^{[\cdot]} = N^{-1}\sum_{i=1}^{N}\widehat{\beta}_j^{[-i]}$. Substitute (2.12) into (2.11) to compute the confidence interval.

Second, the bootstrap method proceeds as follows. The bth bootstrap sample $\{T_i^{\{b\}}, \boldsymbol{X}_i^{\{b\}}, Y_i^{\{b\}}\}_{i=1}^N$ is resampled with replacement from the original sample $\{T_i, \boldsymbol{X}_i, Y_i\}_{i=1}^N$ with the uniform probability. The bth bootstrap replicate, denoted by $\widehat{\boldsymbol{\beta}}^{\{b\}} = (\widehat{\beta}_0^{\{b\}}, \widehat{\beta}_1^{\{b\}}, ..., \widehat{\beta}_{p-1}^{\{b\}})^\top$, is defined as the point estimator for $\boldsymbol{\beta}^*$ computed on the bth bootstrap sample. Repeat B times to get $\{\widehat{\boldsymbol{\beta}}^{\{b\}}\}_{b=1}^B$. The bootstrapped standard error of estimated β_j^* is given by

$$\widehat{SE}_j^{boot} = \left\{\frac{1}{B}\sum_{b=1}^{B}\left(\widehat{\beta}_j^{\{b\}} - \widehat{\beta}_j^{\{\cdot\}}\right)^2\right\}^{\frac{1}{2}}, \tag{2.13}$$

where $\widehat{\beta}_j^{\{\cdot\}} = B^{-1}\sum_{b=1}^{B}\widehat{\beta}_j^{\{b\}}$. Substitute (2.13) into (2.11) to compute the confidence interval. (An alternative bootstrap approach can be found in [22], Sect. 3.3).

Bootstrapping provides another way to construct a confidence interval. Sort the B bootstrap replicates from the smallest to the largest, and relabel them as $\widehat{\beta}_j^{(1)} \leq \cdots \leq \widehat{\beta}_j^{(B)}$. The 95% bootstrapped equitailed confidence interval for β_j^* is given by

$$\left[\widehat{\beta}_j^{(0.025B)}, \quad \widehat{\beta}_j^{(0.975B)}\right]. \tag{2.14}$$

The entire confidence interval (2.11) can be replaced with (2.14). The former bootstrap approach (2.13) relies on the asymptotic normality result, while the latter approach (2.14) does not. We distinguish them hereafter, calling the former *bootstrap method I* and the latter *bootstrap method II*.

2.5 Model Specification Tests

In previous section, we studied continuous treatment effects by imposing a *univariate generalized parametric* model for the functionals of the potential outcome over the treatment variable. If the parametric model is mis-specified, the estimation is biased. This section studies the question of model specification. We propose a consistent specification test for the general continuous treatment effect model.

Let $m\{Y; g(T; \boldsymbol{\beta}^*)\} = L'(Y - g(T; \boldsymbol{\beta}^*))$, $L'(\cdot)$ is the first-order derivative of the general loss function $L(\cdot)$, and $g(t; \boldsymbol{\beta})$ is a parametric working model which is differentiable with respect to $\boldsymbol{\beta}$. We are interested in testing the null hypothesis:

$$H_0 : \mathbb{P}\left(\mathbb{E}[\pi_0(T, \boldsymbol{X})m\{Y; g(T; \boldsymbol{\beta}^*)\}|T] = 0\right) = 1 \text{ for some} \boldsymbol{\beta}^* \in \Theta , \tag{2.15}$$

against the alternative hypothesis

$$H_1 : \mathbb{P}\left(\mathbb{E}[\pi_0(T, \boldsymbol{X})m\{Y; g(T; \boldsymbol{\beta})\}|T] \neq 0\right) > 0 \text{ for all} \boldsymbol{\beta} \in \Theta .$$

where Θ is a compact set in $\mathbb{R}^p$ for some integer $p \geq 1$. Letting

$$U_i := \pi_0(T_i, \boldsymbol{X}_i)m\{Y_i; g(T_i; \boldsymbol{\beta}^*)\} , \tag{2.16}$$

the null hypothesis H_0 is equivalent to $\mathbb{P}\{\mathbb{E}(U_i|T_i) = 0\} = 1$. A popular technique for testing such a conditional moment model is to convert it to an unconditional one. Note that $\mathbb{P}\{\mathbb{E}(U_i|T_i) = 0\} = 1$ if and only if $\mathbb{E}\{U_i M(T_i)\} = 0$ for all bounded and measurable functions $M(\cdot)$. By choosing a proper weight function $\mathscr{H}(\cdot, \cdot)$, $\mathbb{E}(U_i|T_i) = 0$ is a.s. equivalent to

$$\mathbb{E}\left\{U_i \mathscr{H}(T_i, t)\right\} = 0 \text{ for all } t \in \mathcal{T} . \tag{2.17}$$

Popular choices of such a weight function are the logistic function $\mathscr{H}(T_i, t) = 1/\{1 + \exp(c - t \cdot T_i)\}$ with $c \neq 0$, cosine-sine function $\mathscr{H}(T_i, t) = \cos(t \cdot T_i) +$

$\sin(t \cdot T_i)$ and the indicator function $\mathscr{H}(T_i, t) = \mathbb{1}(T_i \le t)$. Now, letting

$$J_N^0(t) = \frac{1}{\sqrt{N}} \sum_{i=1}^{N} U_i \mathscr{H}(T_i, t), \tag{2.18}$$

the sample analogue of $\mathbb{E}\{U_i \mathscr{H}(T_i, t)\}$ multiplied by $\sqrt{N}$, one can test H_0 using the Cramer-von Mises (CM)-type statistic

$$CM_N^0 = \int \{J_N^0(t)\}^2 \widehat{F}_T(dt) = \frac{1}{N} \sum_{i=1}^{N} \{J_N^0(T_i)\}^2, \tag{2.19}$$

or the Kolmogorov-Smirnov (KS)-type statistic

$$KS_N^0 = \sup_{t \in \mathcal{T}} |J_N^0(t)|, \tag{2.20}$$

where $\widehat{F}_T(\cdot)$ is the empirical distribution of $T_1, ..., T_N$. However, both $\pi_0(T, \boldsymbol{X})$ and $\boldsymbol{\theta}^*$ are unknown in practice so that the U_i's are unavailable. We must replace the U_i's with some estimates, which is studied in the following section.

Using the estimated the weights $\widehat{\pi}_K(T, \boldsymbol{X})$ of (2.8), we estimate $\boldsymbol{\beta}^*$ by

$$\widehat{\boldsymbol{\beta}} := \arg\min_{\boldsymbol{\beta} \in \Theta} \|M_N(\boldsymbol{\beta}, \widehat{\pi}_K)\|, \tag{2.21}$$

where $\|\cdot\|$ is the Euclidean norm, and

$$M_N(\boldsymbol{\beta}, \pi) := \frac{1}{N} \sum_{i=1}^{N} \pi(T_i, \boldsymbol{X}_i) m\{Y_i; g(T_i; \boldsymbol{\beta})\} w(T_i; \boldsymbol{\beta}),$$

where $w(T; \boldsymbol{\beta})$ (which may possibly not involve $\boldsymbol{\beta}$) is a prespecified q-dimensional vector with $q \ge p$ such that, under H_0, $\boldsymbol{\beta}^*$ is identified or over-identified. Examples of such vectors include $w(T; \boldsymbol{\beta}) = (1, T, ..., T^{q-1})^\top$ or $w(T; \boldsymbol{\beta}) = \nabla_{\boldsymbol{\beta}} g(T; \boldsymbol{\beta})$, where "$\nabla_{\boldsymbol{\beta}}''$ denotes the derivative with respect to $\boldsymbol{\beta}$.

With the estimators $\{\widehat{\pi}_K(T_i, \boldsymbol{X}_i)\}_{i=1}^N$ of $\{\pi_0(T_i, \boldsymbol{X}_i)\}_{i=1}^N$ and $\widehat{\boldsymbol{\beta}}$ of $\boldsymbol{\beta}^*$, we estimate U_i by $\widehat{U}_i = \widehat{\pi}_K(T_i, \boldsymbol{X}_i) m\{Y_i; g(T_i; \widehat{\boldsymbol{\beta}})\}$, for $i = 1, \ldots, N$. Replacing the U_i's in (2.18) by the $\widehat{U}_i$'s, we have a feasible test statistic for H_0 based on

$$\widehat{J}_N(t) = \frac{1}{\sqrt{N}} \sum_{i=1}^{N} \widehat{U}_i \mathscr{H}(T_i, t),$$

the corresponding estimators of the Cramer-von Mises (CM)-type statistic in (2.19) and the Kolmogorov-Smirnov (KS)-type statistic in (2.20) are, respectively,

$$\widehat{CM}_N = \frac{1}{N}\sum_{i=1}^{N}\{\widehat{J}_N(T_i)\}^2 \quad \text{and} \quad \widehat{KS}_N = \sup_{t\in\mathcal{T}} |\widehat{J}_N(t)|\,, \tag{2.22}$$

where the supremum is calculated as the maximum value over a discretization of $\mathcal{T}$ in practice.

2.5.1 Asymptotic Properties Under Null Hypothesis

To establish the asymptotic properties of $\widehat{J}_N(\cdot)$ and $\widehat{CM}_N$, the following additional assumptions are imposed.

Assumption 8 Under H_0, (i) $\boldsymbol{\beta}^*$ is an interior point of Θ, where Θ is a compact set in $\mathbb{R}^p$; (ii) $\|M_N(\widehat{\boldsymbol{\beta}}, \widehat{\pi}_K)\| = \inf_{\boldsymbol{\beta}\in\Theta_\delta} \|M_N(\boldsymbol{\beta}, \widehat{\pi}_K)\| + o_P(N^{-1/2})$, where $\Theta_\delta := \{\boldsymbol{\beta} \in \Theta : \|\boldsymbol{\beta} - \boldsymbol{\beta}^*\| \le \delta\}$.

Assumption 9 Let $\eta(T, \boldsymbol{X}, Y; t)$ be defined in (2.24), $Var\{\eta(T, \boldsymbol{X}, Y; t)\} < \infty$ for all $t \in \mathcal{T}$.

Assumption 10 (i) $g(t; \boldsymbol{\beta})$ is twice continuously differentiable in $\boldsymbol{\beta} \in \Theta$;
(ii) $\mathbb{E}[m\{Y; g(T; \boldsymbol{\beta}^*)\}|T = t, \boldsymbol{X} = \boldsymbol{x}]$ is continuously differentiable in $(t, \boldsymbol{x})$;
(iii) $\mathbb{E}\left[\pi_0(T, \boldsymbol{X})m\{Y; g(T; \boldsymbol{\beta})\}w(T; \boldsymbol{\beta})|T = t, \boldsymbol{X} = \boldsymbol{x}\right]$ is differentiable w.r.t. $\boldsymbol{\beta}$ and $\nabla_{\boldsymbol{\beta}}\mathbb{E}[\pi_0(T, \boldsymbol{X})m\{Y; g(T; \boldsymbol{\beta})\}w(T; \boldsymbol{\beta})]\big|_{\boldsymbol{\beta}=\boldsymbol{\beta}^*}$ is of full (column) rank.

Assumption 11 (i) $\mathbb{E}\left[\sup_{\boldsymbol{\beta}\in\Theta} |m\{Y; g(T; \boldsymbol{\beta})\}|^{2+\delta}\right] < \infty$ for some $\delta > 0$; (ii) The function class $\left\{m\{Y; g(T; \boldsymbol{\beta})\} : \boldsymbol{\beta} \in \Theta\right\}$ satisfies:

$$\mathbb{E}\left[\sup_{\boldsymbol{\beta}_1:\|\boldsymbol{\beta}_1-\boldsymbol{\beta}\|<\delta} \left|m\{Y; g(T; \boldsymbol{\beta}_1)\} - m\{Y; g(T; \boldsymbol{\beta})\}\right|^2\right]^{1/2} \le C\cdot\delta$$

for any $\boldsymbol{\beta} \in \Theta$ and any small $\delta > 0$ and for some finite positive constant C.

Assumption 8 is essentially stating that the estimating equation is a.s. approximately satisfied; see [23]. Assumption 9 is needed to bound the asymptotic variance of the test statistic. Assumption 10 (i) and (ii) impose sufficient regularity conditions on both the link function g and residual function m. Assumption 10 (iii) ensures that the variance of the test statistic is finite. Assumption 11 is a stochastic equicontinuity condition, which is needed to establish the weak convergence of our test statistic; see [21].

To aid presentation of the asymptotic properties of the test statistic, define the following quantities:

$$\phi(T_i, \boldsymbol{X}_i; t) := \pi_0(T_i, \boldsymbol{X}_i) \cdot \mathscr{H}(T_i, t) \cdot \mathbb{E}[m\left\{Y_i; g(T_i; \boldsymbol{\beta}^*)\right\} | T_i, \boldsymbol{X}_i] \\ - \mathbb{E}[\pi_0(T_i, \boldsymbol{X}_i) m\left\{Y_i; g(T_i; \boldsymbol{\beta}^*)\right\} \cdot \mathscr{H}(T_i, t) | \boldsymbol{X}_i],$$

and

$$\begin{aligned}\psi(T_i, \boldsymbol{X}_i, Y_i; t) :=& \mathbb{E}\left[\pi_0(T_i, \boldsymbol{X}_i) \cdot \frac{\partial}{\partial g}\mathbb{E}[m\left\{Y_i; g(T_i; \boldsymbol{\beta}^*)\right\} | T_i, \boldsymbol{X}_i] \cdot \nabla_{\boldsymbol{\beta}} g(T_i; \boldsymbol{\beta}^*)^\top \mathscr{H}(T_i, t)\right] \\ &\times \left\{\mathbb{E}\left[\pi_0(T_i, \boldsymbol{X}_i) \cdot \frac{\partial}{\partial g}\mathbb{E}[m\left\{Y_i; g(T_i; \boldsymbol{\beta}^*)\right\} | T_i, \boldsymbol{X}_i] \cdot \nabla_{\boldsymbol{\theta}} g(T_i; \boldsymbol{\beta}^*) w(T_i; \boldsymbol{\beta}^*)^\top\right]\right. \\ &\left.\cdot \mathbb{E}\left[\pi_0(T_i, \boldsymbol{X}_i) \cdot \frac{\partial}{\partial g}\mathbb{E}[m\left\{Y_i; g(T_i; \boldsymbol{\beta}^*)\right\} | T_i, \boldsymbol{X}_i] \cdot w(T_i; \boldsymbol{\beta}^*) \nabla_{\boldsymbol{\theta}}^\top g(T_i; \boldsymbol{\beta}^*)\right]\right\}^{-1} \\ &\times \mathbb{E}\left[\pi_0(T_i, \boldsymbol{X}_i) \cdot \frac{\partial}{\partial g}\mathbb{E}[m\left\{Y_i; g(T_i; \boldsymbol{\beta}^*)\right\} | T_i, \boldsymbol{X}_i] \cdot \nabla_{\boldsymbol{\beta}} g(T_i; \boldsymbol{\beta}^*) w(T_i; \boldsymbol{\beta}^*)^\top\right] \\ &\times \left\{\pi_0(T_i, \boldsymbol{X}_i) m\left\{Y_i; g(T_i; \boldsymbol{\beta}^*)\right\} w(T_i; \boldsymbol{\beta}^*)\right. \\ &- \pi_0(T_i, \boldsymbol{X}_i) w(T_i; \boldsymbol{\beta}^*) \cdot \mathbb{E}[m\left\{Y_i; g(T_i; \boldsymbol{\beta}^*)\right\} | T_i, \boldsymbol{X}_i] \\ &\left.+ \mathbb{E}[\pi_0(T_i, \boldsymbol{X}_i) w(T_i; \boldsymbol{\beta}^*) m\left\{Y_i; g(T_i; \boldsymbol{\beta}^*)\right\} | \boldsymbol{X}_i]\right\},\end{aligned} \tag{2.23}$$

and

$$\eta(T_i, \boldsymbol{X}_i, Y_i; t) := U_i \mathscr{H}(T_i, t) - \phi(T_i, \boldsymbol{X}_i; t) - \psi(T_i, \boldsymbol{X}_i, Y_i; t). \tag{2.24}$$

The next theorem establishes the weak convergence of $\widehat{J}_N(\cdot)$ and $\widehat{CM}_N$ under H_0.

Theorem 2.6 *Suppose that Assumptions 1–4 and Assumptions 8–11; then, under* H_0,

$$(i) \quad \widehat{J}_N(t) = \frac{1}{\sqrt{N}} \sum_{i=1}^N \eta(T_i, \boldsymbol{X}_i, Y_i; t) + o_P(1), \ \textit{for all } t \in \mathcal{T},$$

$$(ii) \quad \widehat{J}_N(\cdot) \textit{ converges weakly to } J_\infty(\cdot) \textit{ in } L_2\{\mathcal{T}, dF_T(t)\},$$

where J_∞ *is a Gaussian process with zero mean and covariance function given by*

$$\Sigma(t, t') = \mathbb{E}\left\{\eta(T_i, \boldsymbol{X}_i, Y_i; t)\eta(T_i, \boldsymbol{X}_i, Y_i; t')\right\}.$$

Furthermore,

$$(iii) \quad \widehat{CM}_N \textit{ converges to } \int \{J_\infty(t)\}^2 dF_T(t) \textit{ in distribution},$$

$$(iv) \quad \widehat{KS}_N \textit{ converges to } \sup_{t \in \mathcal{T}} |J_\infty(t)| \textit{ in distribution}.$$

The proof of Theorem 2.6 is the supplementary file of [24]. It can be shown that $\int\{J_\infty(t)\}^2 dF_T(t)$ can be written as an infinite sum of weighted (independent) χ_1^2 random variables with weights depending on the unknown distribution of $(T_i, \boldsymbol{X}_i, Y_i)$.

Hence, it is difficult to obtain the exact critical values. We suggest a simulation method to approximate the critical values for the null limiting distribution of $\widehat{CM}_N$ in the next section.

The effect of the vector $w(T; \boldsymbol{\theta})$ on the asymptotic property of our test statistic is reflected in the term $\psi(T_i, \boldsymbol{X}_i, Y_i, t)$. It is unclear which choice of $w(T; \boldsymbol{\theta})$ would minimize the variance of $\Sigma(t, t)$. A common choice is $w(T; \boldsymbol{\theta}) = \nabla_{\boldsymbol{\theta}} g(T; \boldsymbol{\theta})$. Then, the second and fourth terms of $\psi(T_i, \boldsymbol{X}_i, Y_i, t)$ in (2.23) are canceled out, which also simplifies the calculation approximating the null limiting distribution in practice.

2.5.2 *Approximation for the Null Limiting Distribution*

We know from Theorem 2.6 that $\widehat{CM}_N$ converges in distribTheoution to $\int \{J_\infty(t)\}^2 \, dF_T(t)$. Obtaining the exact critical values is difficult and we here propose a simulation method to approximate the null limiting distribution. Specifically, we first generate B sets of N independent standard normal random variables $w_{1,b}, \ldots, w_{N,b}$, for $b = 1, \ldots, B$ and B a large enough integer. Then we define

$$\widehat{J}^*_{N,b}(t) = \frac{1}{\sqrt{N}} \sum_{i=1}^{N} w_{i,b} \widehat{\eta}(T_i, \boldsymbol{X}_i, Y_i; t), \tag{2.25}$$

where $\widehat{\eta}(T_i, \boldsymbol{X}_i, Y_i; t) = \widehat{U}_i \mathscr{H}(T_i, t) - \widehat{\phi}(T_i, \boldsymbol{X}_i; t) - \widehat{\psi}(T_i, \boldsymbol{X}_i, Y_i; t)$, with $\widehat{\phi}(T_i, \boldsymbol{X}_i; t)$ and $\widehat{\psi}(T_i, \boldsymbol{X}_i, Y_i; t)$, respectively, some consistent nonparametric plug-in estimators of $\phi(T_i, \boldsymbol{X}_i; t)$ and $\psi(T_i, \boldsymbol{X}_i, Y_i; t)$ defined above in Theorem 2.6, for example, the additive penalized spline estimator or the series estimator used in [25].

It is easy to see that $\mathbb{E}^*\{w_{i,b}\widehat{\eta}(T_i, \boldsymbol{X}_i, Y_i; t)\} = 0$ and $\mathbb{E}^*\{w_{i,b}^2 \widehat{\eta}(T_i, \boldsymbol{X}_i, Y_i; t)\widehat{\eta}(T_i, \boldsymbol{X}_i, Y_i; t')\} = \widehat{\eta}(T_i, \boldsymbol{X}_i, Y_i; t)\widehat{\eta}(T_i, \boldsymbol{X}_i, Y_i; t')$, for $i = 1, \ldots, N$, $b = 1, \ldots, B$ and all $t, t' \in \mathcal{T}$, where $\mathbb{E}^*\{\cdot\}$ is the conditional expectation given the data $(T_i, \boldsymbol{X}_i, Y_i)_{i=1}^N$. Because $\widehat{\eta}$ is a consistent estimator of η, then $\widehat{J}^*_{N,b}(\cdot)$ has the same asymptotic behavior as $\widehat{J}_N(\cdot)$ for $b = 1, \ldots, B$. Then, we can approximate the limiting distributions of $\widehat{CM}_N$ and $\widehat{KS}_N$ under H_0, respectively, by

$$\widehat{CM}^*_{N,b} = \frac{1}{N} \sum_{i=1}^{N} \left\{\widehat{J}^*_{N,b}(T_i)\right\}^2 \quad \text{and} \quad \widehat{KS}^*_{N,b} = \sup_{t \in \mathcal{T}} \left|\widehat{J}^*_{N,b}(t)\right|,$$

for $b = 1, \ldots, B$. That is, we can approximate the p-value for the CM-type statistic by $B^{-1} \sum_{b=1}^{B} \mathbb{1}(\widehat{CM}^*_{N,b} \geq \widehat{CM}_N)$ and that for the KS-type statistic by $B^{-1} \sum_{b=1}^{B} \mathbb{1}(\widehat{KS}^*_{N,b} \geq \widehat{KS}_N)$.

2.6 Nonparametric Estimation of ATE

In this section, we consider the nonparametric estimation of the dose-response curve $\theta_t := \mathbb{E}[Y(t)]$ and the average treatment effects (ATE), which is defined by $\theta_{t_1,t_0} := \mathbb{E}[Y(t_1) - Y(t_0)]$ for $t_1 \neq t_0$.

We begin with estimation of θ_t. Note that, for all $t \in \mathcal{T}$ and under Assumption 1, we can rewrite θ_t as

$$\theta_t := \mathbb{E}[Y(t)] = \mathbb{E}\left[\pi_0(T, \boldsymbol{X})Y | T = t\right].$$

With $\pi_0(T, \boldsymbol{X})$ replaced by $\widehat{\pi}_K(T, \boldsymbol{X})$, we estimate θ_t by regressing $\widehat{\pi}_K(T, \boldsymbol{X})Y$ on $u_{K_1}(t)$, thus

$$\widehat{\theta}_t := \left[\sum_{i=1}^{N} \widehat{\pi}_K(T_i, \boldsymbol{X}_i)Y_i u_{K_1}(T_i)^\top\right]\left[\sum_{i=1}^{N} u_{K_1}(T_i)u_{K_1}(T_i)^\top\right]^{-1} u_{K_1}(t).$$

To aid presentation of the asymptotic properties of $\widehat{\theta}_t$, define the following quantities:

$$\begin{aligned}
\Phi_{K_1\times K_1} &:= \mathbb{E}[u_{K_1}(T)u_{K_1}^\top(T)], \\
b_{K_1}(T_i, \boldsymbol{X}_i, Y_i) &:= \pi_0(T_i, \boldsymbol{X}_i)Y_i \cdot u_{K_1}(T_i) - \mathbb{E}\left[\pi_0(T_i, \boldsymbol{X}_i)Y_i \cdot u_{K_1}(T_i)|T_i, \boldsymbol{X}_i\right] \\
&\quad + \mathbb{E}\left[\pi_0(T_i, \boldsymbol{X}_i)Y_i \cdot u_{K_1}(T_i)|\boldsymbol{X}_i\right] - \mathbb{E}\left[\pi_0(T_i, \boldsymbol{X}_i)Y_i \cdot u_{K_1}(T_i)\right], \\
V_t &:= \mathbb{E}\left[\left\{u_{K_1}^\top(t)\Phi_{K_1\times K_1}^{-1} b_{K_1}(T_i, \boldsymbol{X}_i, Y_i)\right\}^2\right] \\
&= u_{K_1}^\top(t) \cdot \Phi_{K_1\times K_1}^{-1} \cdot \mathbb{E}\left[b_{K_1}(T_i, \boldsymbol{X}_i, Y_i)b_{K_1}^\top(T_i, \boldsymbol{X}_i, Y_i)\right] \cdot \Phi_{K_1\times K_1}^{-1} \cdot u_{K_1}(t).
\end{aligned}$$

Theorem 2.7 *Suppose* $\sup_{t\in\mathcal{T}} |\theta_t - (\gamma^*)^\top u_{K_1}(t)| = O(K_1^{-\tilde{\alpha}})$ *holds for some* $\tilde{\alpha} > 0$ *and* $\gamma^* \in \mathbb{R}^{K_1}$, $\lambda_{\min}\left\{\mathbb{E}\left[b_{K_1}(T, \boldsymbol{X}, Y)b_{K_1}^\top(T, \boldsymbol{X}, Y)\right]\right\} \geq \underline{c} > 0$, *and Assumptions 1–4 hold. Then:*

1. *(Consistency)*

$$\int_{\mathcal{T}} |\widehat{\theta}_t - \theta_t|^2 dF_T(t) = O_p\left(\zeta(K)^2\left\{\frac{K}{N} + K^{-2\alpha}\right\} + K_1^{-2\tilde{\alpha}}\right).$$

$$\sup_{t\in\mathcal{T}} |\hat{\theta}_t - \theta_t| = O_p\left(\zeta_1(K_1)\left\{\zeta(K)\left(\sqrt{\frac{K}{N}} + K^{-\alpha}\right) + K_1^{-\tilde{\alpha}}\right\}\right).$$

2. *(Asymptotic Normality) suppose Assumption 6' and* $\sqrt{N}K_1^{-\tilde{\alpha}} \to 0$ *hold. Then for any fixed* $t \in \mathcal{T}$,

$$\sqrt{N}V_t^{-1/2}\left[\hat{\theta}_t - \theta_t\right] \xrightarrow{d} N(0, 1).$$

The proof of Theorem 2.7 can be found in the supplemental file [12]. The proposed estimation procedure can also be used to estimate the average treatment effects (ATE) which is defined by

$$\theta_{t_1,t_0} := \mathbb{E}[Y^*(t_1) - Y^*(t_0)] = \theta_{t_1} - \theta_{t_0} \text{ for } t_1 \neq t_0.$$

The estimator of θ_{t_1,t_0} is defined by $\widehat{\theta}_{t_1,t_0} := \widehat{\theta}_{t_1} - \widehat{\theta}_{t_0}$. Let

$$\begin{aligned} V_{t_1,t_0} &:= \mathbb{E}\left[\left\{u_{K_1}^\top(t_1)\Phi_{K_1\times K_1}^{-1} b_{K_1}(T_i, \boldsymbol{X}_i, Y_i) - u_{K_1}^\top(t_0)\Phi_{K_1\times K_1}^{-1} b_{K_1}(T_i, \boldsymbol{X}_i, Y_i)\right\}^2\right] \\ &= \left\{u_{K_1}(t_1) - u_{K_1}(t_0)\right\}^\top \Phi_{K_1\times K_1}^{-1} \mathbb{E}\left[b_{K_1}(T_i, \boldsymbol{X}_i, Y_i) b_{K_1}^\top(T_i, \boldsymbol{X}_i, Y_i)\right] \Phi_{K_1\times K_1}^{-1} \left\{u_{K_1}(t_1) - u_{K_1}(t_0)\right\}. \end{aligned}$$

Similar to prove Theorem 2.7, we have the following corollary:

Corollary 2.1 *Suppose* $\sup_{t\in\mathcal{T}} |\theta_t - (\gamma^*)^\top u_{K_1}(t)| = O(K_1^{-\tilde{\alpha}})$ *holds for some* $\tilde{\alpha} > 0$ *and* $\gamma^* \in \mathbb{R}^{K_1}$, $\lambda_{\min}\left\{\mathbb{E}\left[b_{K_1}(T, \boldsymbol{X}, Y) b_{K_1}^\top(T, \boldsymbol{X}, Y)\right]\right\} \geq \underline{c} > 0$, *Assumptions 1–6' hold, and* $\sqrt{N}K_1^{-\tilde{\alpha}} \to 0$. *Then*

$$\sqrt{N} V_{t_1,t_0}^{-1/2} \left[\widehat{\theta}_{t_1,t_0} - \theta_{t_1,t_0}\right] \xrightarrow{d} N(0, 1).$$

Feasible versions of the above CLT's are implemented using plug-in sieve estimation of the unknown quantities. For example, V_t can be estimated by

$$\widehat{V}_t = \frac{1}{N}\sum_{i=1}^{N} \left\{u_{K_1}^\top(t)\widehat{\Phi}_{K_1\times K_1}^{-1} \widehat{b}_{K_1}(T_i, \boldsymbol{X}_i, Y_i)\right\}^2,$$

where $\widehat{\Phi}_{K_1\times K_1} := N^{-1}\sum_{i=1}^{N} u_{K_1}(T_i) u_{K_1}^\top(T_i)$,

$$\begin{aligned} \widehat{b}_{K_1}(T_i, \boldsymbol{X}_i, Y_i) : &= \widehat{\pi}_K(T_i, \boldsymbol{X}_i) Y_i \cdot u_{K_1}(T_i) - \widehat{\mathbb{E}}\left[\widehat{\pi}_K(T_i, \boldsymbol{X}_i) Y_i \cdot u_{K_1}(T_i) | T_i, \boldsymbol{X}_i\right] \\ &+ \widehat{\mathbb{E}}[\widehat{\pi}_K(T_i, \boldsymbol{X}_i) Y_i \cdot u_{K_1}(T_i) | \boldsymbol{X}_i] - \widehat{\mathbb{E}}[\widehat{\pi}_K(T_i, \boldsymbol{X}_i) Y_i \cdot u_{K_1}(T_i)] \end{aligned}$$

is the plug-in estimates of $b_{K_1}(T_i, \boldsymbol{X}_i, Y_i)$, and $\widehat{\mathbb{E}}[\widehat{\pi}_K(T, \boldsymbol{X}) Y u_{K_1}(T) | T, \boldsymbol{X}]$ is the least square regression of $\widehat{\pi}_K(T, \boldsymbol{X}) Y u_{K_1}(T)$ on a sieve basis $w_{K_0}(T, \boldsymbol{X})$, and $\widehat{\mathbb{E}}[\widehat{\pi}_K(T, \boldsymbol{X}) Y\ u_{K_1}(T) | \boldsymbol{X}]$ is the least square regression of $\widehat{\pi}_K(T, \boldsymbol{X}) Y u_{K_1}(T)$ on a sieve basis $v_{K_0}(\boldsymbol{X})$.

2.7 Nonparametric Estimation of Distributional and Quantile Treatment Effects

In this section, we estimate the counterfactual distribution function $F_t(\cdot) := P(Y(t) \leq y)$ and its quantile function $q_t(u) := \inf\{z : F_t(z) \geq u\}$ for $u \in [0, 1]$ for all treatment status.

Under Assumption 1, $F_t(y)$ is identified as

$$F_t(y) = \mathbb{E}\{\pi_0(T, \boldsymbol{X})I(Y \leq y)|T = t\}. \tag{2.26}$$

We notice that $\mathbb{E}\{\pi_0(T, \boldsymbol{X})|T = t\} = 1$. For normalization purpose (explained below), we write

$$F_t(y) = \frac{\mathbb{E}\{\pi_0(T, \boldsymbol{X})I(Y \leq y)|T = t\}}{\mathbb{E}\{\pi_0(T, \boldsymbol{X})|T = t\}}.$$

The counterfactual distribution $F_t(y)$ is now estimated by replacing the true weights with the estimated ones:

$$\widehat{F}_{t,h}(y) := \frac{\sum_{i=1}^{N} \widehat{\pi}_K(T_i, \boldsymbol{X}_i)I(Y_i \leq y)K\left(\frac{T_i-t}{h}\right)}{\sum_{i=1}^{N} \widehat{\pi}_K(T_i, \boldsymbol{X}_i)K\left(\frac{T_i-t}{h}\right)},$$

and the quantile function of $Y^*(t)$ is estimated by inverting the estimated distribution function:

$$\widehat{q}_{t,h}(u) := \inf\left\{y : \widehat{F}_{t,h}(y) \geq u\right\}, \quad u \in [0, 1].$$

To establish the asymptotic properties of the estimated distribution and quantile functions, the following additional conditions shall be maintained throughout the paper.

Assumption 12 For any $t \in \mathcal{T}$, the potential outcome $Y(t)$ has a compact support.

Assumption 13 (i) For any given $(t, \boldsymbol{x}) \in \mathcal{T} \times \mathcal{X}$, the conditional distribution function $F_{Y|T,X}(y|t, \boldsymbol{x})$ is continuous in $y \in \mathcal{Y}$; (ii) For any $y \in \mathcal{Y}$, $F_{Y|T,X}(y|t, \boldsymbol{x})$ is continuously differentiable in $(t, x) \in \mathcal{T} \times \mathcal{X}$; (iii) The density function $f_T(t)$ is third-order continuously differentiable; (iv) The function $F_t(y)$ is continuous in y and third-order continuously differentiable in t.

Assumption 14 $K(\cdot)$ is a univariate kernel function, symmetric around origin and satisfying (i) $\int K(u)du = 1$; (ii) $\int u^2K(u)du = \kappa_{21} \in (0, \infty)$; (iii) $\int K^2(u)du = \kappa_{02} < \infty$; (iv) As $N \to \infty$, $Nh \to \infty$ and $h \to 0$.

Assumption 12 is not needed in our derivation of large sample properties of the estimated distribution functions, but it is needed in the derivation of large sample properties of the estimated quantile functions (also see Assumption 3.1 of [25]). Assumption 14 is the standard assumption commonly imposed in the kernel regression literature (see [26]).

Theorem 2.8 *For any fixed $t \in \mathcal{T}$, suppose that Assumptions 1–4 and 12–14 are satisfied and $Nh^5 \to 0$, then we have the following linear representation:*

$$\sup_{y\in\mathcal{Y}}\left|\sqrt{Nh}\left\{\widehat{F}_{t,h}(y)-F_t(y)\right\}-\sqrt{\frac{h}{N}}\sum_{i=1}^{N}\psi_{t,h}(Y_i,T_i,\boldsymbol{X}_i;y)\right|=o_P(1),$$

where

$$\psi_{t,h}(Y_i,T_i,\boldsymbol{X}_i;y):=\frac{\pi_0(T_i,\boldsymbol{X}_i)}{p_{t,h}}K\left(\frac{T_i-t}{h}\right)\left\{I(Y_i\le y)-F_{Y|T,X}(y|T_i,\boldsymbol{X}_i)\right\}$$

and $p_{t,h} := \mathbb{E}\left[K(\frac{T-t}{h})\right]$. Furthermore,

$$\sqrt{Nh}\left\{\widehat{F}_{t,h}(\cdot)-F_t(\cdot)\right\}\Rightarrow\Psi_t(\cdot).$$

where "$\Rightarrow$" denotes weakly convergence, and $\Psi_t(\cdot)$ is a mean zero Gaussian process with covariance function $\Omega_t(y_1, y_2)$ which is defined by

$$\begin{aligned}\Omega_t(y_1,y_2)&:=\lim_{h\to0}h\cdot\mathbb{E}\left[\psi_{t,h}(Y_i,T_i,\boldsymbol{X}_i;y_1)\psi_{t,h}(Y_i,T_i,\boldsymbol{X}_i;y_2)\right]\\&=\frac{\kappa_{02}}{f_T(t)}\cdot\mathbb{E}\Big[\pi_0(T_i,\boldsymbol{X}_i)^2\left\{I(Y_i\le y_1)-F_{Y|T,X}(y_1|T_i,\boldsymbol{X}_i)\right\}\\&\qquad\times\left\{I(Y_i\le y_2)-F_{Y|T,X}(y_2|T_i,\boldsymbol{X}_i)\right\}\Big|T_i=t\Big],\end{aligned}$$

and $\kappa_{ij} := \int u^i K^j(u)du$.

The proof of Theorem 2.8 is presented in the supplemental material of [27].

Given that the quantile map is Hadamard differentiable, the asymptotic properties of quantile function can be derived by applying the functional delta method. By applying Theorem 20.8 and Lemma 21.3 of [28], we obtain

$$\sqrt{Nh}\left\{\widehat{q}_{t,h}(u)-q_t(u)\right\}=-\frac{1}{f_{Y(t)}(q_t(u))}\cdot\sqrt{Nh}\left\{\widehat{F}_{t,h}(q_t(u))-F_t(q_t(u))\right\}+o_p(1).$$

The following result is an application of Theorem 2.8 whose proof is presented in the supplemental material of [27].

Theorem 2.9 *Suppose that Assumptions 1–4 and 12–14 are satisfied and $Nh^5 \to 0$. Then for any fixed $t \in \mathcal{T}$, we have:*

$$\sqrt{Nh}\left\{\widehat{q}_{t,h}(\cdot)-q_t(\cdot)\right\}\Rightarrow\mathcal{Q}_t(\cdot),$$

where $\mathcal{Q}_t(\cdot)$ is a mean zero Gaussian process with covariance function

$$\begin{aligned}\Gamma_t(u_1,u_2) &= \frac{1}{f_{Y(t)}(q_t(u_1))f_{Y(t)}(q_t(u_2))}\\ &\times \frac{\kappa_{02}}{f_T(t)}\cdot\mathbb{E}\Big[\pi_0(T,\boldsymbol{X})^2\left\{I(Y\le q_t(u_1))-F_{Y|T,X}(q_t(u_1)|T,\boldsymbol{X})\right\}\\ &\times\left\{I(Y\le q_t(u_2))-F_{Y|T,X}(q_t(u_2)|T,\boldsymbol{X})\right\}\Big|T=t\Big],\end{aligned}$$

where $(u_1,u_2)\in[0,1]\times[0,1]$.

2.7.1 Consistent Variances

The results of Theorems 2.8 and 2.9 are not sufficient for statistical inference. To conduct statistical inference, we must have consistent variance estimates available. We estimate the variance functions $\Omega_t(y,y)$ and $\Gamma_t(y,y)$ by

$$\widehat{\Omega}_t(y,y)=\frac{\kappa_{02}}{\widehat{f_T}(t)}\cdot\frac{\sum_{i=1}^N\widehat{\pi}_K(T_i,\boldsymbol{X}_i)^2\left\{I(Y_i\le y)-\widehat{F}_{Y|T,X}(y|T_i,\boldsymbol{X}_i)\right\}^2K\left(\frac{T_i-t}{h}\right)}{\sum_{i=1}^N K\left(\frac{T_i-t}{h}\right)}$$

and

$$\widehat{\Gamma}_t(u,u)=\frac{1}{\widehat{f}_{Y^*(t)}(\widehat{q}_t(u_1))^2}\frac{\kappa_{02}}{\widehat{f}_T(t)}\cdot\frac{\sum_{i=1}^N\widehat{\pi}_K(T_i,\boldsymbol{X}_i)^2\left\{I(Y_i\le\widehat{q}_t(u_1))-\widehat{F}_{Y|T,X}(\widehat{q}_t(u_1)|T_i,\boldsymbol{X}_i)\right\}^2K\left(\frac{T_i-t}{h}\right)}{\sum_{i=1}^N K\left(\frac{T_i-t}{h}\right)},$$

where

$$\widehat{f_T}(t):=\frac{1}{Nh_T}\sum_{i=1}^N K\left(\frac{T_i-t}{h_T}\right),\quad \widehat{f}_{Y^*(t)}(y):=\partial_t\widehat{F}_t(y),$$

$$\widehat{F}_{Y|T,X}(y|t,\boldsymbol{x}):=\frac{\sum_{i=1}^N I(Y_i\le y)K\left(\frac{T_i-t}{h_T}\right)\prod_{j=1}^r K\left(\frac{X_i^{(j)}-x^{(j)}}{h_X}\right)}{\sum_{i=1}^N K\left(\frac{T_i-t}{h_T}\right)\prod_{j=1}^r K\left(\frac{X_i^{(j)}-x^{(j)}}{h_X}\right)}.$$

By [26], the kernel regression estimators are consistent if the following condition holds:

Assumption 15 As $N\to\infty$, $Nh_T\to\infty$, $Nh_X^r\to\infty$, $h_T\to0$, and $h_X\to0$.

Under this additional condition, we establish:

Theorem 2.10 *Suppose that Assumptions 1–4 and 12–15 are satisfied. Then we have* $\widehat{\Omega}_t(y,y)\xrightarrow{p}\Omega_t(y,y)$ *and* $\widehat{\Gamma}_t(y,y)\xrightarrow{p}\Gamma_t(y,y)$.

The results of Theorem 2.8–2.10 can be used to construct tests for the distributional effects, a subject we now turn to in the next section.

2.8 Testing Distributional Effects

Detecting evidence of the treatment effect is the primary goal of the program evaluation literature. Although the existing literature is mostly concerned with comparing some moments of the counterfactual distributions (e.g., means and quantiles), the best detection should compare the entire distributions. There are three approaches to conduct the comparison. Below we describe each of the three approaches.

2.8.1 *Uniform Confidence Band*

The first approach is the difference test based on the difference between the counterfactual distributions at two treatment levels $t_1 \neq t_0$. The null and alternative hypothesis are, respectively, given by

$$\begin{aligned} H_0 &: F_{t_1}(y) = F_{t_0}(y) \text{ for all } y \in \mathcal{Y}; \\ H_1 &: F_{t_1}(y) \neq F_{t_0}(y) \text{ for some } y \in \mathcal{Y}. \end{aligned}$$

Obviously, the null hypothesis is equivalent to $F_{t_1}(y) - F_{t_0}(y) = 0$ for all $y \in \mathcal{Y}$, and the corresponding test statistic is $\sqrt{Nh}\left\{\widehat{F}_{t_1,h}(y) - \widehat{F}_{t_0,h}(y)\right\}$ which converges to a Gaussian process uniformly by Theorem 2.8. With the consistent variance provided by Theorem 2.10, we can construct a uniform confidence band for a given level of significance. If 0 is outside the confidence band, then the null hypothesis is rejected. If the null is rejected, we can plot the difference together with the confidence band against y. The plot could reveal the range of y with negative difference and the range of y with positive difference. In other words, we could infer the population benefited from (or harmed by) changing treatment status from t_0 to t_1.

2.8.2 *Mann-Whitney Test*

Despite the difference test can reveal the distributional difference at all outcome levels, it cannot tell how the difference varies with treatment. This is because the difference test depends on the triplet (y, t_0, t_1), and it is difficult to view the four-dimensional graph. An alternative test is the Mann-Whitney test based on the Mann-Whitney indicator $\theta_{t_1,t_0} = \int_{\mathcal{Y}} F_{t_1}(y) dF_{t_0}(y)$. The null hypothesis is equivalent to $\theta_{t_1,t_0} = 1/2$. Moreover, $\theta_{t_1,t_0} > 1/2$ if $F_{t_1}(y) > F_{t_0}(y)$ for all y and $\theta_{t_1,t_0} < 1/2$ if $F_{t_1}(y) < F_{t_0}(y)$ for all y. The advantage of this indicator is that it reduces comparison of two distributions to a single parameter. Moreover, the value of this parameter may reveal the stochastic dominance of potential outcomes. By plotting θ_{t_1,t_0} against (t_1, t_0) we may be able to infer the range of treatment and the possible stochastic dominant outcomes. The Mann-Whitney test statistic is defined by

$$\widehat{\theta}_{t_1,t_0,h} = \int_{\mathcal{Y}} \widehat{F}_{t_1,h}(y)d\widehat{F}_{t_0,h}(y),$$

whose asymptotic distribution is given by

Theorem 2.11 *Suppose that Assumptions 1–4 and 12–15 are satisfied and $Nh^5 \to 0$. We have*

$$\sqrt{Nh}\{\widehat{\theta}_{t_1,t_0,h} - \theta_{t_1,t_0}\} \xrightarrow{d} \mathcal{N}(0, V_{t_1,t_0}),$$

where

$$V_{t_1,t_0} = \frac{\kappa_{02}}{f_T(t_1)} \cdot \mathbb{E}\left[\pi_0(T_i, \boldsymbol{X}_i)^2 \left\{\int_{\mathcal{Y}} \{I(Y_i \le y) - F_{Y|T,X}(y|T_i, \boldsymbol{X}_i)\}dF_{t_0}(y)\right\}^2 \middle| T = t_1\right]$$
$$+ \frac{\kappa_{02}}{f_T(t_0)} \cdot \mathbb{E}\left[\pi_0(T_i, \boldsymbol{X}_i)^2 \left\{\int_{\mathcal{Y}} \{I(Y_i \le y) - F_{Y|T,X}(y|T_i, \boldsymbol{X}_i)\}dF_{t_1}(y)\right\}^2 \middle| T = t_0\right].$$

The proof of Theorem 2.11 is presented in Appendix of [27]. The asymptotic variance in Theorem 2.11 is estimated by

$$\widehat{V}_{t_1,t_0} := \frac{\kappa_{02}}{\widehat{f}_T(t_1)} \cdot \frac{\sum_{i=1}^N \widehat{\pi}_K(T_i, \boldsymbol{X}_i)^2 \left\{\int_{\mathcal{Y}}\{I(Y_i \le y) - \widehat{F}_{Y|T,X}(y|T_i, \boldsymbol{X}_i)\}d\widehat{F}_{t_0,h}(y)\right\}^2 K\left(\frac{T_i - t_1}{h_T}\right)}{\sum_{i=1}^N K\left(\frac{T_i - t_1}{h_T}\right)}$$
$$+ \frac{\kappa_{02}}{\widehat{f}_T(t_0)} \cdot \frac{\sum_{i=1}^N \widehat{\pi}_K(T_i, \boldsymbol{X}_i)^2 \left\{\int_{\mathcal{Y}}\{I(Y_i \le y) - \widehat{F}_{Y|T,X}(y|T_i, \boldsymbol{X}_i)\}d\widehat{F}_{t_1,h}(y)\right\}^2 K\left(\frac{T_i - t_0}{h_T}\right)}{\sum_{i=1}^N K\left(\frac{T_i - t_0}{h_T}\right)},$$

where:

$$\widehat{f}_T(t) := \frac{1}{Nh_T}\sum_{i=1}^N K\left(\frac{T_i - t}{h_T}\right),$$

$$\widehat{F}_{Y|T,X}(y|t, \boldsymbol{x}) := \frac{\sum_{i=1}^N I(Y_i \le y)K\left(\frac{T_i - t}{h_T}\right)\prod_{j=1}^r K\left(\frac{X_i^{(j)} - x^{(j)}}{h_X}\right)}{\sum_{i=1}^N K\left(\frac{T_i - t}{h_T}\right)\prod_{j=1}^r K\left(\frac{X_i^{(j)} - x^{(j)}}{h_X}\right)}.$$

Thus, the null hypothesis H_0 is rejected if

$$\sqrt{\frac{Nh}{\widehat{V}_{t_1,t_0}}}\left|\widehat{\theta}_{t_1,t_0,h} - 1/2\right| > z_{1-\alpha/2},$$

where $z_{1-\alpha/2}$ is the $1 - \alpha/2$ quantile of the standard normal distribution.

2.8.3 *Stochastic Dominance Test*

The Mann-Whitney test is useful for detecting the distributional difference. But it cannot ascertain the stochastic dominance when $\theta_{t_1,t_0} > (<) 1/2$. This is because $\theta_{t_1,t_0} > (<) 1/2$ does not necessarily mean $F_{t_1}(y)$ dominates (is dominated by) $F_{t_0}(y)$. To ascertain the dominance, we consider a stochastic dominance test, which was first introduced to econometrics by [29] and further studied by [25, 30–34]. In this subsection, we consider the following null and alternative hypothesis:

$$H_0 : F_{t_1}(y) \le F_{t_0}(y) \text{ for all } y \in \mathcal{Y};$$
$$H_1 : F_{t_1}(y) > F_{t_0}(y) \text{ for some } y \in \mathcal{Y}.$$

Clearly the Kolmogorov-Smirnov (KS) statistic

$$\widehat{S}_h = \sqrt{Nh} \sup_{y \in \mathcal{Y}} \left(\widehat{F}_{t_1,h}(y) - \widehat{F}_{t_0,h}(y) \right)$$

cannot be too positive if the null is true. Thus, for a small positive value c (to be determined), the decision rule is

$$\text{Reject } H_0 \text{ if } \widehat{S}_h > c.$$

The difficulty with KS test is that the null hypothesis is an inequality and the asymptotic distribution of the test statistic depends on the true distributions which are not known under the null. To overcome this difficulty, we follow the literature by adopting the least favorable configuration (LFC). The LFC finds an upper bound that is equal to the KS statistic when the two distributions are identical. The asymptotic distribution of the upper bound is known so that the upper bound can be used to determine the critical value. Applying this idea to our problem, we find that under the null hypothesis

$$\begin{aligned} \widehat{F}_{t_1,h}(y) - \widehat{F}_{t_0,h}(y) =& \left\{ (\widehat{F}_{t_1,h}(y) - \widehat{F}_{t_0,h}(y)) - (F_{t_1}(y) - F_{t_0}(y)) \right\} + (F_{t_1}(y) - F_{t_0}(y)) \\ \le& (\widehat{F}_{t_1,h}(y) - \widehat{F}_{t_0,h}(y)) - (F_{t_1}(y) - F_{t_0}(y)). \end{aligned}$$

Hence,

$$\begin{aligned} \widehat{S}_h =& \sqrt{Nh} \sup_{y \in \mathcal{Y}} \left\{ \widehat{F}_{t_1,h}(y) - \widehat{F}_{t_0,h}(y) \right\} \\ \le& \sqrt{Nh} \sup_{y \in \mathcal{Y}} \left\{ (\widehat{F}_{t_1,h}(y) - \widehat{F}_{t_0,h}(y)) - (F_{t_1}(y) - F_{t_0}(y)) \right\}. \end{aligned}$$

Applying Theorem 2.8, the upper bound $\sqrt{Nh} \sup_{y \in \mathcal{Y}} \left\{ (\widehat{F}_{t_1,h}(y) - \widehat{F}_{t_0,h}(y)) - (F_{t_1}(y) - F_{t_0}(y)) \right\}$ is asymptotically equivalent to $\sup_{y \in \mathcal{Y}} \left(\Psi_{t_1}(y) - \Psi_{t_0}(y) \right)$, which shall be used to determine the critical value.

The distribution of $\sup_{y\in\mathcal{Y}}\left(\Psi_{t_1}(y)-\Psi_{t_0}(y)\right)$ is still too complicate to compute. To compute the critical value, we follow [25] by approximating its distribution. By Theorem 2.8, for any fixed $t \in \mathcal{T}$, we have

$$\sup_{y\in\mathcal{Y}}\left|\sqrt{Nh}\left\{\widehat{F}_{t,h}(y)-F_t(y)\right\}-\sqrt{\frac{h}{N}}\cdot\sum_{i=1}^{N}\psi_{t,h}(Y_i,T_i,\boldsymbol{X}_i;y)\right|=o_P(1).$$

Let $\{U_i\}_{i=1}^N$ be i.i.d. random variables with mean zero and variance one and independent of the sample $\{T_j,\boldsymbol{X}_j,Y_j\}_{j=1}^N$. We approximate the distribution of $\sup_{y\in\mathcal{Y}}\left(\Psi_{t_1}(y)-\Psi_{t_0}(y)\right)$ by the distribution of

$$\overline{S}_u := \sup_{y\in\mathcal{Y}}\left(\Psi^u_{t_1,h}(y)-\Psi^u_{t_0,h}(y)\right),$$

where

$$\Psi^u_{t,h}(y) := \sqrt{\frac{h}{N}}\cdot\sum_{j=1}^{N}U_j\widehat{\psi}_{t,h}(Y_j,T_j,\boldsymbol{X}_j;y),$$

$$\widehat{\psi}_{t,h}(Y_j,T_j,\boldsymbol{X}_j;y)=\frac{\widehat{\pi}_K(T_j,\boldsymbol{X}_j)}{\widehat{p}_{t,h}}K\left(\frac{T_j-t}{h}\right)\left\{I(Y_j\le y)-\widehat{F}_{Y|T,X}(y|T_j,\boldsymbol{X}_j)\right\},$$

$$\widehat{p}_{t,h} := \frac{1}{N}\sum_{i=1}^{N}K\left(\frac{T_i-t}{h}\right).$$

Employing the same arguments as those in [25], we show that $\Psi_t^u(\cdot)\Rightarrow\Psi_t(\cdot)$ conditional on the sample path $\{T_i,\boldsymbol{X}_i,Y_i : i=1,\ldots,N\}$ with probability approaching to 1. Given the significant level α_0, the simulated critical value $\hat{c}$ is defined to be the $(1-\alpha_0)$-th quantile of $\overline{S}_u$:

$$\hat{c} := \sup\{q : P_u\left(\overline{S}_u\le q\right)\le 1-\alpha_0\}.$$

The following theorem is established with the proof similar to that of Theorem 5.1 in [25].

Theorem 2.12 *Suppose that Assumptions 1–4 and 12–15 are satisfied. Consider the decision rule: reject* H_0 *when* $\widehat{S}_h>\widehat{c}$. *We can show that*

1. *if* H_0 *is true,* $\limsup P(\text{reject } H_0)=\limsup P(\widehat{S}_h>\hat{c})\le\alpha_0$ *where the equality holds when* $F_{t_0}(y)=F_{t_1}(y)$ *for all* $y\in\mathcal{Y}$*;*
2. *if* H_0 *is false,* $\lim_N P(\text{reject } H_0)=1$.

Evidently, the critical value determined by the upper bound is larger than necessary and consequently the KS test is conservative in Type I error.

2.9 Empirical Study: Presidential Campaign Data

We revisit the U.S. presidential campaign data analyzed by [6, 35]. The motivation of the original study, [35], is well summarized in Sect. 2 of [6]:

> Urban and Niebler [35] explored the potential causal link between advertising and campaign contributions. Presidential campaigns ordinarily focus their advertising efforts on competitive states, but if political advertising drives more donations, then it may be worthwhile for candidates to also advertise in non-competitive states. The authors exploit the fact that media markets sometimes cross state boundaries. This means that candidates may inadvertently advertise in non-competitive states when they purchase advertisements for media markets that mainly serve competitive states. By restricting their analysis to non-competitive states, the authors attempt to isolate the effect of advertising from that of other campaigning, which do not incur these media market spillovers.

The treatment of interest, the number of political advertisements aired in each zip code, can be regarded as a continuous variable since it takes a range of values from 0 to 22379 across $N = 16265$ zip codes. Restricting themselves to a binary treatment framework, [35] compared 5230 zip codes that received more than 1000 advertisements and 11035 zip codes that received less than 1000 advertisements. Their empirical results suggest that advertising in non-competitive states had a significant causal effect on the level of campaign contributions.

Fong et al. [6] used the continuous treatment model, taking advantage of their proposed CBGPS method. Their empirical results suggest, contrary to [35], that advertising in non-competitive states did *not* have a significant causal effect on the level of campaign contributions (cf. [6], Table 2).

Using the generalized optimization estimator, we analyze the impact of advertisements on contributions based on both binary and continuous treatment models. Let Y_i and T_i be the log of the campaign contribution and political advertisement in zip code $i \in \{1, \dots, N\}$, respectively. Stack eight covariates as

$$X = \begin{bmatrix} \log(\text{Population}) \\ \%\text{Over 65} \\ \log(\text{Income} + 1) \\ \%\text{Hispanic} \\ \%\text{Black} \\ \log(\text{Population Density} + 1) \\ \%\text{College Graduates} \\ \text{Can Commute} \end{bmatrix}. \tag{2.27}$$

Subscript i is omitted for brevity, but (2.27) is defined for each zip code. The definition of each covariate is almost self-explanatory (see [6], Sect. 5 for more details). The log-transformation is implemented for Y, T, and some of the covariates in order to stabilize computation. Urban and Niebler [35] made the data publicly available at the American Journal of Political Science (AJPS) Dataverse archive. See Sect. 2.9.1 for the binary treatment model and Sect. 2.9.2 for the continuous treatment model.

2.9.1 Binary Treatment Model

We dichotomize the treatment variable (i.e., the log-advertisement) as $D = \mathbf{1}(T > 4)$. This is equivalent to dichotomizing the advertisement at 100, and 7137 zip codes out of $N = 16265$ are above the cut-off level. The potential outcome model is written as

$$\mathbb{E}[Y(d)] = \beta_1 + \beta_2 \times d.$$

Then, the stabilized weight reduces to

$$\pi_0(D, \boldsymbol{X}) = D \times \frac{P(D=1)}{P(D=1 \mid \boldsymbol{X})} + (1-D) \times \frac{P(D=0)}{P(D=0 \mid \boldsymbol{X})}.$$

The parameters of interest, $\boldsymbol{\beta} = (\beta_1, \beta_2)^\top$, are identified as

$$\beta_1 = \mathbb{E}[Y(0)] = \frac{\mathbb{E}\left[(1-D)\pi_0(D, \boldsymbol{X})Y\right]}{\mathbb{E}[(1-D)\pi_0(D, \boldsymbol{X})]},$$
$$\beta_2 = \mathbb{E}[Y(1) - Y(0)] = \frac{\mathbb{E}\left[D\pi_0(D, \boldsymbol{X})Y\right]}{\mathbb{E}[D\pi_0(D, \boldsymbol{X})]} - \frac{\mathbb{E}\left[(1-D)\pi_0(D, \boldsymbol{X})Y\right]}{\mathbb{E}[(1-D)\pi_0(D, \boldsymbol{X})]}.$$

The covariate-balancing equation of propensity score becomes

$$\frac{\mathbb{E}\left[D\pi(D, \boldsymbol{X})v(\boldsymbol{X})\right]}{\mathbb{E}[D]} = \mathbb{E}[v(\boldsymbol{X})] = \frac{\mathbb{E}\left[(1-D)\pi(D, \boldsymbol{X})v(\boldsymbol{X})\right]}{\mathbb{E}[1-D]}.$$

Our proposed estimator of stabilized weights becomes

$$\hat{\pi}_K(D_i, \boldsymbol{X}_i) = D_i \rho'\left(\hat{\lambda}_{1K}^\top v_K(\boldsymbol{X}_i)\right) + (1-D_i)\rho'\left(\hat{\lambda}_{2K}^\top v_K(\boldsymbol{X}_i)\right),$$

where

$$\hat{\lambda}_{1K} = \arg\max_{\lambda_1} \left\{ \frac{\sum_{i=1}^N D_i \rho\left(\lambda_1^\top v_K(\boldsymbol{X}_i)\right)}{\sum_{i=1}^N D_i} - \frac{1}{N}\sum_{i=1}^N \lambda_1^\top v_K(\boldsymbol{X}_i) \right\},$$
$$\hat{\lambda}_{2K} = \arg\max_{\lambda_2} \left\{ \frac{\sum_{i=1}^N (1-D_i) \rho\left(\lambda_2^\top v_K(\boldsymbol{X}_i)\right)}{\sum_{i=1}^N (1-D_i)} - \frac{1}{N}\sum_{i=1}^N \lambda_2^\top v_K(\boldsymbol{X}_i) \right\}.$$

Finally, the generalized optimization estimator for $\boldsymbol{\beta}$ is given by

$$\hat{\beta}_1 = \frac{\sum_{i=1}^N (1-D_i)\hat{\pi}_K(D_i, \boldsymbol{X}_i)Y_i}{\sum_{i=1}^N (1-D_i)\hat{\pi}_K(D_i, \boldsymbol{X}_i)}, \qquad \hat{\beta}_2 = \frac{\sum_{i=1}^N D_i\hat{\pi}_K(D_i, \boldsymbol{X}_i)Y_i}{\sum_{i=1}^N D_i\hat{\pi}_K(D_i, \boldsymbol{X}_i)} - \hat{\beta}_1.$$

The sieve basis function is specified as $v_K(\boldsymbol{X}) = (1, \boldsymbol{X}^\top)^\top$ with $K = 9$, where the covariates are given in (2.27). The exponential tilting function $\rho(w) = -e^{-w-1}$ is used. As in the simulation study in Sect. 7.3.5, 95% confidence intervals for β_1 and β_2 are computed via the bootstrap method II with $B = 1000$ iterations; recall (2.14).

Our empirical results are as follows. First, $\hat{\beta}_1 = 1.227$ and the bootstrapped confidence interval is [1.198, 1.257]. Second, $\hat{\beta}_2 = 0.061$ and the bootstrapped confidence interval is [0.003, 0.076]. The latter result indicates that advertising in non-competitive states has a significantly positive causal effect on the level of campaign contributions at the 5% level, which is a consistent result with [35].

2.9.2 Continuous Treatment Model

The procedure for the continuous treatment model is described in detail in Sect. 7.3.5, hence we refrain from repeating it here. The link function is specified as $g(T, \boldsymbol{\beta}) = \beta_1 + \beta_2 T + \beta_3 T^2$, where $\boldsymbol{\beta} = (\beta_1, \beta_2, \beta_3)^\top$. The sieve basis functions are specified as $u_{K_1}(T) = (1, T, T^2)^\top$ with $K_1 = 3$ and $v_{K_2}(\boldsymbol{X}) = (1, \boldsymbol{X}^\top)^\top$ with $K_2 = 9$, where the covariates are given in (2.27). The exponential tilting function $\rho(w) = -e^{-w-1}$ is used. 95% confidence intervals for $\boldsymbol{\beta}$ are computed via the bootstrap method II with $B = 1000$ iterations.

Our empirical results are as follows. First, $\hat{\beta}_1 = 1.100$ and the bootstrapped confidence interval is [0.909, 1.320]. Second, $\hat{\beta}_2 = 0.140$ and the confidence interval is [−0.025, 0.232]. Third, $\hat{\beta}_3 = -0.015$ and the confidence interval is [−0.025, 0.001]. The latter two results suggest that advertising in non-competitive states does not have a significant causal effect on the level of campaign contributions, which is a consistent result with [6].

The binary and continuous approaches lead to the opposite conclusions; the former finds the marginally significant impact of advertisements on campaign contributions at the 5% level, while the latter finds the marginally insignificant impact. These results suggest that the causal effect should be small if it exists at all. The binary model involves only one sieve basis function $v_9(\boldsymbol{X})$, while the continuous model involves two sieve basis functions $u_3(T)$ and $v_9(\boldsymbol{X})$. The latter requires the joint estimation of a relatively large-dimensional parameter matrix $\Lambda_{3\times 9}$; recall (2.9). This numerical complexity might be a reason why a significant causal effect is not detected under the continuous model.

References

1. Hirano, K., Imbens, G.W.: The propensity score with continuous treatments. Appl. Bayesian Model. Causal Infer. Incomplete-Data Perspect. **226164**, 73–84 (2004)
2. Robins, J.M., Hernán, M.A., Brumback, B.: Marginal structural models and causal inference in epidemiology. Epidemiology **11**, 550–560 (2000)

3. Buchinsky, M.: Quantile regression, box-cox transformation model, and the u.s. wage structure, 1963–1987. J. Econometrics **65**(1), 109–154 (1995)
4. Hirano, K., Imbens, G.W., Ridder, G.: Efficient estimation of average treatment effects using the estimated propensity score. Econometrica **71**(4), 1161–1189 (2003)
5. Imai, K., van Dyk, D.A.: Causal inference with general treatment regimes: generalizing the propensity score. J. Am. Stat. Assoc. **99**, 854–866 (2004)
6. Fong, C., Hazlett, C., Imai, K.: Covariate balancing propensity score for a continuous treatment: application to the efficacy of political advertisements. Ann. Appl. Stat. **12**, 156–177 (2018)
7. Bickel, P.J., Klaassen, C.A.J., Ritov, Y., Wellner, J.A.: Efficient and Adaptive Estimation for Semiparametric Models. The Johns Hopkins University Press (1993)
8. Ai, C., Linton, O., Motegi, K., Zhang, Z.: A unified framework for efficient estimation of general treatment models. Quant. Econ. **12**(3), 779–816 (2021)
9. Hahn, J.: On the role of the propensity score in efficient semiparametric estimation of average treatment effects. Econometrica **66**, 315–331 (1998)
10. Cattaneo, M.D.: Efficient semiparametric estimation of multi-valued treatment effects under ignorability. J. Econometrics **155**(2), 138–154 (2010)
11. Firpo, S.: Efficient semiparametric estimation of quantile treatment effects. Econometrica **75**(1), 259–276 (2007)
12. Ai, C., Linton, O., Motegi, K., Zhang, Z.: Supplemental material for a unified framework for efficient estimation of general treatment models. Chinese University of Hong Kong, Shenzhen, Technical report (2020)
13. Athey, S., Imbens, G., Pham, T., Wager, S.: Estimating average treatment effects: supplementary analyses and remaining challenges. Am. Econ. Rev. **107**(5), 278–81 (2017)
14. Chernozhukov, V., Chetverikov, D., Demirer, M., Duflo, E., Hansen, C., Newey, W., Robins, J.: Double/debiased machine learning for treatment and structural parameters. Economet. J. **21**(1), C1–C68 (2018)
15. Newey, W.K.: Convergence rates and asymptotic normality for series estimators. J. Econometrics **79**(1), 147–168 (1997)
16. Chen, X.: Large sample sieve estimation of semi-nonparametric models. Handb. Econometrics **6**(B), 5549–5632 (2007)
17. Kitamura, Y., Stutzer, M.: An information-theoretic alternative to generalized method of moments estimation. Econometrica: J. Econometric Soc. 861–874 (1997)
18. Imbens, G., Johnson, P., Spady, R.H.: Information theoretic approaches to inference in moment condition models. Econometrica **66**(2), 333–357 (1998)
19. Tseng, P., Bertsekas, D.P.: Relaxation methods for problems with strictly convex costs and linear constraints. Math. Oper. Res. **16**(3), 462–481 (1991)
20. Andrews, D.W.K.: Asymptotic normality of series estimators for nonparametric and semiparametric regression models. Econometrica **59**(2), 307–345 (1991)
21. Andrews, D.W.K.: Empirical process methods in econometrics. In: Engle, R.F., McFadden, D.L. (eds.) Handbook of Econometrics, vol. 4, Chap. 37, pp. 2247–2294. Citeseer (1994)
22. Chen, X., Linton, O., Van Keilegom, I.: Estimation of semiparametric models when the criterion function is not smooth. Econometrica **71**(5), 1591–1608 (2003)
23. Chen, X., Linton, O., Van Keilegom, I.: Estimation of semiparametric models when the criterion function is not smooth. Econometrica **71**(5), 1591–1608 (2003)
24. Huang, W., Linton, O., Zhang, Z.: A unified framework for specification tests of continuous treatment effect models. J. Bus. Econ. Stat. **40**(4), 1817–1830 (2022)
25. Donald, S.G., Hsu, Y.C.: Estimation and inference for distribution functions and quantile functions in treatment effect models. J. Econometrics **178**(3), 383–397 (2014)
26. Li, Q., Racine, J.S.: Nonparametric Econometrics: Theory and Practice. Princeton University Press (2007)
27. Ai, C., Linton, O., Zhang, Z.: Estimation and inference for the counterfactual distribution and quantile functions in continuous treatment models. J. Econometrics **228**(1), 39–61 (2022)
28. van der Vaart, A.W.: Asymptotic Statistics. Cambridge University Press (1998)

29. McFadden, D.: Testing for stochastic dominance. In: Studies in the Economics of Uncertainty, pp. 113–134. Springer, Berlin (1989)
30. Anderson, G.: Nonparametric tests of stochastic dominance in income distributions. Econometrica: J. Econometric Soc. 1183–1193 (1996)
31. Davidson, R., Duclos, J.Y.: Statistical inference for stochastic dominance and for the measurement of poverty and inequality. Econometrica **68**(6), 1435–1464 (2000)
32. Barrett, G.F., Donald, S.G.: Consistent tests for stochastic dominance. Econometrica **71**(1), 71–104 (2003)
33. Linton, O., Maasoumi, E., Whang, Y.J.: Consistent testing for stochastic dominance under general sampling schemes. Rev. Econ. Stud. **72**(3), 735–765 (2005)
34. Linton, O., Song, K., Whang, Y.J.: An improved bootstrap test of stochastic dominance. J. Econometrics **154**(2), 186–202 (2010)
35. Urban, C., Niebler, S.: Dollars on the sidewalk: Should u.s. presidential candidates advertise in uncontested states? Am. J. Polit. Sci. **58**(2), 322–336 (2014)

Chapter 3
Causal Inference with Measurement Errors

3.1 Basic Framework

In many empirical studies, it may not be available to observe the accurate data. For instance, [1] indicated that over 75% of the variance in fat intake data from the Epidemiologic Study Cohort in NHANES-I stems from measurement error. This chapter focuses on continuous treatment data measured with classical error, where researchers only observe the sum of the treatment and a random error, rather than the actual treatment received.

Let T be a continuously valued treatment with the probability density function $f_T(t)$ and support $\mathcal{T} \subset \mathbb{R}$. $Y(t)$ denotes the potential outcome if being treated at level t for $t \in \mathcal{T}$. The observed the outcome is $Y := Y(T)$. In addition, we observe a vector of confounders $\boldsymbol{X} \in \mathbb{R}^r$, where r is a positive integer. In this chapter, we consider the scenario that the treatment is measured with classical error, i.e., instead of observing the true treatment T, we only observe the contaminated treatment S satisfying

$$S = T + U, \tag{3.1}$$

where U denotes the measurement error and it is independent of T, $\boldsymbol{X}$, and $\{Y(t)\}_{t\in\mathcal{T}}$. We assume that the characteristic function of U, denoted by ϕ_U, is known. See Remark 3.1 for more discussion on the situation of unknown ϕ_U. In this chapter, we aim to develop a nonparametric method for estimating the unconditional ADRF, denoted as $\mu(t) := \mathbb{E}\{Y(t)\}$, based on an independent and identically distributed (i.i.d.) sample $\{S_i, \boldsymbol{X}_i, Y_i\}_{i=1}^N$.

We impose the following assumption to identify $\mu(t)$, which is a standard condition in most of causal inference literature (e.g., [2–6]).

Assumption 1 We assume

(i) (Unconfoundedness) Given $\boldsymbol{X}$, T is independent of $Y(t)$ for every $t \in \mathcal{T}$;
(ii) (No Interference) The outcome of one individual is not affected by the treatment status of the other individuals;

Z. Zhang et al., *Big Data in Economics and Management*, Statistics and Big Data 1,
https://doi.org/10.1007/978-981-95-3125-7_3

(iii) (Consistency) $Y = Y(t)$ when $T = t$;
(iv) (Positivity) The conditional probability density function of T given $\boldsymbol{X}$, denoted by $f_{T|X}$, satisfies $f_{T|X}(t|\boldsymbol{X}) > 0$ a.s. for all $t \in \mathcal{T}$.

Under Assumption 1, we can identify $\mu(t)$ as follows:

$$
\begin{aligned}
\mu(t) &= \mathbb{E}[\mathbb{E}\{Y(t)|\boldsymbol{X}\}] = \mathbb{E}[\mathbb{E}\{Y(t)|\boldsymbol{X}, T = t\}] \quad \text{(using Assumption 36 (i) and (ii))} \\
&= \mathbb{E}\{\mathbb{E}(Y|\boldsymbol{X}, T = t)\} \quad \text{(using Assumption 36 (iii))} \\
&= \int_{\mathcal{X}} \int_{\mathcal{Y}} \frac{f_T(t)}{f_{T|X}(t|x)} y f_{Y|X,T}(y|x,t) f_{X|T}(x|t) dy dx \quad \text{(using Assumption 36 (iv))} \\
&= \mathbb{E}\{\pi_0(t, \boldsymbol{X}) Y | T = t\},
\end{aligned} \tag{3.2}
$$

where $\pi_0(t, x)$ is the *stabilized weights* defined by

$$
\pi_0(t, x) := \frac{f_T(t)}{f_{T|X}(t|x)} \tag{3.3}
$$

and f_T is the density function of T.

If T can be observed and $\pi_0(T, \boldsymbol{X})$ is known, we can estimate $\mu(t)$ with the Nadaraya-Watson estimator:

$$
\mu_{NW}(t) := \frac{\sum_{i=1}^{N} \pi_0(t, \boldsymbol{X}_i) Y_i L\{(t - T_i)/h\}}{\sum_{i=1}^{N} L\{(t - T_i)/h\}}, \quad t \in \mathcal{T}, \tag{3.4}
$$

where $L(\cdot)$ is the ordinary kernel function and h is the bandwidth. However, the challenges are (i) we cannot directly observe T due to the measurement error; and (ii) the stabilized weight $\pi_0(t, \boldsymbol{x})$ is unknown in practice. We will address these problems in the next section.

Remark 3.1 The measurement error's characteristic function, denoted as ϕ_U, may remain elusive in certain empirical applications. Numerous approaches have been put forward in literature to reliably approximate it. For instance, in the study by [7], it was postulated that the data originating from the error distribution could be observed, leading to a nonparametric estimation of ϕ_U based on the error data. In certain scenarios, the replication of error-laden observations enables the estimation of the density function using these duplicates, as discussed in [8]. Parametric methods come into play when certain parameters of ϕ_U are known, as exemplified in [9]. Additionally, a data-independent nonparametric method is also accessible, as demonstrated in [10]. Once a reliable estimator of ϕ_U is derived, our proposed method can be straightforwardly applied.

In practical situations, it is customary for the assumptions (a) that the error distribution is known up to identifiable parameters based on previous studies and (b) that repeated error-contaminated data are available to be satisfied. Under assumption (a), we typically acquire estimators for the parameters and, consequently, estimators for $\phi_U(t)$ that are uniformly consistent across the real line, where N_U represents the sample size of previous studies. As long as $N = O(N_U)$ and considering the slower convergence rate of our proposed estimator $\widehat{\mu}(t)$ compared to $N^{-1/2}$, the asymptotic behavior remains unaffected by the estimation of ϕ_U.

Assuming (b), we can observe that

$$S_{jk} = T_j + U_{jk}, \quad k = 1, 2, \quad j = 1, \ldots, N,$$

where the U_{jk}'s are i.i.d.. It is important to note that

$$\mathbb{E}[\exp\{it(S_{j1} - S_{j2})\}] = \mathbb{E}\{\exp(itU_{j1})\}\mathbb{E}\{\exp(-itU_{j2})\} = |\phi_U(t)|^2.$$

The approach proposed by [11] estimates $\phi_U(t)$ using $\widehat{\phi_U}(t) = \left|N^{-1}\sum_{j=1}^{N} \cos\{it(S_{j1} - S_{j2})\}\right|^{1/2}$.Their study demonstrates that the deconvolution kernel local constant estimator employing $\widehat{\phi_U}(t)$ and the one utilizing $\phi_U(t)$ exhibit similar asymptotic behavior under certain regularity conditions. Specifically, in the case of ordinary smoothness for U (as defined in (3.18)), sufficient smoothness of f_T relative to the density of U is required. Conversely, if U exhibits supersmoothness as described in (3.19), the deconvolution kernel local constant estimator achieves an optimal convergence rate logarithmic in N, resulting in negligible estimation error when approximating ϕ_U. This conclusion can be extended to our specific scenario.

3.2 Estimation Method

To address the challenge posed by the unobservable $L(t - T_i)/h$'s in empirical scenarios, we employ the *deconvolution* kernel approach (e.g., [12, 13]). This technique is common in nonparametric regression settings where covariates are subject to classical errors, as exemplified in Eq. (3.1). The fundamental idea is as follows: the density of S is a result of convolving the densities of T and U, indicating that $\phi_S(w) = \phi_T(w)\phi_U(w)$, where ϕ_S and ϕ_T denote the characteristic functions of S and T, respectively. We assume that U satisfies $\phi_U(w) \neq 0$ for all $w \in \mathbb{R}$. By utilizing the Fourier inversion theorem, when $|\phi_T|$ is integrable, we obtain the following expression:

$$f_T(t) = \frac{1}{2\pi}\int_{-\infty}^{\infty} \exp(-iwt)\frac{\phi_S(w)}{\phi_U(w)}dw. \tag{3.5}$$

This inspired [12] to propose an estimation method for f_T using $\widehat{f}T, h(t) := (Nh)^{-1}\sum^N i = 1L_U(t - S_i)/h$, where

$$L_U(v) := \frac{1}{2\pi}\int_{-\infty}^{\infty} \exp(-iwv)\frac{\phi_L(w)}{\phi_U(w/h)}dw, \tag{3.6}$$

and ϕ_L represents the Fourier transform of the kernel L. The purpose of the kernel L is to prevent $\widehat{f}_{T,h}$ from becoming excessively large in its tails.

Expanding on this concept, [13] introduced a consistent errors-in-variables regression estimator by replacing the $L(t - T_i)/h$ terms in (3.4) with the $L_U(t - S_i)/h$ terms. In our specific scenario, an errors-in-variables estimator of $\mu(t)$ can be formulated as follows:

$$\widetilde{\mu}(t) := \frac{\sum_{i=1}^{N} \pi_0(t, \boldsymbol{X}_i) Y_i L_U\{(t - S_i)/h\}}{\sum_{i=1}^{N} L_U\{(t - S_i)/h\}}. \tag{3.7}$$

Note that $U \perp (T, \boldsymbol{X}, Y)$, we have

$$\begin{aligned}
&\mathbb{E}\big[L_U\{(t - S)/h\}|T, \boldsymbol{X}, Y\big] = \mathbb{E}\big[L_U\{(t - S)/h\}|T\big] \\
=&\frac{1}{2\pi}\int_{-\infty}^{\infty} \exp\{-iw(t - T)/h\}\mathbb{E}[\exp(iwU/h)]\frac{\phi_L(w)}{\phi_U(w/h)}\,dw \\
=&\frac{1}{2\pi}\int_{-\infty}^{\infty} \exp\{-iw(t - T)/h\}\phi_L(w)\,dw = L\{(t - T)/h\},
\end{aligned} \tag{3.8}$$

where the last equation is derived from the Fourier inversion theorem. Exploiting this property, $\widetilde{\mu}(t)$ exhibits the same asymptotic bias as $\mu_{NW}(t)$, which diminishes as $h \to 0$. To establish its consistency with $\mu(t)$, it suffices to demonstrate that its asymptotic variance diminishes to zero as $N \to \infty$, which can be achieved through a straightforward extension of the proof outlined in [13].

However, a challenge arises from the fact that π_0 is typically unknown in practice. We will now elucidate the process of estimating $\pi_0(t, \boldsymbol{X})$ using the error-contaminated data $(S_i, \boldsymbol{X}_i, Y_i)$, $i = 1, \ldots, N$.

3.2.1 Estimating $\pi_0(t, X)$

Observing Eq. (3.3), a direct approach to estimate π_0 involves estimating f_T and $f_{T|X}$ and subsequently computing their ratio. However, this ratio estimator is susceptible to inaccuracies, particularly when $f_{T|X}$ takes low values. Even slight errors in its estimation can result in significant errors in the ratio estimator. To address this issue, similar to the methodology in the error-free treatment effect literature, we consider π_0 as a unified entity and estimate it directly to mitigate this sensitivity. Specifically, we employ a nonparametric estimation approach by progressively expanding the set of equations for estimation.

When the T_i's are fully observable, in Chap. 2, we have derived the moment equation as follows:

$$\mathbb{E}\{\pi_0(T, \boldsymbol{X})u(\boldsymbol{X})v(T)\} = \mathbb{E}\{v(T)\}\mathbb{E}\{u(\boldsymbol{X})\} \tag{3.9}$$

holds for any integrable functions $u(\boldsymbol{X})$ and $v(T)$, and it identifies $\pi_0(\cdot,\cdot)$. In our analysis, we estimated the function $\pi_0(\cdot,\cdot) : \mathcal{T} \times \mathcal{X} \to \mathbb{R}$ by maximizing a generalized empirical likelihood, subject to the constraints of the sample version of (3.9). Estimating $\pi_0(\cdot,\cdot)$ in this scenario is challenging because the constraints based on the moment equation are not computable when T is unobservable. Additionally, accurately estimating the moment $\mathbb{E}[v(T)]$ for a general function $v(\cdot)$ from contaminated data $S_i{}_{i=1}^N$ poses significant challenges, and deriving its theoretical properties is difficult, if not impossible. An example can be found in the work of [14]. They explored the estimation of the absolute moment $\mathbb{E}[|T|^q]$, where T is subject to ordinary smooth error (as defined in (3.18)). The properties of the estimator varied depending on the value of q. If q was an even integer, achieving $\sqrt{N}$-consistency required a stringent condition $\mathbb{E}[T^{2q}] + \mathbb{E}[U^{2q}] < \infty$. When q was an odd integer, $\sqrt{N}$-consistency was possible if and only if the distribution of the measurement error was suitably "rough," as indicated by the convergence rate of ϕ_U to zero in its tails. However, for $q > 0$ not being a positive integer, $\sqrt{N}$-consistency was generally unattainable. For other forms of $v(T)$ or when dealing with supersmooth error (as defined in (3.19)), the consistent nonparametric estimation of $\mathbb{E}[v(T)]$, as well as understanding its theoretical behavior, remains an open problem to the best of our knowledge.

To improve the stability of estimating π_0, an alternative approach is proposed, involving an expanding set of equations that can identify π_0 from error-contaminated data without the need to estimate $\mathbb{E}[v(T)]$. Instead of directly estimating the function $\pi_0(\cdot,\cdot) : \mathcal{T} \times \mathcal{X} \to \mathbb{R}$, the focus shifts to estimating its projection $\pi_0(t,\cdot) : \mathcal{X} \to \mathbb{R}$ for each fixed $t \in \mathcal{T}$. It is found that

$$\mathbb{E}\left\{\pi_0(t,\boldsymbol{X})u(\boldsymbol{X})|T=t\right\} = \int_{\mathcal{X}} \frac{f_T(t)}{f_{T|X}(t|x)} u(x) f_{X|T}(x|t)dx = \mathbb{E}\{u(\boldsymbol{X})\} \quad (3.10)$$

holds for any integrable function $u(\boldsymbol{X})$. Indeed, while Eq. (3.10) still involves the unobservable variable T, the term $\mathbb{E}\left\{\pi_0(t,\boldsymbol{X})u(\boldsymbol{X})|T=t\right\}$ can be estimated using the deconvolution kernel introduced in Eq. (3.8) based on the observable variables $(S,\boldsymbol{X})$. In the following theorem, it is demonstrated that the corresponding moment condition can identify the function $\pi_0(t,\cdot)$ from the observed data $(S,\boldsymbol{X})$ for every fixed $t \in \mathcal{T}$.

Theorem 3.3 *Let $L_U(\cdot)$ be the deconvolution kernel function defined in* (3.6). *For every fixed $t \in \mathcal{T}$ and any integrable function $u(\boldsymbol{X})$,*

$$\lim_{h_0 \to 0} \frac{\mathbb{E}\left[\pi(t,\boldsymbol{X})u(\boldsymbol{X})L_U\{(t-S)/h_0\}\right]}{\mathbb{E}\left[L_U\{(t-S)/h_0\}\right]} = \mathbb{E}[u(\boldsymbol{X})] \quad (3.11)$$

holds if and only if $\pi(t,\boldsymbol{X}) = \pi_0(t,\boldsymbol{X})$ a.s.

The proof of Theorem 3.3 can be found in the supplementary file of [15]. It establishes a method for estimating the weighting function by solving a sample analogue of Eq. (3.11) for any integrable function $u(x)$, where h_0 approaches 0 as the sample size tends to infinity. However, in practice, solving this infinite set of equations becomes impractical when dealing with a finite sample of observations. To overcome this challenge, an approximation approach is employed by considering a series of finite-dimensional sieves to approximate the infinite-dimensional function space of $u(\boldsymbol{x})$. Specifically, let $u_K(x)$ denote a set of predetermined basis functions with a dimension of K (such as power series, B-splines, or trigonometric polynomials). The function $u_K(x)$ serves as an approximation sieve that effectively approximates any suitable function $u(x)$ as K approaches infinity (for more details on sieve approximation, refer to [16]).

By using the approximation sieve, $\pi_0(t, \boldsymbol{X})$ can also be approximated within a subspace of the original function space, satisfying

$$\lim_{h_0 \to 0} \frac{\mathbb{E}[\pi(t, \boldsymbol{X}) u_K(\boldsymbol{X}) L_U\{(t-S)/h_0\}]}{\mathbb{E}[L_U\{(t-S)/h_0\}]} = \mathbb{E}\{u_K(\boldsymbol{X})\}. \tag{3.12}$$

Equation (3.12), as $K \to \infty$, asymptotically identifies $\pi_0(t, \boldsymbol{X})$. It is worth noting that for any increasing and globally concave function $\rho(v)$,

$$\pi^*(t, \boldsymbol{X}) := \rho' \left\{ \lambda_t^{*\top} u_K(\boldsymbol{X}) \right\} \tag{3.13}$$

solves (3.12), where $\rho'(\cdot)$ is the derivative of $\rho(\cdot)$, $\lambda_t^* := \operatorname{argmax}_{\lambda \in \mathbb{R}^K} G_t^*(\lambda)$ and $G_t^*(\lambda)$ is a strictly concave function defined by

$$G_t^*(\lambda) := \lim_{h_0 \to 0} \frac{\mathbb{E}\left[\rho\{\lambda^\top u_K(\boldsymbol{X})\} L_U\{(t-S)/h_0\}\right]}{\mathbb{E}[L_U\{(t-S)/h_0\}]} - \lambda^\top \mathbb{E}\{u_K(\boldsymbol{X})\}.$$

Certainly, by utilizing the first-order condition $\nabla G_t^{(}\lambda_t^{)} = 0$, it is observed that (3.12) holds with $\pi(t, \boldsymbol{X}) = \pi^*(t, \boldsymbol{X})$. As a result, we can expect the estimator of $\pi_0(t, \boldsymbol{X})$ to be defined as the empirical counterpart of (3.13). Therefore, for any fixed $t \in \mathcal{T}$, a proposed approach for estimating $\pi_0(t, \boldsymbol{X})$ is given by

$$\widehat{\pi}(t, \boldsymbol{X}) = \rho' \left\{ \widehat{\lambda}_t^\top u_K(\boldsymbol{X}) \right\} \tag{3.14}$$

with $\widehat{\lambda}_t = \operatorname{argmax}_{\lambda \in \mathbb{R}^K} \widehat{G}_t(\lambda)$ and

$$\widehat{G}_t(\lambda) := \frac{\sum_{i=1}^N \rho\{\lambda^\top u_K(\boldsymbol{X}_i)\} L_U\{(t-S_i)/h_0\}}{\sum_{i=1}^N L_U\{(t-S_i)/h_0\}} - \lambda^\top \left\{ \frac{1}{N} \sum_{i=1}^N u_K(\boldsymbol{X}_i) \right\}. \tag{3.15}$$

Indeed, in a finite sample, it is possible for some of the deconvolution kernel $L_U(t-S_i)/h_0$'s to take negative values, which can lead to a non-strictly concave empirical objective function $\widehat{G}_t(\cdot)$. However, as the sample size tends to infinity ($N \to \infty$) and the bandwidth parameter h_0 tends to zero ($h_0 \to 0$), the empirical objective function $\widehat{G}_t(\cdot)$ converges in probability to the true objective function $G_t^*(\cdot)$, which is strictly concave. As a result, with probability approaching one, $\widehat{G}t(\cdot)$

becomes strictly concave, ensuring the existence of a unique solution $\widehat{\lambda}t$. Remark 3.2 in Sect. 3.4 presents an efficient and robust method to solve this maximization problem using finite samples.

Our estimator $\widehat{\pi}(t, \mathbf{X}_i)$ can be interpreted as a localized generalized empirical likelihood estimator. Specifically, it is obtained as the solution to the following maximization problem for local generalized empirical likelihood, for every fixed $t \in \mathcal{T}$:

$$\begin{cases} \max_{\{\pi_i\}_{i=1}^N} & -\frac{\sum_{i=1}^N D(\pi_i)L_U(\{t-S_i\}/h_0)}{\sum_{i=1}^N L_U(\{t-S_i\}/h_0)} \\ \text{subject to} \ \frac{\sum_{i=1}^N \pi_i u_K(\boldsymbol{X}_i)L_U(\{t-S_i\}/h_0)}{\sum_{i=1}^N L_U(\{t-S_i\}/h_0)} & = \frac{1}{N}\sum_{i=1}^N u_K(\boldsymbol{X}_i), \end{cases} \tag{3.16}$$

where $D(v)$ is a distance measure from v to 1 for $v \in \mathbb{R}$, which is continuously differentiable and satisfies that $D(1) = 0$ and

$$\rho(-v) = D\{(D')^{-1}(v)\} - v \cdot (D')^{-1}(v).$$

The objective of Eq. (3.16) is to minimize a certain distance measure between the target weight $N^{-1}\pi_i$ and the observed frequencies N^{-1} within a local vicinity of $T_i = t$, while adhering to the sample version of the moment constraint (2.6).

To facilitate subsequent discussions, in acknowledge of the equivalence between the dual formulation (3.14) and the primal problem (3.16), we denote the estimator in terms of $\rho(v)$ in the following discussions. For example: $\rho(v) = -\exp(-v-1)$ corresponds to exponential tilting [5, 17, 18], $\rho(v) = \log(1+v)$ corresponds to empirical likelihood [19], $\rho(v) = -(1-v)^2/2$ corresponds to continuous updating of the generalized method of moments [20], $\rho(v) = v - \exp(-v)$ corresponds to the inverse logistic. By substituting $\pi_0(t, \mathbf{X}_i)$ in Eq. (3.7) with $\widehat{\pi}(t, \mathbf{X}_i)$, we obtain an estimate of $\mu(t)$ as follows:

$$\widehat{\mu}(t) := \frac{\sum_{i=1}^N \widehat{\pi}(t, X_i)Y_i L_U\{(t-S_i)/h\}}{\sum_{i=1}^N L_U\{(t-S_i)/h\}}. \tag{3.17}$$

3.3 Large Sample Properties

In this section, the focus is on establishing the convergence rates of $\widehat{\pi}(t, \cdot)$ in L_∞ and L_2 norms for every fixed $t \in \mathcal{T}$. Subsequently, the asymptotic behavior of the proposed ADRF estimator $\widehat{\mu}(t)$ is explored. It is important to note that $\widehat{\mu}(t)$ (or $\widehat{\pi}(t, \cdot)$) acts as a nonparametric estimator, and its asymptotic behavior is influenced by both asymptotic bias and variance. The asymptotic bias is defined as the expectation of the limiting distribution of $\widehat{\mu}(t) - \mu(t)$ (or $\widehat{\pi}(t, \cdot) - \pi_0(t, \cdot)$), while the variance pertains to their respective variances.

Based on Eq. (3.8), it will be demonstrated that the asymptotic biases of the two estimators remain consistent with their counterparts in the error-free scenario. This consistency relies on several conditions related to the smoothness of π_0 and μ, the density of T, and the approximation error associated with the sieve basis u_K. The following conditions are particularly required:

Assumption 2 The kernel function $L(\cdot)$ is an even function such that $\int_{-\infty}^{\infty} L(u)\,du = 1$ and has finite moments of order 3.

Assumption 3 We assume

(i) the support $\mathcal{X}$ of $\boldsymbol{X}$ is a compact subset of $\mathbb{R}^r$. The support $\mathcal{T}$ of the treatment variable T is a compact subset of $\mathbb{R}$.
(ii) (Strict Positivity) there exist a positive constant η_{min} such that $f_{T|X}(t|\boldsymbol{x}) \geq \eta_{\min} > 0$, for all $\boldsymbol{x} \in \mathcal{X}$.

Assumption 4 (i) The densities $f_T(t)$, $f_{T|X}(t|\boldsymbol{X})$ and $f_{T|Y,X}(t|Y, \boldsymbol{X})$ are third-order continuously differentiable w.r.t. t almost surely. (ii) The derivatives of $f_{T|X}(t|\boldsymbol{X})$ and $f_{T|Y,X}(t|Y, \boldsymbol{X})$, denoted by $\{\partial_t^d f_{T|X}(t|\boldsymbol{X}),\ \partial_t^d f_{T|Y,X}(t|Y, \boldsymbol{X})$ for $d = 0, 1, 2, 3\}$, are integrable almost surely in t.

Assumption 5 For every $t \in \mathcal{T}$, (i) the function $\pi_0(t, \boldsymbol{x})$ is s-times continuously differentiable w.r.t. $\boldsymbol{x} \in \mathcal{X}$, where $s > r/2$ is an integer; (ii) there exist $\lambda_t \in \mathbb{R}^K$ and a positive constant $\alpha > 0$ such that $\sup_{\boldsymbol{x}\in\mathcal{X}} \left|(\rho')^{-1}\{\pi_0(t, \boldsymbol{x})\} - \lambda_t^\top u_K(\boldsymbol{x})\right| = O(K^{-\alpha})$.

Assumption 6 (i) For every K, the eigenvalues of $\mathbb{E}\left[u_K(\boldsymbol{X})u_K(\boldsymbol{X})^\top|T = t\right]$ are bounded away from zero and infinity, and twice differentiable w.r.t. t for $t \in \mathcal{T}$. (ii) There is a sequence of constants $\zeta(K)$ satisfying $\sup_{\boldsymbol{x}\in\mathcal{X}} \|u_K(\boldsymbol{x})\| \leq \zeta(K)$, such that $\zeta(K)\{K^{-\alpha} + h_0^2 + h^2\} \to 0$ as $N \to \infty$, where $\|\cdot\|$ denotes the Euclidean norm.

Assumption 7 For every $t \in \mathcal{T}$, there exist $\gamma_t \in \mathbb{R}^K$ and a positive constant $\ell > 0$ such that $\sup_{\boldsymbol{x}\in\mathcal{X}} \left|m(t, \boldsymbol{x}) - \gamma_t^\top u_K(\boldsymbol{x})\right| = O(K^{-\ell})$, where $m(t, \boldsymbol{x}) = \mathbb{E}[Y|T = t, \boldsymbol{X} = \boldsymbol{x}]$.

Assumption 8 $R_1^{2+\delta}(t) := \mathbb{E}\big[|\pi_0(t, \boldsymbol{X})Y - \mu(t)|^{2+\delta}|T = t\big]$, $R_2^{2+\delta}(t) := \mathbb{E}\big[|\pi_0(t, \boldsymbol{X})m(t, \boldsymbol{X}) - \mu(t)\big|^{2+\delta}|T = t\big]$ and $R_3^{2+\delta}(t) := \mathbb{E}\big[|\pi_0(t, \boldsymbol{X})\{Y - m(t, \boldsymbol{X})\}\big|^{2+\delta}|T = t\big]$ are bounded for some $\delta > 0$, for all $t \in \mathcal{T}$.

Assumption 3 (i) imposes constraints on the boundedness of both the covariates $\boldsymbol{X}$ and the treatment T, which is a common requirement in nonparametric regression studies. However, this condition can be relaxed by constraining the tail distributions of $\boldsymbol{X}$ and T, as discussed in [21]. Assumption 3 (ii) is a stringent positivity requirement that mandates that each subject has a nonzero probability of receiving every treatment level, regardless of their covariate values. This condition is commonly found in the literature, particularly in the absence of measurement error. It is often used when no restrictions are placed on the potential outcome distribution. Relaxing this condition is possible by imposing other smoothness conditions on the potential outcome distribution, as suggested in [22], or by considering different target parameters. For example, [23] investigated the estimation of a stochastic intervention causal parameter under a weaker positivity condition. They defined the parameter as $\mathbb{E}[\mathbb{E}[Y|T+a(\boldsymbol{X}), \boldsymbol{X}]]$ and imposed the condition $\sup_{t\in\mathcal{T}} f_{T|X}(t-a(\boldsymbol{X})|\boldsymbol{X})/f_{T|X}(t|\boldsymbol{X}) < \infty$ almost everywhere, where $a(\boldsymbol{X})$ is a user-specified intervention function. Similarly, [24] explored the estimation of a conditional causal dose-response curve defined by $\mathbb{E}[\mathbb{E}[Y|T=t, \boldsymbol{X}]|\boldsymbol{Z}]$ under a weaker positivity condition. They used the condition $\sup_{t\in\mathcal{T}} b(t, \boldsymbol{Z})/f_{T|X}(t|\boldsymbol{X}) < \infty$ almost everywhere, where $\boldsymbol{Z} \subset \boldsymbol{X}$ is a subset of observed covariates and $b(t, \boldsymbol{Z})$ is a user-specified weight function. Although Assumption 3 (ii) may not be the mildest condition, it is retained in this paper due to its technical advantages, particularly in the presence of measurement error.

Assumption 4 is introduced to ensure the smoothness required for nonparametric estimation. Under Assumption 1, the parameter of interest $\mu(t) = \mathbb{E}[Y^*(t)]$ can be alternatively expressed as $\mu(t) = \mathbb{E}[\pi_0(t, \boldsymbol{X})Y|T=t] = \mathbb{E}\left[f_{T|Y,X}(t|Y, \boldsymbol{X})Y/f_{T|X}(t|\boldsymbol{X})\right]$. It is important to note that Assumption 4 (i) implies that the function $t \mapsto \frac{f_{T|Y,X}(t|Y,\boldsymbol{X})}{f_{T|X}(t|\boldsymbol{X})}$ is third-order continuously differentiable almost surely. Moreover, by leveraging Leibniz's integral rule and Assumption 4 (ii), it can be concluded that the target parameter $t \mapsto \mu(t)$ is also third-order continuously differentiable.

Assumption 5 (i) is introduced to control the complexity of the function class $\pi_0(t, \boldsymbol{x}), \boldsymbol{x} \in \mathcal{X}$ based on the uniform entropy integral. This assumption ensures that the function class forms a Donsker class, which is important for the application of empirical process theory ([25], Corollary 2.7.2). By controlling the complexity, it allows for effective nonparametric estimation. Assumption 5 (ii) requires the sieve approximation error of $\rho'^{-1}\{\pi_0(t, \boldsymbol{x})\}$ to diminish at a polynomial rate. This condition holds for various sieve basis functions. Specifically, if $\boldsymbol{X}$ is discrete, the condition can be satisfied with $\alpha = +\infty$. On the other hand, if $\boldsymbol{X}$ is continuous and $u_K(\boldsymbol{x})$ is a power series or a B-spline, the condition can be satisfied with $\alpha = s/r$, where s represents the smoothness of the approximand and r signifies the dimension of $\boldsymbol{X}$. Assumption 7 imposes a similar sieve approximation error condition for $m(t, \boldsymbol{x})$.

Assumption 6 (i) is introduced to preclude near multicollinearity among the approximating basis functions. This condition is commonly encountered in sieve regression literature. Assumption 6 (ii) is met with $\zeta(K) = O(K)$ if $u_K(\boldsymbol{x})$ is a power series and with $\zeta(K) = O(\sqrt{K})$ if u_K is a B-spline [26]. Assumption 8 stipulates boundedness conditions on the moment of the response variable, standard in errors-in-variables problems (e.g., [8, 13]). These conditions are essential for deriving the asymptotic distribution of the proposed estimator using the Lyapunov central limit theorem.

The choice of U distribution and the decay rates of h_0 and h have an impact on the asymptotic variance of our estimator, which is different from the error-free case. We examine two types of U: ordinary smooth and supersmooth cases, standard in errors-in-variables literature (see, e.g., [8, 13, 27], among others, for more details).

An ordinary smooth error of order $\beta \geq 1$ satisfies

$$\lim_{t\to\infty} t^{\beta}\phi_U(t) = c \quad \text{and} \quad \lim_{t\to\infty} t^{\beta+1}\phi_U^{(1)}(t) = -c\beta\,, \tag{3.18}$$

for some constant $c > 0$. A supersmooth error of order $\beta \geq 1$ satisfies

$$d_0|t|^{\beta_0}\exp(-|t|^\beta/\gamma) \leq |\phi_U(t)| \leq d_1|t|^{\beta_1}\exp(-|t|^\beta/\gamma) \quad \text{as} \quad |t| \to \infty\,, \tag{3.19}$$

for some positive constants d_0, d_1, γ and some constants β_0 and β_1. Examples of ordinary smooth errors in the literature include Laplace errors, Gamma errors, and their convolutions. Cauchy errors, Gaussian errors, and their convolutions are categorized as supersmooth errors. The parameter β describes the decay rate of the characteristic function $\phi_U(t)$ as $t \to \infty$, corresponding to the smoothness of the error distribution. For instance, the Cauchy distribution corresponds to $\beta = 1$, Laplace and Gaussian distributions correspond to $\beta = 2$, and in the case of the Gamma distribution, β is related to both the shape and scale parameters.

The inverse Fourier transform representation (3.5) involves dividing by ϕ_U, and it is natural to expect better estimation results when $|\phi_U|$ is larger (corresponding to smaller β values). Indeed, as documented in the literature (see, e.g., [8, 13, 27, 28], among others), for both ordinary smooth and supersmooth cases, a higher order β leads to a more challenging deconvolution, resulting in slower convergence of the variance of a deconvolution kernel estimator. This inherent difficulty arises in nonparametric estimation when dealing with errors in variables [1, 13].

Such an impact on the convergence rate of our estimator is demonstrated in the following theorems. Depending on the distribution type of U, we need to impose different conditions on L in order to derive the asymptotic variance. The conditions are as follows:

Assumption O (*Ordinary Smooth Case*) $\|\phi_L\|_\infty < \infty$, $\int_{-\infty}^{\infty}|t|^{\beta+1}\{|\phi_L(t)| + |\partial_t\phi_L(t)|\}\,dt < \infty$ and $\int_{-\infty}^{\infty}|t^\beta\phi_L(t)|^2\,dt < \infty$.

Assumption S (*Supersmooth Case*) $\phi_L(t)$ is support on $[-1, 1]$ and bounded.

These assumptions are about the prespecified kernel function and can be readily satisfied. For instance, a kernel function whose Fourier transform is $\phi_L(u) = (1 - u^2)^3 \cdot \mathbb{1}_{[-1,1]}(u)$ fulfills these conditions (see, e.g., [8, 13]). In the subsequent two sections, the large sample properties of $\widehat{\pi}(t, \cdot)$ and $\widehat{\mu}(t)$ under the two types of U are established.

3.3.1 Asymptotics for the Ordinary Smooth Error

To establish the large sample properties of $\widehat{\mu}(t)$, we initially demonstrate the consistency of the estimated weight function $\widehat{\pi}(t, \cdot)$. We then compute its convergence rates under both the L_∞ norm and the L_2 norm.

Theorem 3.4 *Suppose that the error U is ordinary smooth of order β satisfying (3.18) and that Assumption O holds. Under Assumptions 2–6 and $\zeta(K)\sqrt{K\big/\left(Nh_0^{1+2\beta}\right)} \to 0$ as $N \to \infty$, for every fixed $t \in \mathcal{T}$, then*

$$\sup_{x\in\mathcal{X}}|\widehat{\pi}(t,\boldsymbol{x})-\pi_0(t,\boldsymbol{x})| = O_p\left(\zeta(K)\left\{K^{-\alpha}+h_0^2\right\}+\zeta(K)\sqrt{\frac{K}{Nh_0^{1+2\beta}}}\right),$$

$$\int_{\mathcal{X}}|\widehat{\pi}(t,\boldsymbol{x})-\pi_0(t,\boldsymbol{x})|^2 dF_X(\boldsymbol{x}) = O_p\left(\{K^{-2\alpha}+h_0^4\}+\frac{K}{Nh_0^{1+2\beta}}\right),$$

$$\frac{1}{N}\sum_{i=1}^{N}|\widehat{\pi}(t,\boldsymbol{X}_i)-\pi_0(t,\boldsymbol{X}_i)|^2 = O_p\left(\{K^{-2\alpha}+h_0^4\}+\frac{K}{Nh_0^{1+2\beta}}\right).$$

The first part of the rates, $\zeta(K)\left\{K^{-\alpha}+h_0^2\right\}$ and $K^{-2\alpha}+h_0^4$, represents the rates of the asymptotic bias. On the other hand, $\zeta(K)\sqrt{K/Nh_0^{1+2\beta}}$ and $K/Nh_0^{1+2\beta}$ correspond to the asymptotic variance.

Next, we establish the asymptotic linear expansion and asymptotic normality of $\widehat{\mu}(t)-\mu(t)$. To facilitate the presentation, we introduce the following definitions. For $i=1,\ldots,N$, $\eta_{h,h_0}(S_i,\boldsymbol{X}_i,Y_i;t) := \phi_h(S_i,\boldsymbol{X}_i,Y_i;t)+\psi_{h_0}(S_i,\boldsymbol{X}_i,Y_i;t)$, where

$$\begin{aligned}
\phi_h(S_i,\boldsymbol{X}_i,Y_i;t) &:= \left[\pi_0(t,\boldsymbol{X}_i)Y_iL_{U,h}(t-S_i)-\mathbb{E}\{\pi_0(t,\boldsymbol{X})YL_{U,h}(t-S)\}\right]\\
&\quad -\mu(t)\left[L_{U,h}(t-S_i)-\mathbb{E}\{L_{U,h}(t-S)\}\right],\\
\psi_{h_0}(S_i,\boldsymbol{X}_i,Y_i;t) &:= \mu(t)\left[L_{U,h_0}(t-S_i)-\mathbb{E}\{L_{U,h_0}(t-S)\}\right]\\
&\quad -\left[m(t,\boldsymbol{X}_i)\pi_0(t,\boldsymbol{X}_i)L_{U,h_0}(t-S_i)-\mathbb{E}\{m(t,\boldsymbol{X})\pi_0(t,\boldsymbol{X})L_{U,h_0}(t-S)\}\right],
\end{aligned}$$

with $L_{U,h}(v) := h^{-1}L_U(v/h)$. The population mean of both ϕ_h and ψ_{h_0} are zero. Let "$*$" denote the convolution operator, we define

$$V_j := f_T^{-2}(t)(R_j^2 f_T) * f_U(t)\cdot C\,,\ \text{for } j=1,2\,,$$

where $C := \int_{-\infty}^{\infty}J^2(v)\,dv = (2\pi c^2)^{-1}\int|w|^{2\beta}\phi_L^2(w)\,dw$, with c defined in (3.18), $J(v) := (2\pi c)^{-1}\int_{-\infty}^{\infty}\exp(-iwv)\phi_L(w)w^{\beta}\,dw$ and R_1^2, R_2^2 defined in Assumption 8. Moreover, let $(R_1R_2)(t) := \mathbb{E}\left[\{\pi_0(t,\boldsymbol{X})Y-\mu(t)\}\{\mu(t)-\pi_0(t,\boldsymbol{X})m(t,\boldsymbol{X})\}|T=t\right]$ and $v_h(t) := \mathbb{E}\{L_{U,h}^2(t-S)\}$.

Theorem 3.5 *Suppose that the error U is ordinary smooth of order β satisfying* (3.18) *and that Assumption O and Assumptions 1–8 as well as the following condition hold:*

$$\frac{(K^{-\ell}+h_0^2)\cdot(K^{-\alpha}+h_0^2)}{h^2}+\frac{(h\wedge h_0)^{1/2+\beta}}{h_0^{1+2\beta}}\frac{K}{\sqrt{N}}\to 0,$$

where $(h\wedge h_0) = h\mathbb{1}\{h=O(h_0)\}+h_0\mathbb{1}\{h_0=o(h)\}$. *Then, for every fixed* $t\in\mathcal{T}$,

$$
\begin{aligned}
\widehat{\mu}(t)-\mu(t) =& \frac{\kappa_{21}}{2}\left[\frac{f_T(t)\Phi_1(t)-\mu(t)\partial_t^2 f_T(t)}{f_T(t)}\right]\cdot h^2+o(h^2)\\
&+\frac{\kappa_{21}}{2}\left[\frac{\mu(t)\partial_t^2 f_T(t)-f_T(t)\Phi_2(t)}{f_T(t)}\right]\cdot h_0^2+o(h_0^2)\\
&+\sum_{i=1}^{N}\frac{\eta_{h,h_0}(S_i,\boldsymbol{X}_i,Y_i;t)}{N\cdot f_T(t)}+o_P\left\{\frac{1}{\sqrt{N(h\wedge h_0)^{1+2\beta}}}\right\},
\end{aligned}
\tag{3.20}
$$

where $\kappa_{ij} := \int u^i L^j(u)du$, $\Phi_1(t) := \mathbb{E}\big[\{Y\partial_t^2 f_{T|Y,X}(t|Y,\boldsymbol{X})\}/\{f_{T|X}(t|\boldsymbol{X})\}\big]$, *and* $\Phi_2(t) := \mathbb{E}\big[\{m(t,\boldsymbol{X})\partial_t^2 f_{T|X}(t|\boldsymbol{X})\}/\{f_{T|X}(t|\boldsymbol{X})\}\big]$. *Furthermore,*

(a) if $h=o(h_0)$, *then* $\sqrt{h^{1+2\beta}/N}\sum_{i=1}^N \eta_{h,h_0}(S_i,\boldsymbol{X}_i,Y_i;t)/f_T(t)\overset{d}{\to}N(0,V_1)$;

(b) if $h_0=o(h)$, *then* $\sqrt{h_0^{1+2\beta}/N}\sum_{i=1}^N \eta_{h,h_0}(S_i,\boldsymbol{X}_i,Y_i;t)/f_T(t)\overset{d}{\to}N(0,V_2)$;

(c) if $h_0=\tilde{c}h$ *for a constant* $\tilde{c}>0$, *then* $\sqrt{h^{1+2\beta}/N}\sum_{i=1}^N \eta_{h,h_0}(S_i,\boldsymbol{X}_i,Y_i;t)/f_T(t)\overset{d}{\to}N(0,V_3)$, *where*

$$
\begin{aligned}
V_3 :=& \frac{(R_1^2 f_T)*f_U(t)}{f_T^2(t)}\cdot\int_{-\infty}^{\infty}J^2(v)\,dv+\frac{(R_2^2 f_T)*f_U(t)}{\tilde{c}^{(2+2\beta)}f_T^2(t)}\cdot\int_{-\infty}^{\infty}J^2(v/\tilde{c})\,dv\\
&+\frac{2\{(R_1R_2)f_T\}*f_U(t)}{\tilde{c}^{(1+\beta)}f_T^2(t)}\cdot\int_{-\infty}^{\infty}J(v)J(v/\tilde{c})\,dv\,.
\end{aligned}
$$

In particular, when $\tilde{c}=1$, V_3 *reduces to* $f_T^{-2}(t)(R_3^2 f_T)*f_U(t)\cdot C$ *with* R_3^2 *defined in Assumption 8.*

The proof of Theorem 3.5 can be found in the Appendix of [15]. It shows that under certain conditions, such as sufficient smoothness of $\pi_0(t,\cdot)$ and $m(t,\cdot)$ or sufficiently fast growth of K, and rapid decay of h_0 satisfying $(K^{-\ell}+h_0^2)\cdot(K^{-\alpha}+h_0^2)=o(h^2)$, the error resulting from the sieve approximation becomes asymptotically negligible. In this scenario, $\widehat{\mu}(t)-\mu(t)$ achieves the optimal convergence rate of $N^{-2/(2\beta+5)}$ when $h_0\asymp h\asymp N^{-1/(2\beta+5)}$. To satisfy these conditions, it is necessary to have $K=o(h^{-2})$, $\alpha+\ell>1$, and $\alpha>1/2$ when using a spline basis, or $\alpha>1$ when using a power series (detailed derivation can be found in Appendix A.5 of [15]).

The convergence rate $N^{-2/(2\beta+5)}$ mentioned above is optimal for all possible nonparametric regression estimators when the regressors are affected by ordinary smooth errors, as demonstrated in [13]. It is worth noting that for the error-free local constant estimator, the convergence rates of the asymptotic bias and variance are h^2 and $(Nh)^{-1/2}$, respectively (see, for example, [29]). Our proposed estimator $\widehat{\mu}(t)$ shares the same rate of asymptotic bias as the error-free case, but the asymptotic variance degenerates by $h^{-\beta}$ due to the ordinary smoothness of the error distribution.

Additionally, as for asymptotic normality, we give the asymptotic linear expansion of $\widehat{\mu}(t)-\mu(t)$ in (3.20), facilitating statistical inference. Estimating the closed-form asymptotic variances V_1, V_2, and V_3 can be challenging in the context of measurement error (see, e.g., [30] Appendix C). However, leveraging our linear expansion in

(3.20), to estimate the asymptotic variance, we merely require consistent estimators of $\phi_h(S, \boldsymbol{X}, Y; t)$ and $\psi_{h_0}(S, \boldsymbol{X}, Y; t)$. For instance, $\pi_0(t, \boldsymbol{X})$ and $\mu(t)$ can be, respectively, estimated using our $\widehat{\pi}(t, \boldsymbol{X})$ and $\widehat{\mu}(t)$, and $\mathbb{E}[Y|T = t, \boldsymbol{X}]$ can be estimated using the method proposed by [31]. Subsequently, a pointwise confidence interval for $\mu(t)$ can be achieved by employing the undersmoothing technique. Alternative confidence intervals based on bias-correction (see, e.g., [32, 33]) are also viable but need a more precise estimation of the asymptotic bias and corresponding adjustments of the variance estimation with theoretical justification.

3.3.2 Asymptotics for the Supersmooth Error

The asymptotic properties of our estimator for the supersmooth case are showed in the next two theorems.

Theorem 3.6 *Suppose that the error U is supersmooth of order β satisfying* (3.19) *and Assumption S holds. Under Assumptions 2–6 and $\zeta^2(K)K \cdot (Nh_0)^{-1} \cdot \exp(2h_0^{-\beta}/\gamma) \to 0$ as $N \to \infty$, for every fixed $t \in \mathcal{T}$, then*

$$\sup_{\boldsymbol{x}\in\mathcal{X}} |\widehat{\pi}(t, \boldsymbol{x}) - \pi_0(t, \boldsymbol{x})| = O_p\left(\zeta(K)\cdot\left[\left\{K^{-\alpha} + h_0^2\right\} + \frac{\exp\left(h_0^{-\beta}/\gamma\right)}{\sqrt{h_0}}\cdot\sqrt{\frac{K}{N}}\right]\right),$$

$$\int_{\mathcal{X}} |\widehat{\pi}(t, \boldsymbol{x}) - \pi_0(t, \boldsymbol{x})|^2 dF_X(\boldsymbol{x}) = O_p\left(\{K^{-2\alpha} + h_0^4\} + \frac{\exp\left(2h_0^{-\beta}/\gamma\right)}{h_0}\cdot\frac{K}{N}\right),$$

$$\frac{1}{N}\sum_{i=1}^{N} |\widehat{\pi}(t, \boldsymbol{X}_i) - \pi_0(t, \boldsymbol{X}_i)|^2 = O_p\left(\{K^{-2\alpha} + h_0^4\} + \frac{\exp\left(2h_0^{-\beta}/\gamma\right)}{h_0}\cdot\frac{K}{N}\right).$$

We provide the proof of Theorem 3.6 in Appendix of [15]. We compare these results with those in Theorem 3.4 and find that the asymptotic bias remains unchanged from the ordinary smooth case. However, in the errors-in-variables context, the rate of the asymptotic variance becomes significantly slower as expected. This phenomenon has been well-documented in the literature, as demonstrated by [8, 13] in the errors-in-variables setting.

Theorem 3.7 *Suppose that the error U is supersmooth of order β satisfying* (3.19) *and that Assumption S and Assumptions 1–8 hold. Letting $e(h) := h^{1/2}\exp(-h^{-\beta}/\gamma)$, we have $v_h(t) = \mathbb{E}\{L_{U,h}^2(t - S)\} = O\{e(h)^{-2}\}$. If, as $h \to 0$, $v_h(t) \to \infty$ and*

$$\frac{(K^{-\ell} + h_0^2)\cdot(K^{-\alpha} + h_0^2)}{h^2} + \frac{K}{\{e(h)\wedge e(h_0)\}\sqrt{N}} \to 0 \text{ as } N \to \infty,$$

then, for every fixed $t \in \mathcal{T}$,

$$\begin{aligned}\widehat{\mu}(t) - \mu(t) =& \frac{\kappa_{21}}{2}\left[\frac{f_T(t)\Phi_1(t) - \mu(t)\partial_t^2 f_T(t)}{f_T(t)}\right] \cdot h^2 + o(h^2) \\ &+ \frac{\kappa_{21}}{2}\left[\frac{\mu(t)\partial_t^2 f_T(t) - f_T(t)\Phi_2(t)}{f_T(t)}\right] \cdot h_0^2 + o(h_0^2) \\ &+ \sum_{i=1}^{N} \frac{\eta_{h,h_0}(S_i, \boldsymbol{X}_i, Y_i; t)}{N \cdot f_T(t)} \cdot \{1 + o_P(1)\},\end{aligned} \tag{3.21}$$

where κ_{21}, $\Phi_1(t)$, $\Phi_2(t)$ *and* η_{h,h_0} *are defined as those in Theorem 3.5 and*

$$\{Nf_T(t)\}^{-1} \sum_{i=1}^{N} \eta_{h,h_0}(S_i, \boldsymbol{X}_i, Y_i; t) = O_p\{N^{-1/2}\{e(h) \wedge e(h_0)\}^{-1}\}.$$

Moreover, if $\upsilon_h(t) \geq d_1 f_S(t) h^{d_3} \exp(2h^{-\beta}/\gamma - d_2 h^{-d_4\beta})$ *for some constants* $d_1, d_2 > 0$, $1 > d_4 > 0$ *and* d_3, *we have*

$$[var\{\eta_{h,h_0}(S_i, \boldsymbol{X}_i, Y_i; t)\}]^{-1/2} \cdot \frac{1}{\sqrt{N}} \sum_{i=1}^{N} \left\{\eta_{h,h_0}(S_i, \boldsymbol{X}_i, Y_i; t)\right\} \xrightarrow{D} N(0, 1).$$

The proof of Theorem 3.7 can be found in Appendix of [15]. Similar to the case of ordinary smoothness, provided that $\pi_0(t, \cdot)$ exhibits sufficient smoothness or K increases rapidly, and h_0 diminishes swiftly, the approximation error of our estimator $\widehat{\mu}(t)$ via the sieve method becomes asymptotically negligible. The primary bias term remains unchanged from that of the ordinary smooth case. However, the asymptotic variance is influenced by the measurement error U. The convergence rate of the variance for bandwidth $b = h$ or h_0, denoted as $\{Nb\exp(-2b^{-\beta}/\gamma)\}^{-1/2}$, is degraded by $\exp(b^{-\beta}/\gamma)$ compared to the rate $(Nb)^{-1/2}$ for the error-free scenario, due to the supersmoothness of the error distribution.

From the theorem, when $h \asymp h_0$ and $\min(h, h_0) = d(\log N)^{-1/\beta}$ for a constant $d > (2/\gamma)^{1/\beta}$, it is observed that the rate of variance, $N^{-1}\{e(h) \wedge e(h_0)\}^{-2} = o(h^4 + h_0^4)$, becomes negligible compared to the asymptotic bias, and the convergence rate of $\widehat{\mu}(t) - \mu(t)$ is $(\log N)^{-2/\beta}$. This result aligns with findings in the literature on nonparametric regression with measurement error (cf. [13, 30, 34, 34] and others). It demonstrates that our estimator achieves the optimal convergence rate for all feasible nonparametric regression estimators when the regressors are measured with supersmooth errors, as demonstrated in [13].

It is worth noting that deriving an explicit expression and exact convergence rate of $\text{var}\{\eta_{h,h_0}(S, \boldsymbol{X}, Y; t)\}$ under supersmooth error is extremely challenging (if not impossible) without additional assumptions. To establish the asymptotic distribution of $\widehat{\mu}$ using Lyapunov central limit theorem, it is necessary to have a lower bound on the second moment of the deconvolution kernel. Specifically, we need $\upsilon_h(t) \geq d_1 f_S(t) h^{d_3} \exp(2h^{-\beta}/\gamma - d_2 h^{-d_4\beta})$. This constraint is commonly imposed in the measurement error literature as demonstrated by [30, 34] among others.

Fan [34] demonstrated that this lower bound holds under some mild conditions on ϕ_U and ϕ_L (e.g., (3.19) and Assumption S hold, $\phi_L(t) > c_L(1-t)^3$ for $t \in [1-\epsilon, 1)$ for some $c_L, \epsilon > 0$, and the real part $R_U(t)$ and the imaginary part $I_U(t)$ of ϕ_U satisfy $R_U(t) = o\{I_U(t)\}$ or $I_U(t) = o\{R_U(t)\}$ as $t \to \infty$). In these mentioned assumptions, we do not preclude the use of commonly used kernel functions like L defined below Assumption S and error distributions such as Gaussian, Cauchy, and Gaussian mixture.

For statistical inference, our explicit asymptotic linear expansion of $\widehat{\mu}(t) - \mu(t)$ in (3.21) is particularly valuable in the supersmooth error case. This is because deriving an explicit expression for the asymptotic variance is challenging in this setting. Most of the existing literature only provides information about the convergence rate; see [9, 13, 27], among others.

3.4 Select the Smoothing Parameters

In this section, we address the selection of the three smoothing parameters K, h_0, and h required for computing our estimator $\widehat{\mu}(t)$ (see (3.14), (3.15), and (3.17)). In the beginning, we establish some preliminary concepts.

3.4.1 *Preliminaries*

In nonparametric regression, the choice of smoothing parameters is often made by minimizing cross-validation (CV) criteria or by approximating the asymptotic bias and variance of the estimator.

However, in errors-in-variables regression, as highlighted by [1, 27], approximating the asymptotic bias and variance of the estimator can be extremely challenging, if not impossible. Unfortunately, the CV criteria are also not computable. To illustrate this, let's consider the case where K and h_0 in Eqs. (3.14) and (3.15) are given. We can adapt the CV criteria to select the smoothing parameter h.

$$CV(h) = \sum_{i=1}^{N} \left\{\widehat{\pi}(T_i, X_i)Y_i - \widehat{\mu}^{-i}(T_i)\right\}^2 w(T_i), \tag{3.22}$$

where w is a weight function preventing CV from becoming excessively large due to unreliable data points from the tails of the distributions of T, and $\widehat{\mu}^{-i}$ denotes the estimator obtained as in (3.17), but without incorporating observations from the individual i. However, it is important to note that Eq. (3.22) cannot be computed in errors-in-variables regression problems due to the unobservability of the T_i values.

To address this issue, [35] proposed a combination of the CV and SIMEX methods (e.g. [36, 37]). The approach involves generating additional sets of contaminated data in the simulation step. Specifically, generate $S^*_{i,d} = S_i + U^*_{i,d}$ and $S^{**}_{i,d} = S^*_{i,d} + U^{**}_{i,d}$, for $i = 1, \ldots, N$ and $d = 1, \ldots, D$ with D a large number, where the $U^*_{i,d}$'s and $U^{**}_{i,d}$'s are i.i.d. as U in (3.1). By replacing the original S_i values with $S^*_{i,d}$ and $S^{**}_{i,d}$ in Eqs. (3.14) and (3.17), two sets of estimates $(\widehat{\pi}^*_d, \widehat{\mu}d^*)$ and $(\widehat{\pi}^{**}_d, \widehat{\mu}^{**}_d)$ are obtained for $d = 1, \ldots, D$. The authors then proposed deriving two CV-type bandwidths, $\widehat{h}^*$ and $\widehat{h}^{**}$, by minimizing $\sum_{d=1}^{D} CV^*_d(h)/D$ and $\sum_{d=1}^{D} CV^{**}_d(h)/D$, respectively, where

$$CV^*_d(h) = \sum_{i=1}^{N} \left\{\widehat{\pi}^*_d(S_i, \boldsymbol{X}_i)Y_i - \widehat{\mu}^{*,-i}_d(S_i)\right\}^2 w(S_i),$$

$$CV^{**}_d(h) = \sum_{i=1}^{N} \left\{\widehat{\pi}^{**}_d(S^*_{i,d}, \boldsymbol{X}_i)Y_i - \widehat{\mu}^{**,-i}_d(S^*_{i,d})\right\}^2 w(S^*_{i,d}),$$

for $d = 1, \ldots, D$, where $\widehat{\mu}^{*,-i}_d$ and $\widehat{\mu}^{**,-i}_d$ are obtained, respectively, as $\widehat{\mu}^*_d$ and $\widehat{\mu}^{**}_d$, but without the observations from individual i.

The $S^{**}_{i,d}$'s represent the contaminated versions of the S^*_i's, similar to how the S^*_i's relate to the S_i's, and the S_i's relate to the T_i's. It is expected that there is a relationship between $\widehat{h}^*$ and the desired bandwidth h similar to the relationship between $\widehat{h}^{**}$ and $\widehat{h}^*$. Consequently, the authors proposed an extrapolation step to estimate the bandwidth h. In particular, they assumed that $h/\widehat{h}^* \approx \widehat{h}^*/\widehat{h}^{**}$ and employed a linear back-extrapolation procedure. In our specific context, this procedure would yield an estimator for the bandwidth

$$\widehat{h}_{DH} = (\widehat{h}^*)^2/\widehat{h}^{**}. \tag{3.23}$$

3.4.2 Two-Step Procedure and Local Constant Extrapolation

It is crucial to consider that we have two additional smoothing parameters, namely K and h_0. We can approach this by considering two potential strategies: either extending the SIMEX method to simultaneously select all three parameters or adopting a two-step procedure. The first approach, which involves simultaneously selecting K, h_0, and h using the SIMEX method, can be computationally demanding and may exhibit instability in practical settings. Hence, we opt for the second approach, which is a two-step procedure.

Referring to Theorems 3.5 and 3.7, we can observe that our estimator achieves an optimal rate when $h \asymp h_0$, striking a balance between the bias rate $h^2 + h_0^2$ and the standard deviation $\sqrt{v_h(t)/N + v_{h_0}(t)/N}$. Furthermore, it is worth noting that the plug-in bandwidth h_{PI} for the kernel deconvolution estimator with bandwidth h of the density of T, as proposed by [38], minimizes the asymptotic mean squared error (MSE) of the estimator. In this case, the bias is of rate h^2 and the standard deviation is $\sqrt{v_h(t)/N}$. Therefore, our goal is to have $h \asymp h_0 \asymp h_{PI}$ and $K = o(h^{-2})$ to ensure that the condition of K satisfies all the requirements in our theorems when $h \asymp h_0$.

First, set $h_0 = h_{PI}$. Then, To determine the value of K, we can utilize the fact that $\mathbb{E}\pi_0(t, \boldsymbol{X})\exp(\boldsymbol{X})|T = t = \mathbb{E}\exp(\boldsymbol{X})$ from Eq. (3.11). We propose selecting $K = \lfloor \tilde{c} h_{PI}^{-2} \log(h_{PI} + 1) \rfloor$ such that $K \geq 2$, where the constant $\tilde{c}$ minimizes the following generalized CV criterion [39]:

$$\int_T \left| \frac{\sum_{i=1}^N \widehat{\pi}(t, \boldsymbol{X}_i)\exp(\boldsymbol{X}_i) L_U\{(t - S_i)/h_{PI}\}}{\sum_{i=1}^N L_U\{(t - S_i)/h_{PI}\}} - \frac{\sum_{i=1}^N \exp(\boldsymbol{X}_i)}{N} \right|^2 \Big/ (1 - K/N)^2 \, dt.$$

It is important to note that while the selection of K and h_0 may not directly minimize the error for the final estimator $\widehat{\mu}(t)$, it ensures the optimal convergence rate of $\widehat{\mu}(t)$ given the chosen bandwidth h. This selection is particularly effective when employing a B-spline basis. However, when using a polynomial sieve basis, an additional consideration is needed. Specifically, the smoothing parameter α, as defined in Assumption 5, must exceed 1 to achieve the desired convergence rate.

In the second step of the procedure, one option is to directly use $\widehat{h}_{DH}$ from Eq. (3.23). However, in our numerical investigations, we observed that the linear back-extrapolation occasionally led to highly unstable outcomes. Increasing the number of simulations D can reduce variability, and previous works such as [35] have used $D = 20$. However, even with $D = 40$, we still encountered some unacceptable results. This is not entirely unexpected as determining the appropriate extrapolation function for practical use remains uncertain [1]. Therefore, we propose a novel extrapolation procedure to address these challenges

In our proposed extrapolation procedure, instead of parametrically extrapolating from $\widehat{h}^*$ and $\widehat{h}^{**}$, we approximate the relationship between the h_d^*'s and h_d^{**}'s using a local constant estimator, as detailed in [29]. Here, $h_d^* = c_d^* h_{PI}$ and $h_d^{**} = c_d^{**} h_{PI}$ with the constants c_d^* and c_d^{**} minimizing the respective cross-validation criteria $CV_d^*(h)$ and $CV_d^{**}(h)$ for $d = 1, \ldots, D$. By employing this approximated relationship as the extrapolant function, we determine the bandwidth h as follows:

$$\widehat{h} = \frac{\sum_{d=1}^D h_d^* \cdot \varphi\{(\widehat{h}^* - h_d^{**})/b\}}{\sum_{d=1}^D \varphi\{(\widehat{h}^* - h_d^{**})/b\}}, \tag{3.24}$$

where φ represents the Gaussian kernel function. The bandwidth b can be determined using leave-one-out cross-validation, which is a commonly used approach in bandwidth selection. Local constant estimation has been extensively studied and widely applied, demonstrating fast and stable performance in many cases. Our simulation study indicated that employing $D = 35$ is sufficient for achieving satisfactory performance.

Remark 3.2 In Eq. (3.15), it is worth recalling that some deconvolution kernels $L_U(t - S_i)/h_0$ may result in negative values, which can lead to a violation of the strict concavity of $\widehat{G}t(\lambda)$ in finite samples. To address this issue, one option is to set $h_0 = hPI$ and truncate these negative values of $L_U(t - S_i)/h_0$ to 0. Our simulations have demonstrated that this approach yields satisfactory performance in terms of accuracy and reliability.

3.5 Real Data Example

To illustrate the practical value of our data-driven $\widehat{\text{CM}}$ estimator using Epidemiologic Study Cohort data from NHANES-I, our objective is to estimate the causal effect of the long-term log-transformed daily saturated fat intake on the risk of breast cancer. The dataset consists of 3,145 women aged between 25 and 50. The analysis of this data was conducted by [1] using a logistic regression calibration method, accessible in https://carroll.stat.tamu.edu/data-and-documentation. Within the dataset, the daily saturated fat intake was measured using a single 24-hour recall, and the log-transformation was calculated as $\log(5 + \text{saturated fat})$. Previous nutrition studies have suggested that measurement error results in more than 75% of the variance in these data . According to [1], a classical measurement error model (i.e., Eq. (3.1)) is suitable to be assumed, with a Gaussian measurement error U on the data. The outcome variable Y indicates whether an individual has breast cancer (taking the value 1) or not (taking the value 0). The covariates $\boldsymbol{X}$ include age, poverty index ratio, body mass index, alcohol use (yes or no), family history of breast cancer, age at menarche (a binary variable indicating age ≤ 12), menopausal status (pre or post), and race. It is assumed that these covariates were measured without appreciable error.

Initially, we analyze the data under our estimator with the assumption of a Gaussian measurement error with an error variance ratio of $\text{var}(U)/\text{var}(S) = 0.75$. This ratio corresponds to $\text{var}(U)/\text{var}(T) = 3$. It is important to note that estimation of error variances from other nutrition studies may be unreliable, as highlighted by [40]. Hence, we also examine the situations where $\text{var}(U)/\text{var}(S) = 0.43$, $\text{var}(U)/\text{var}(S) = 0.17$, and $\text{var}(U) = 0$ (i.e., $\text{var}(U)/\text{var}(T) = 0.75$, $\text{var}(U)/\text{var}(T) = 0.2$, and the error-free case).

In Fig. 3.1, the estimated curves are plotted using the smoothing parameters selected according to the procedure outlined in Sect. 3.4. Additionally, a 95% undersmoothing pointwise confidence band is included for two scenarios: $\text{var}(U)/\text{var}(S) = 0.43$ and $\text{var}(U)/\text{var}(S) = 0.75$. For the case where $\text{var}(U)/\text{var}(S) = 0.43$, the reliable range of t is slightly shorter compared to $\text{var}(U)/\text{var}(S) = 0.17$. For the case

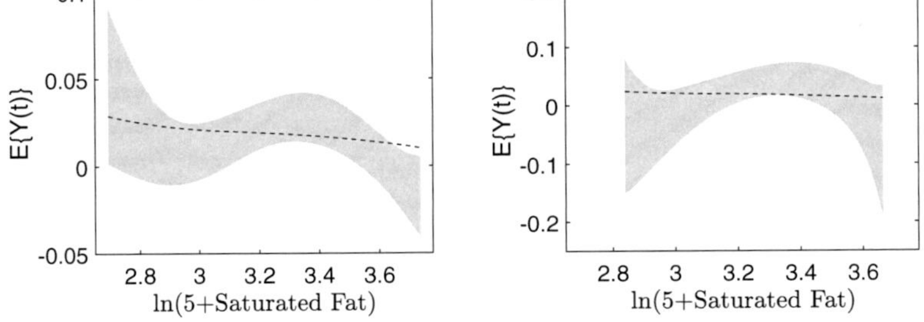

Fig. 3.1 Estimation of the treatment effect of the log-saturated fat intake on the risk of breast cancer and a 95% pointwise confidence band for a Gaussian error of $\text{var}(U)/\text{var}(S) = 0.43$ (left) and $\text{var}(U)/\text{var}(S) = 0.75$ (right)

where $\text{var}(U)/\text{var}(S) = 0.75$, the reliable range of t is notably shorter. However, despite the differences in the reliable ranges, the confidence band suggests a similar trend between $t = 3$ and $t = 3.6$ in the cases with $\text{var}(U)/\text{var}(S) = 0.17$ and 0.43. Specifically, there is a slight increase in the estimated curves before $t = 3.4$, followed by a significant decrease thereafter.

In general, the estimated risk of breast cancer demonstrates a decreasing trend across the range of transformed saturated fat intake. These findings align with the conclusions drawn by [1], who conducted a multivariate logistic regression calibration analysis and observed a significant and negative coefficient for the log-transformed saturated fat intake in relation to the risk of breast cancer. However, we have to raise caution when interpreting these results. There are a couple of factors that should be taken into consideration. First, there may be misclassification in the breast cancer data, which can introduce uncertainty in the analysis. Additionally, it is important to note that follow-up data for breast cancer cases with high fat intakes might be lacking, which could potentially impact the conclusions. We refer to [1].

References

1. Carroll, R.J., Rupper, D., Stefanski, L.A., Crainiceanu, C.M.: Measurement Error in Nonlinear Models: A Modern Perspective. Chapman & Hall/CRC (2006)
2. Rubin, D.B.: Comment: Neyman (1923) and causal inference in experiments and observational studies. Stat. Sci. **5**(4), 472–480 (1990)
3. Kennedy, E.H., Ma, Z., McHugh, M.D., Small, D.S.: Non-parametric methods for doubly robust estimation of continuous treatment effects. J. R. Stat. Soc. Ser. B (Statistical Methodology) **79**, 1229–1245 (2017)
4. Ai, C., Linton, O., Motegi, K., Zhang, Z.: A unified framework for efficient estimation of general treatment models. Quant. Econ. **12**(3), 779–816 (2021)
5. Ai, C., Linton, O., Zhang, Z.: Estimation and inference for the counterfactual distribution and quantile functions in continuous treatment models. J. Econometrics **228**(1), 39–61 (2022)
6. D'Amour, A., Ding, P., Feller, A., Lei, L., Sekhon, J.: Overlap in observational studies with high-dimensional covariates. J. Econometrics **221**(2), 644–654 (2021)
7. Diggle, P.J., Hall, P.: A fourier approach to nonparametric deconvolution of a density estimate. J. R. Stat. Soc. Ser. B (Methodol.) **55**, 523–531 (1993)
8. Delaigle, A., Fan, J., Carroll, R.J.: A design-adaptive local polynomial estimator for the errors-in-variables problem. J. Am. Stat. Assoc. **104**(485), 348–359 (2009)
9. Meister, A.: Density estimation with normal measurement error with unknown variance. Statistica Sinica 195–211 (2006)
10. Delaigle, A., Hall, P.: Methodology for non-parametric deconvolution when the error distribution is unknown. J. R. Stat. Soc. Ser. B: Stat. Methodol. 231–252 (2016)
11. Delaigle, A., Hall, P., Meister, A., et al.: On deconvolution with repeated measurements. Ann. Stat. **36**, 665–685 (2008)
12. Stefanski, L., Carroll, R.J.: Deconvoluting kernel density estimators. Statistics **2**, 169–184 (1990)
13. Fan, J., Truong, Y.K.: Nonparametric regression with errors in variables. Ann. Stat. **21**, 1900–1925 (1993)
14. Hall, P., Lahiri, S.N.: Estimation of distributions, moments and quantiles in deconvolution problems. Ann. Stat. **36**(5), 2110–2134 (2008)

15. Huang, W., Zhang, Z.: Nonparametric estimation of the continuous treatment effect with measurement error. J. R. Stat. Soc. Ser. B Stat Methodol. **85**(2), 474–496 (2023)
16. Chen, X.: Large sample sieve estiamtion of semi-nonparametric models. Handb. Econ. **6**, 5549–5632 (2007)
17. Kitamura, Y., Stutzer, M.: An information-theoretic alternative to generalized method of moments estimation. Econometrica: J. Econometric Soc. 861–874 (1997)
18. Imbens, G., Johnson, P., Spady, R.H.: Information theoretic approaches to inference in moment condition models. Econometrica **66**(2), 333–357 (1998)
19. Owen, A.B.: Empirical Likelihood. Chapman and Hall/CRC (2001)
20. Hansen, L.: Large sample properties of generalized method of moments estimators. Econometrica **50**, 1029–1054 (1982)
21. Chen, X., Hong, H., Tarozzi, A.: Semiparametric efficiency in gmm models with auxiliary data. Ann. Stat. **36**(2), 808–843 (2008)
22. Ma, X., Wang, J.: Robust inference using inverse probability weighting. J. Am. Stat. Assoc. **115**(532), 1851–1860 (2020)
23. Muñoz, I.D., Van Der Laan, M.: Population intervention causal effects based on stochastic interventions. Biometrics **68**(2), 541–549 (2012)
24. Díaz, I., van der Laan, M.J.: Targeted data adaptive estimation of the causal dose-response curve. J. Causal Infer. **1**(2), 171–192 (2013)
25. Van Der Vaart, A.W., van der Vaart, A., van der Vaart, A.W., Wellner, J.: Weak Convergence and Empirical Processes: With Applications to Statistics. Springer Science & Business Media (1996)
26. Newey, W.K.: Convergence rates and asymptotic normality for series estimators. J. Econometrics **79**(1), 147–168 (1997)
27. Meister, A.: Deconvolution Problems in Nonparametric Statistics. Springer, Berlin Heidelberg (2009)
28. Fan, J.: On the optimal rates of convergence for nonparametric deconvolution problems. Ann. Stat. 1257–1272 (1991)
29. Fan, J., Gijbels, I.: Local Polynomial Modelling and Its Applications. Chapman & Hall/CRC (1996)
30. Delaigle, A., Hall, P., Wishart, J.: Confidence bands in nonparametric errors-in-variables regression. J. R. Stat. Soc. Ser. B (Statistical Methodology) **77**, 149–169 (2015)
31. Liang, H.: Asymptotic normality of parametric part in partially linear models with measurement error in the nonparametric part. J. Stat. Plann. Infer. **86**(1), 51–62 (2000)
32. Calonico, S., Cattaneo, M.D., Farrell, M.H.: On the effect of bias estimation on coverage accuracy in nonparametric inference. J. Am. Stat. Assoc. **113**(522), 767–779 (2018)
33. Takatsu, K., Westling, T.: Debiased inference for a covariate-adjusted regression function (2022). arXiv:2210.06448
34. Fan, J.: Asymptotic normality for deconvolution kernel density estimators. Sankhyā: Indian J. Stat. Ser. A 97–110 (1991)
35. Delaigle, A., Hall, P.: Using simex for smoothing-parameter choice in errors-in-variables problems. J. Am. Stat. Assoc. **103**, 280–287 (2008)
36. Cook, J., Stefanski, L.: Simulation-extrapolation estimation in parametric measurement error models. J. Am. Stat. Assoc. **89**, 1314–1328 (1994)
37. Stefanski, L., Cook, J.: Simulation-extrapolation: the measurement error jackknife. J. Am. Stat. Assoc. **90**, 1247–1256 (1995)
38. Delaigle, A., Gijbels, I.: Estimation of integrated squared density derivatives from a contaminated sample. J. R. Stat. Soc. Ser. B (Statistical Methodology) **64**, 869–886 (2002)
39. Craven, P., Wahba, G.: Smoothing noisy data with spline functions. Numer. Math. **31**(4), 377–403 (1978)
40. Delaigle, A., Gijbels, I.: Bootstrap bandwidth selection in kernel density estimation from a contaminated sample. Ann. Inst. Stat. Math. **56**(1), 19–47 (2004)

Part II
Financial Model Computing and Decisions

In the second part, we introduce efficient computing algorithms for high-dimensional data in econometrics.

With the development in science and technology, high-dimensional data have been widely used in the study of econometrics. We focus on the fitting algorithms for quantile regression and portfolio allocation models in this part. Specifically, we introduce parallel computing algorithms based on feature split, which resolves the computational burden caused by high-dimensional data.

We introduce the computing algorithms in detail and also introduce the theoretical properties of the algorithms. Then we conduct extensive simulation studies to show the efficiency of the new algorithms and also apply them in real data examples. By comparing the parallel algorithms with existing method, we notice the advantage of the new algorithm in handling high-dimensional data in terms of both computing accuracy and efficiency.

Chapter 4
Efficient Computing for High-Dimensional Econometric Models

4.1 Introduction

With the development of science and technology, high-dimensional data are widely applied in the research of econometrics. Designing efficient computing algorithm for high-dimensional econometric models become more and more important recently. Portfolio optimization and quantile regression (QR) are two fundamental pillars in econometrics and finance, respectively. This chapter focuses on introducing efficient computing algorithms for these two models.

Mean-variance portfolio theory, proposed by [1], holds a central position in econometrics research, offering analytical solutions dependent on assets' mean excess returns and covariance matrix inverses. However, the optimality of strategies based on traditional estimations such as sample means and covariance matrices is compromised by estimation errors, particularly in high-dimensional settings [2, 3]. High-dimensional portfolio optimization has thus emerged as a prominent research area, and estimation of covariance matrices has attracted significant attention. Various methods have been proposed, including factor-based risk estimators [4], nonparametric eigenvalue-regularized estimators [5], robust covariance estimators [6], and a Kronecker product model [7]. More references for the estimation of high-dimensional covariance matrices include but not limited to [8–14]. The estimates in these aforementioned works converge to the true covariance matrix under certain regularity conditions, but it is still unclear whether the derived portfolio allocation provides the optimal strategy. In recent literature, constrained portfolio optimization, such as no-short-selling constraints [15] and norm constraints [16–18], has been explored to mitigate risks and accommodate investors' preferences. Pun and Wong [18] investigated a constrained ℓ_1 minimization approach and named the new estimator as the linear programming optimal (LPO) estimator and its corresponding estimated optimal portfolio as the LPO portfolio. The conventional approach to obtain the LPO

Z. Zhang et al., *Big Data in Economics and Management*, Statistics and Big Data 1,
https://doi.org/10.1007/978-981-95-3125-7_4

estimator is through linear programming (LP). However, as the volume and dimension of data used for statistical learning increase, the efficiency of LP is severely affected.

Quantile regression (QR), a powerful tool for analyzing data with heterogeneous effects, has seen extensive applications in econometrics since its inception [19]. Recent advancements include theories for high-dimensional quantile regression [20], inference frameworks for auction settings [21], and robust estimation methods [22]. Other recent studies of QR include, but not limited to [23–29]. Penalized quantile regression (PQR) has become pivotal in variable selection for QR, especially in high-dimensional contexts [30, 31]. Belloni et al. [32] derived a nice error bound for PQR with the Lasso penalty (ℓ_1-QR for short). Wang et al. [33] studied the PQR with folded concave penalty such as the smoothed clip absolute deviation (SCAD) penalty [34] and minimax concave penalty (MCP) [35], and further established the oracle property for PQR in the ultrahigh dimension setting under mild conditions. However, efficient numerical solutions for high-dimensional PQR remain elusive due to the nonsmoothness and possible nonconvexity of the objective function. Sherwood and Maidman [36] developed an R-package *rqPen* for ℓ_1-QR, and the algorithm is similar to the ℓ_1-QR introduced in [37]. Peng and Wang [38] developed an iterative coordinate descent algorithm (QICD) for solving PQR with nonconvex penalty. Gu et al. [39] introduced fast alternating direction method of multiplier (ADMM) [40] for PQR in high dimension. With the advent of big data, it is of crucial importance to study numerical algorithms for PQR in ultrahigh dimensions and/or a large data size. Yu et al. [41] and Fan et al. [42] developed parallel algorithms for PQR based on **sample-splitting** ADMM. By sample-splitting, it means by its name that the algorithm partitions the data across samples. Ultrahigh dimensionality adds another challenge in minimizing the objective function of ultrahigh-dimensional PQR.

To address the challenges in high-dimensional portfolio optimization and PQR, in this chapter we propose an efficient and parallelizable method in ultrahigh dimensions based on three-block ADMM. Using this method, we introduce a parallel ADMM approach for solving LPO problems with high-dimensional portfolio through **asset splitting**, and develop a **feature-splitting** algorithm for PQR to tackle the simultaneous challenges of nonsmoothness, nonconvexity and ultrahigh dimensionality. Using related techniques in [43], we establish the rate of convergence of the proposed algorithms and the theoretical convergence guarantee, and address the convergence uncertainty. The compatibility of the proposed three-block ADMM algorithms with parallel computing alleviates the storage and scalability limitations of a single machine in the large-scale data processing. The proposed three-block ADMM algorithms also enjoy numerical efficiency over the directly extended two-block ADMM. It is worthy to note that the newly proposed algorithms can be directly applied to both portfolio optimization and PQR with various penalties including the ℓ_1 and the SCAD penalties by local linear approximation to the penalties [44]. Based on theories developed in [45, 46], the proposed algorithms are able to obtain portfolio and PQR estimate with strong oracle properties in ultrahigh dimension. Simulation studies and empirical analyses demonstrate the convergence and performance of the proposed algorithms and estimators in both synthetic and real-world datasets.

The rest of the chapter is organized as follows. In Sect. 4.2, we first propose asset-splitting algorithm for LPO portfolio with ℓ_1 penalty, and then further propose the two-step procedure for LPO estimator with SCAD penalty and establish its strong oracle property. Parallel to Sect. 4.2, in Sect. 4.3 we give feature-splitting methods for both PQR-Lasso and PQR-SCAD. In Sect. 4.4, we conduct a numerical study to assess the performance of the proposed methods. All technical proofs are given in the Appendix.

Throughout the chapter, we adopt the following notations. For a matrix $\mathbf{m}M = (m_{ij})_{s\times t}$, denote $\|\mathbf{m}M\|_{\max} = \max_{(i,j)} |m_{ij}|$, $\|\mathbf{m}M\|_{\min} = \min_{(i,j)} |m_{ij}|$, and $\lambda_{\min}(\mathbf{m}M)$ and $\lambda_{\max}(\mathbf{m}M)$ as the smallest and largest eigenvalues of $\mathbf{m}M$, respectively. $\mathbf{X}_A$ denotes the sub-matrix of $\mathbf{X}$ with the columns indexed by A. $\mathbf{m}M \succ 0$ indicates that $\mathbf{m}M$ is positive definite. For a positive semidefinite operator or matrix $\mathbf{m}M$, $\|\mathbf{x}\|^2_{\mathbf{m}M} = \mathbf{x}^T\mathbf{m}M\mathbf{x}$.

4.2 Asset-Splitting Algorithm for Portfolio Selection

4.2.1 A Constrained ℓ_1 Minimization Approach

Portfolio allocation garners significant attention in financial econometrics and quantitative finance research. In his seminal work, [1] introduced the mean-variance portfolio theory, which formulates portfolio allocation as an optimization problem based on excess returns

$$\min_{\mathbf{w}} \mathbf{w}^T\boldsymbol{\Sigma}\mathbf{w} \quad \text{subject to } \mathbf{w}^T\boldsymbol{\mu} \geq \mu_0, \tag{4.1}$$

where $\boldsymbol{\mu} = (\mu_1, \ldots, \mu_p)^T$ is the mean vector of the excess return of p assets, $\boldsymbol{\Sigma}$ is the covariance matrix of the excess returns of the assets, and μ_0 is the expected return of the portfolio. The optimal weights $\mathbf{w}_{\text{opt}}$ of the optimization problem (4.1) can be derived as

$$\mathbf{w}_{\text{opt}} = \frac{\mu_0\boldsymbol{\Sigma}^{-1}\boldsymbol{\mu}}{\boldsymbol{\mu}^T\boldsymbol{\Sigma}^{-1}\boldsymbol{\mu}} = \frac{\mu_0\boldsymbol{\Omega}\boldsymbol{\mu}}{\boldsymbol{\mu}^T\boldsymbol{\Omega}\boldsymbol{\mu}}, \tag{4.2}$$

where $\boldsymbol{\Omega} = \boldsymbol{\Sigma}^{-1}$ represents the precision matrix of the excess returns of the assets. According to (4.2), the optimal weights $\mathbf{w}_{\text{opt}}$ are directly related to $\boldsymbol{\Omega}\boldsymbol{\mu}$. Consequently, we define $\boldsymbol{\beta} = \boldsymbol{\Omega}\boldsymbol{\mu}$ as the effective parameter for executing the optimization problem (4.1).

Inspired by [18, 47] proposed the LPO estimator based on the constrained ℓ_1 minimization approach. This approach is to directly estimate $\boldsymbol{\beta}$ as

$$\min_{\boldsymbol{\beta}\in\mathbb{R}^p} \|\boldsymbol{\beta}\|_1 \quad \text{s.t. } \|\hat{\boldsymbol{\Sigma}}\boldsymbol{\beta} - \mathbf{m}\hat{\mu}\|_\infty \leq \lambda, \tag{4.3}$$

where λ denotes a tuning parameter, $\hat{\boldsymbol{\mu}} = (\hat{\mu}_1, \ldots, \hat{\mu}_p)$ represents the sample mean vector of the excess returns, $\hat{\boldsymbol{\Sigma}}$ is the sample covariance matrix of the excess returns,

$\|\mathbf{m}a\|_1 = \sum_{i=1}^{p} |a_i|$ and $\|\mathbf{m}a\|_\infty = \max_i |a_i|$. Estimating $\boldsymbol{\beta}$ from (4.3) allows us to plug it into (4.2) to compute the estimated optimal weights of the assets in the portfolio. In this framework, the sparsity assumption applies to the product $\boldsymbol{\beta}$ rather than to $\boldsymbol{\mu}$ and $\boldsymbol{\Omega}$. Cai and Liu [47] demonstrated that $\boldsymbol{\beta}$ can be directly and efficiently estimated, even when separate estimations of $\boldsymbol{\mu}$ and/or $\boldsymbol{\Omega}$ pose challenges. Furthermore, [18] elaborated on several notable advantages of the LPO approach.

The minimization problem (4.3) constitutes a convex optimization problem. Traditional algorithms such as linear programming and classical ADMM algorithms may be employed to tackle (4.3) when the portfolio comprises a moderately large number of assets. A succinct overview of linear programming and the ADMM algorithm is provided in the Appendix.

4.2.2 Asset-Splitting ADMM Algorithm

The ultrahigh dimensionality presents a significant computational challenge in solving Problem (4.3). To address this issue, we propose a strategy of asset partitioning and parallel computing on subsets of assets. Assuming that the effective parameter $\boldsymbol{\beta}$ can be partitioned into $K > 1$ groups, $\boldsymbol{\beta} = (\boldsymbol{\beta}_1^T, \ldots, \boldsymbol{\beta}_K^T)^T$ with $\boldsymbol{\beta}_i \in \mathbb{R}^{p_i}$, where $\sum_{i=1}^{K} p_i = p$, and partition matrix $\hat{\boldsymbol{\Sigma}}$ as $(\hat{\boldsymbol{\Sigma}}_1, \ldots \hat{\boldsymbol{\Sigma}}_K)$ with $\hat{\boldsymbol{\Sigma}}_i \in \mathbb{R}^{p \times p_i}$. We then introduce the auxiliary variable $\mathbf{z} = \hat{\boldsymbol{\Sigma}}\boldsymbol{\beta} - \hat{\boldsymbol{\mu}}$ and incorporate the condition $\|\hat{\boldsymbol{\Sigma}}\boldsymbol{\beta} - \hat{\boldsymbol{\mu}}\|_\infty \le \lambda$ into the objective function, leading to:

$$
\begin{aligned}
&\min_{\boldsymbol{\beta},\mathbf{z}} \sum_{i=1}^{K} \|\boldsymbol{\alpha}_i \circ \boldsymbol{\beta}_i\|_1 + \delta_{\mathcal{Z}_0}(\mathbf{z}) \\
&\text{s.t. } \hat{\boldsymbol{\Sigma}}\boldsymbol{\beta} - \mathbf{z} = \hat{\boldsymbol{\mu}}.
\end{aligned}
$$

where $\boldsymbol{\alpha}_i \circ \boldsymbol{\beta}_i = (\alpha_{i1}\beta_{i1}, \ldots, \alpha_{ip_i}\beta_{ip_i})$, $\mathcal{Z}_0 = \{\mathbf{z} : |z_s| \le \lambda\alpha_s, s = 1, \ldots, p\}$ and α_s is the s-th element of $\boldsymbol{\alpha} = (\boldsymbol{\alpha}_i, \ldots, \boldsymbol{\alpha}_K)$, and

$$
\delta_{\mathcal{Z}_0}(\mathbf{z}) = \begin{cases} 0, & \text{if } \mathbf{z} \in \mathcal{Z}_0 \\ +\infty, & \text{otherwise.} \end{cases}
$$

To enforce a multi-block structure on the primal variables, we introduce $\mathbf{m}\omega = (\mathbf{m}\omega_2, \ldots \mathbf{m}\omega_K)$, $\mathbf{m}\omega_i \in \mathcal{R}^p, i = 2, \ldots, K$ into the problem such that $\hat{\boldsymbol{\Sigma}}_i\boldsymbol{\beta}_i = \mathbf{m}\omega_i,\ i = 2, \ldots, K$, then we have

$$
\begin{aligned}
&\min_{\boldsymbol{\beta},\mathbf{z},\mathbf{m}\omega} \sum_{i=1}^{K} \|\boldsymbol{\alpha}_i \circ \boldsymbol{\beta}_i\|_1 + \delta_{\mathcal{Z}_0}(\mathbf{z}) \\
&\text{s.t. } \hat{\boldsymbol{\Sigma}}_1\boldsymbol{\beta}_1 + \mathbf{m}\omega_2 + \ldots + \mathbf{m}\omega_K - \mathbf{z} = \hat{\boldsymbol{\mu}}, \\
&\qquad \hat{\boldsymbol{\Sigma}}_i\boldsymbol{\beta}_i = \mathbf{m}\omega_i,\ i = 2, \ldots, K.
\end{aligned} \tag{4.4}
$$

It's important to highlight that when $\boldsymbol{\alpha} = \mathbf{1}_p$, the weighted ℓ_1 penalized problem simplifies to the ℓ_1 penalized problem. By introducing the third block variable $\boldsymbol{\omega}$, we restructure the original objective function into a 3-block problem. This restructuring enables us to partition $\boldsymbol{\beta}$ into distinct groups, facilitating parallel computation.

The augmented Lagrangian function for (4.4) is

$$\begin{aligned}\mathcal{L}_\phi(\boldsymbol{\beta}, \mathbf{z}, \mathbf{m}\omega; \boldsymbol{\gamma}) = &\sum_{i=1}^{K} \|\boldsymbol{\alpha}_i \circ \boldsymbol{\beta}_i\|_1 + \delta_{\mathcal{Z}_0}(\mathbf{z}) + \boldsymbol{\gamma}_1^T(\hat{\boldsymbol{\Sigma}}_1\boldsymbol{\beta}_1 + \sum_{i=2}^{K} \mathbf{m}\omega_i - \mathbf{z} - \hat{\boldsymbol{\mu}}) \\ &+ \sum_{i=2}^{K} \boldsymbol{\gamma}_i^T(\hat{\boldsymbol{\Sigma}}_i\boldsymbol{\beta}_i - \mathbf{m}\omega_i) + \frac{\phi}{2}\|\hat{\boldsymbol{\Sigma}}_1\boldsymbol{\beta}_1 + \sum_{i=2}^{K} \mathbf{m}\omega_i - \mathbf{z} - \hat{\boldsymbol{\mu}}\|_2^2 \\ &+ \frac{\phi}{2}\sum_{i=2}^{K} \|\hat{\boldsymbol{\Sigma}}_i\boldsymbol{\beta}_i - \mathbf{m}\omega_i\|_2^2. \end{aligned} \tag{4.5}$$

Here, $\boldsymbol{\beta}_1, \ldots, \boldsymbol{\beta}_K$ are made separable in the augmented Lagrangian function through the introduction of slack variables $\boldsymbol{\omega}$. This allows for a straightforward parallelization when updating $\boldsymbol{\beta}$. Considering that $\boldsymbol{\beta}$-subproblems usually entail the most computational expense, the 3-block formulation shows potential for significantly accelerating computation. Inspired by the methodology proposed by [43], we employ the following update scheme.

$$\begin{cases} \boldsymbol{\beta}^{t+1} &= \operatorname{argmin} \mathcal{L}_\phi(\boldsymbol{\beta}, \mathbf{z}^t, \mathbf{m}\omega^t; \gamma^t), \\ \mathbf{m}\omega^{t+\frac{1}{2}} &= \operatorname{argmin} \mathcal{L}_\phi(\boldsymbol{\beta}^{t+1}, \mathbf{z}^t, \mathbf{m}\omega; \gamma^t), \\ \mathbf{z}^{t+1} &= \operatorname{argmin} \mathcal{L}_\phi(\boldsymbol{\beta}^{t+1}, \mathbf{z}, \mathbf{m}\omega^{t+\frac{1}{2}}; \gamma^t), \\ \mathbf{m}\omega^{t+1} &= \operatorname{argmin} \mathcal{L}_\phi(\boldsymbol{\beta}^{t+1}, \mathbf{z}^{t+1}, \mathbf{m}\omega; \gamma^t), \\ \boldsymbol{\gamma}_1^{t+1} &= \boldsymbol{\gamma}_1^t + \theta\phi(\hat{\boldsymbol{\Sigma}}_1\boldsymbol{\beta}_1 + \sum_{i=2}^{K} \mathbf{m}\omega_i^t - \mathbf{z}^t - \hat{\boldsymbol{\mu}}), \\ \boldsymbol{\gamma}_i^{t+1} &= \boldsymbol{\gamma}_i^t + \theta\phi(\hat{\boldsymbol{\Sigma}}_i\boldsymbol{\beta}_i^{t+1} - \mathbf{m}\omega_i^{t+1}), \quad i = 2, \ldots, K. \end{cases} \tag{4.6}$$

Algorithm (4.6) comprises three separable blocks in the objective function, with the third part being linear. It follows a specific block coordinate descent cycle with the order $\boldsymbol{\beta} \to \mathbf{m}\omega \to \mathbf{z} \to \mathbf{m}\omega$ for updating the variable blocks. Consequently, only two updates are required for the third variable block to achieve convergence.

For high-dimensional data, the matrices $\hat{\boldsymbol{\Sigma}}_i^T\hat{\boldsymbol{\Sigma}}_i$ for $i = 1, \ldots, K$ are not necessarily positive definite. Consequently, iterative sequences may not be well defined. To address this issue, we introduce a proximal term to the $\boldsymbol{\beta}$ subproblem.

$$\boldsymbol{\beta}_1^{t+1} = \underset{\boldsymbol{\beta}_1 \in \mathbb{R}^{p_1}}{\operatorname{argmin}} \frac{\phi}{2}\|\hat{\boldsymbol{\Sigma}}_1\boldsymbol{\beta}_1 + \sum_{i=2}^{K} \mathbf{m}\omega_i^t - \mathbf{z}^t - \hat{\boldsymbol{\mu}} + \frac{\boldsymbol{\gamma}_1^t}{\phi}\|_2^2 + \lambda\|\boldsymbol{\alpha}_1 \circ \boldsymbol{\beta}_1\|_1 + \frac{1}{2}\|\boldsymbol{\beta}_1 - \boldsymbol{\beta}_1^t\|_{\mathbf{m}\mathcal{T}_1}^2, \tag{4.7}$$

$$\boldsymbol{\beta}_i^{t+1} = \underset{\boldsymbol{\beta}_i \in \mathbb{R}^{p_i}}{\operatorname{argmin}} \lambda\|\boldsymbol{\alpha}_i \circ \boldsymbol{\beta}_i\|_1 + \frac{\phi}{2}\|\hat{\boldsymbol{\Sigma}}_i\boldsymbol{\beta}_i - \mathbf{m}\omega_i^t + \frac{\boldsymbol{\gamma}_i^t}{\phi}\|_2^2 + \frac{1}{2}\|\boldsymbol{\beta}_i - \boldsymbol{\beta}_i^t\|_{\mathbf{m}\mathcal{T}_i}^2, \quad i = 2, \ldots, K, \tag{4.8}$$

where the proximal operators $\mathbf{m}\mathcal{T}_i \succ \mathbf{m}0, i = 1, \ldots, K$ and $\|\mathbf{X}\|_{\mathbf{m}P}^2 = \mathbf{X}^T \mathbf{m} P \mathbf{X}$. The positive definiteness of $\mathbf{m}\mathcal{T}_i$ will make $\{\boldsymbol{\beta}^t\}$ automatically well defined. One reasonable choice of $\mathbf{m}\mathcal{T}_i = \eta_i \mathbf{m} I_{p_i} - \phi \hat{\boldsymbol{\Sigma}}_i^T \hat{\boldsymbol{\Sigma}}_i$ with $\eta_i > \phi \lambda_{\max}(\hat{\boldsymbol{\Sigma}}_i^T \hat{\boldsymbol{\Sigma}}_i)$. Then we can derive the solution of $\boldsymbol{\beta}-$subproblem as follows

$$\boldsymbol{\beta}_1^{t+1} = \mathrm{S}\Big(\beta_{1j}^t - \frac{\phi}{\eta_1} \hat{\boldsymbol{\Sigma}}_{1j}^T (\hat{\boldsymbol{\Sigma}}_1 \boldsymbol{\beta}_1^t + \sum_{i=2}^K \mathbf{m}\omega_i^t - \mathbf{z}^t - \hat{\boldsymbol{\mu}} + \frac{\boldsymbol{\gamma}_1^t}{\phi}), \frac{\alpha_{1j}}{\eta_1}\Big)_{j=1,\ldots,p_1}, \tag{4.9}$$

$$\boldsymbol{\beta}_i^{t+1} = \mathrm{S}\Big(\beta_{ij}^t - \frac{\phi}{\eta_i} \hat{\boldsymbol{\Sigma}}_{ij}^T (\hat{\boldsymbol{\Sigma}}_i \boldsymbol{\beta}_i^t - \mathbf{m}\omega_i^t + \frac{\boldsymbol{\gamma}_i^t}{\phi}), \frac{\alpha_{ij}}{\eta_i}\Big)_{j=1,\ldots,p_i}, \tag{4.10}$$

where $\mathrm{S}(x, t) = \mathrm{sign}(x)(|x| - t) I(|x| > t)$ is the soft thresholding function. This asset-splitting algorithm is denoted as AS-ADMM throughout this chapter. Subsequently, we obtain the analytical solutions for the $\boldsymbol{\omega}$-subproblem and $\mathbf{z}$-subproblem.

$$\mathbf{m}\omega_i^{t+\frac{1}{2}} = \frac{1}{K}(\hat{\boldsymbol{\mu}} + \mathbf{z}^t + K \hat{\boldsymbol{\Sigma}}_i \boldsymbol{\beta}_i^{t+1} - \hat{\boldsymbol{\Sigma}} \boldsymbol{\beta}^{t+1}),\ i = 2, \ldots, K, \tag{4.11}$$

$$\mathbf{m}\omega_i^{t+1} = \frac{1}{K}(\hat{\boldsymbol{\mu}} + \mathbf{z}^{t+1} + K \hat{\boldsymbol{\Sigma}}_i \boldsymbol{\beta}_i^{t+1} - \hat{\boldsymbol{\Sigma}} \boldsymbol{\beta}^{t+1}),\ i = 2, \ldots, K, \tag{4.12}$$

$$\mathbf{z}^{t+1} = \min\left(\max\left(\hat{\boldsymbol{\Sigma}}_1 \boldsymbol{\beta}_1^{t+1} + \sum_{i=2}^K \mathbf{m}\omega_i^{t+\frac{1}{2}} - \hat{\boldsymbol{\mu}} + \frac{\boldsymbol{\gamma}_1^t}{\phi}, -\lambda \boldsymbol{\alpha}\right), \lambda \boldsymbol{\alpha}\right). \tag{4.13}$$

The update for $\boldsymbol{\gamma}$ in (4.6) is

$$\begin{aligned} \boldsymbol{\gamma}_1^{t+1} &= \boldsymbol{\gamma}_1^t + \theta \phi (\hat{\boldsymbol{\Sigma}}_1 \boldsymbol{\beta}_1^{t+1} + \sum_{i=2}^K \mathbf{m}\omega_i^{t+1} - \mathbf{z}^{t+1} - \hat{\boldsymbol{\mu}}), \\ \boldsymbol{\gamma}_i^{t+1} &= \boldsymbol{\gamma}_i^t + \theta \phi (\hat{\boldsymbol{\Sigma}}_i \boldsymbol{\beta}_i^{t+1} - \mathbf{m}\omega_i^{t+1}), i = 2, \ldots, K. \end{aligned} \tag{4.14}$$

Let the primal residual at the t-th iteration be

$$\epsilon_{pri}^t = \|\hat{\boldsymbol{\Sigma}}_1 \boldsymbol{\beta}_1 + \sum_{i=2}^K \mathbf{m}\omega_i^t - \mathbf{z}^t - \hat{\boldsymbol{\mu}}\|_2 + \|\sum_{i=2}^K (\hat{\boldsymbol{\Sigma}}_i \boldsymbol{\beta}_i - \mathbf{m}\omega_i^t)\|_2, \tag{4.15}$$

we stop the algorithm at iteration t if $|\epsilon_{pri}^t - \epsilon_{pri}^{t-1}| < \epsilon$. We summarize the approach of AS-ADMM in Algorithm 1.

The rate of convergence of Algorithm 1 for solving LPO is given in Theorem 4.1 below.

Theorem 4.1 *Take $\theta \in (0, (1 + \sqrt{5})/2)$, then sequence $(\boldsymbol{\beta}^t, \mathbf{z}^t, \mathbf{m}\boldsymbol{\omega}^t, \boldsymbol{\gamma}^t)$ generated by Algorithm 1 converges in probability to a unique limit $(\bar{\boldsymbol{\beta}}, \bar{\mathbf{z}}, \mathbf{m}\bar{\boldsymbol{\omega}}, \bar{\boldsymbol{\gamma}})$ such that $(\bar{\boldsymbol{\beta}}, \bar{\mathbf{z}}, \mathbf{m}\bar{\boldsymbol{\omega}})$ is the primal optimal of* (4.4) *and $\bar{\boldsymbol{\gamma}}$ is the dual optimal. Moreover, there exists a constant $\mu \in (0, 1)$ such that*

Algorithm 1 AS-ADMM for LPO

Require: $\boldsymbol{\beta}^0$, $\mathbf{m}\omega^0$, $\mathbf{z}^0$, $\boldsymbol{\gamma}^0$, and $\phi > 0, \theta > 0$ are given.
while the stopping criterion is not satisfied, **do**
Compute $\boldsymbol{\beta}^{t+1}$ by

$$\boldsymbol{\beta}_1^{t+1} = \mathrm{S}\Big(\beta_{1j}^t - \tfrac{\phi}{\eta_1}\hat{\boldsymbol{\Sigma}}_{1j}^T(\hat{\boldsymbol{\Sigma}}_1\boldsymbol{\beta}_1 + \textstyle\sum_{i=2}^K \mathbf{m}\omega_i^t - \mathbf{z}^t - \hat{\boldsymbol{\mu}} + \tfrac{\boldsymbol{\gamma}_1^t}{\phi}), \tfrac{\alpha_{1j}}{\eta_1}\Big)_{j=1,\ldots,p_1},$$

$$\boldsymbol{\beta}_i^{t+1} = \mathrm{S}\Big(\beta_{ij}^t - \tfrac{\phi}{\eta_i}\hat{\boldsymbol{\Sigma}}_{ij}^T(\hat{\boldsymbol{\Sigma}}_i\boldsymbol{\beta}_i - \mathbf{m}\omega_i^t + \tfrac{\boldsymbol{\gamma}_i^t}{\phi}), \tfrac{\alpha_{ij}}{\eta_i}\Big)_{j=1,\ldots,p_i}.$$

Compute $\mathbf{m}\omega_i^{t+\frac{1}{2}}$, $\mathbf{z}^{t+1}$ and $\mathbf{m}\omega_i^{t+1}$ by (4.11), (4.13) and (4.12).
Compute $\boldsymbol{\gamma}^{t+1}$ by (4.14).
end while

$$Dist^{t+1} \le \mu Dist^t,$$

where the distance function $Dist^t$ at iteration t is defined as

$$Dist^t = \sum_{i=1}^K \|\hat{\boldsymbol{\Sigma}}_i(\boldsymbol{\beta}_i^t - \bar{\boldsymbol{\beta}}_i)\|_2^2 - \frac{1}{K}\|\sum_{i=1}^K \hat{\boldsymbol{\Sigma}}_i(\boldsymbol{\beta}_i^t - \bar{\boldsymbol{\beta}}_i)\|_2^2 + \sum_{i=1}^K \|\boldsymbol{\beta}_i^t - \bar{\boldsymbol{\beta}}_i\|_{\mathbf{m}\mathcal{T}_i}^2$$
$$+ \|\mathbf{z}^t - \bar{\mathbf{z}}\|_2^2 + \frac{K-1}{K}\|\mathbf{z}^t - \mathbf{z}^{t-1}\|_2^2 + \frac{m_1}{K}\|\sum_{i=1}^K \hat{\boldsymbol{\Sigma}}_i(\boldsymbol{\beta}_i^t - \bar{\boldsymbol{\beta}}_i) + (\mathbf{z}^t - \bar{\mathbf{z}})\|_2^2,$$

where $m_1 = 1 + d_1 - d_1\theta - (1 - d_1)\min\{\theta, \frac{1}{\theta}\}$ and $d_1 \in (0, \frac{1}{2})$.

4.2.3 Two-Step Procedures for LPO

In this section, we introduce the LPO problem (4.3) with a general class of folded concave penalties and verify its oracle properties. We assume throughout this section that the folded concave penalty functions satisfy the following conditions:

(1) $P_\lambda(t)$ is non-decreasing and concave for $t \in [0, \infty)$ with $P_\lambda(0) = 0$,
(2) $P_\lambda(t)$ is differentiable in $(0, \infty)$,
(3) $P_\lambda'(t) \ge a_1\lambda$ for $t \in (0, a_2\lambda]$,
(4) $P_\lambda'(t) = 0$ for $t \in [a\lambda, \infty)$ with $a > 1$.

Here, a_1 and a_2 are two constants associated with each penalty. This class of penalty functions was introduced in [48]. It includes all non-decreasing concave functions of $\|t\|$ with $P_\lambda(0) = 0$ and $\|P_\lambda\|_\infty > 0$, and is closed under the operations of summation and maximization. Fan and Li [48] proposed the SCAD penalty, which is a folded concave penalty with $a_1 = a_2 = 1$, and its derivative being

$$P_\lambda'(|\beta|) = \lambda\left\{I(|\beta| \le \lambda) + \frac{(a\lambda - |\beta|)_+}{(a-1)\lambda}I(|\beta| > \lambda)\right\} \tag{4.16}$$

for some $a > 2$ and $P'_\lambda(0+) = \lambda$. The SCAD penalty possesses both sparsity and unbiasedness properties, coupled with a bounded derivative. This characteristic facilitates the utilization of gradient descent and other gradient-based algorithms to explore suitable solutions within large subspaces. However, the nonconvex nature of the penalty term poses significant computational hurdles due to the presence of multiple local optima. To tackle this challenge, [44] introduced a methodology and theory for devising an efficient one-step sparse estimation procedure in nonconcave penalized likelihood models using a Local Linear Approximation (LLA). The LLA approximates a nonconvex penalized problem by breaking it down into a series of adaptive Lasso problems. Specifically, the LLA applied to the penalty term is expressed as:

$$P_\lambda(|\beta_j|) \approx P_\lambda(|\beta_j^0|) + P'_\lambda(|\beta_j^0|)(|\beta_j| - |\beta_j^0|), \text{ for } \beta_j \approx \beta_j^0, \tag{4.17}$$

where β_j^0 is a given initial value. Consider a folded concave penalized estimation problem,

$$\min_{\boldsymbol{\beta}} L(\boldsymbol{\beta}; \mathbf{y}, \mathbf{X}) + P_\lambda(|\boldsymbol{\beta}|), \tag{4.18}$$

then the LLA algorithm iteratively updates $\boldsymbol{\beta}^{t+1}$ as

$$\boldsymbol{\beta}^{t+1} = \operatorname*{argmin}_{\boldsymbol{\beta}} \left\{ L(\boldsymbol{\beta}; \mathbf{y}, \mathbf{X}) + \sum_{j=1}^{p} P'_\lambda(|\beta_j^t|)|\beta_j| \right\}, \tag{4.19}$$

where $L(\boldsymbol{\beta}; \mathbf{y}, \mathbf{X})$ is a convex loss function. The LLA inherits favorable properties from Lasso in terms of computational efficiency. Fan et al. [46] demonstrated that when the LLA algorithm is initialized with the Lasso solution, it yields an oracle solution with overwhelming probability in high-dimensional linear regression scenarios.

We next study Problem (4.3) with $\|\boldsymbol{\beta}\|_1$ replaced by folded concave penalty $P_\lambda(|\boldsymbol{\beta}|)$. Let $\rho(t, \lambda) = \lambda^{-1} P_\lambda(t)$ and write it as $\rho(t)$ for simplicity, then obtain the LPO with folded concave penalty as follows,

$$\min_{\boldsymbol{\beta} \in \mathbb{R}^p} \rho(|\boldsymbol{\beta}|) \text{ s.t. } |\hat{\boldsymbol{\Sigma}}_j^T \boldsymbol{\beta} - \hat{\mu}_j| \le \lambda \rho'(|\boldsymbol{\beta}_j|), \quad j = 1, \ldots, p, \tag{4.20}$$

where $\hat{\boldsymbol{\Sigma}}_j$ is the j-th column of $\hat{\boldsymbol{\Sigma}}$ and $\hat{\mu}_j$ is the j-th element of $\hat{\mathbf{m}\mu}$. As discussed, the LLA algorithm can redirect problem (4.20) to an adaptive version of problem (4.3), which is essentially a weighted ℓ_1 regularization model,

$$\begin{aligned} &\min_{\boldsymbol{\beta} \in \mathbb{R}^p} \sum_{j=1}^{p} \rho'(|\hat{\beta}_j^0|)|\beta_j| \\ &\text{s.t. } |\hat{\boldsymbol{\Sigma}}_j^T \boldsymbol{\beta} - \hat{\mu}_j| \le \lambda \rho'(|\hat{\beta}_j^0|), \quad j = 1, \ldots, p. \end{aligned} \tag{4.21}$$

Denote the matrix as $S_B = \hat{\boldsymbol{\mu}}\hat{\boldsymbol{\mu}}^T$, then we can express the oracle estimator as

$$\hat{\boldsymbol{\beta}}^{\text{oracle}} = \underset{\beta:\beta_{\mathcal{A}^c}=\mathbf{m}0}{\operatorname{argmin}} \; L(\boldsymbol{\beta}) = \underset{\beta:\beta_{\mathcal{A}^c}=\mathbf{m}0}{\operatorname{argmin}} \; -\frac{\boldsymbol{\beta}^T S_B \boldsymbol{\beta}}{\boldsymbol{\beta}^T \hat{\boldsymbol{\Sigma}} \boldsymbol{\beta}}, \tag{4.22}$$

where the support set $\mathcal{A} = \{j | \beta_j^* \neq 0, 1 \leq j \leq p\}$, $\boldsymbol{\beta}^* = (\beta_1^*, \ldots, \beta_p^*)$ is the true value of $\boldsymbol{\beta}$, the cardinality of $|\mathcal{A}| = q$ and $\mathcal{A}^c$ is the complement of $\mathcal{A}$. We assume that $q\sqrt{\log(qp)/n} \longrightarrow 0$ and $\|\boldsymbol{\beta}^*\|_{\min} \gg q\sqrt{\log(pq)/n}$ to guarantee the consistency of the oracle estimator [49, 50].

In the subsequent analysis, we assume that $\hat{\boldsymbol{\Sigma}}_{\mathcal{A}}^T \boldsymbol{\beta} - \hat{\boldsymbol{\mu}}_{\mathcal{A}} = 0$ possesses regularity with a unique solution, where $\boldsymbol{\Sigma}_{\mathcal{A}}$ denotes the sub-matrix containing the columns of $\boldsymbol{\Sigma}$ indexed by the set $\mathcal{A}$. This assumption essentially stipulates that $\hat{\boldsymbol{\Sigma}}_{\mathcal{A}}$ is of full rank. Zhang et al. [51] established the sign consistency of the standard LPO rule $\hat{\boldsymbol{\beta}}^{\text{LPO}}$ (4.3) under sparsity assumptions, and we utilize $\hat{\boldsymbol{\beta}}^{\text{LPO}}$ to initialize the LLA algorithm. The following regularity conditions are imposed to facilitate the technical proof of the oracle property of the proposed estimator:

(A1) $\|\boldsymbol{\Sigma}_{\mathcal{A}^c,\mathcal{A}} \boldsymbol{\Sigma}_{\mathcal{A},\mathcal{A}}^{-1} \mathbf{m}r\|_{\max} \leq 1 - c, c \in (0, 1), \forall \mathbf{m}r \in \{-1, 0, 1\}^q$.
(A2) $\|\boldsymbol{\Sigma}_{\mathcal{A},\mathcal{A}}^{-1} \mathbf{m}r\|_{\max} \leq l, l > 0, \forall \mathbf{m}r \in \{-1, 0, 1\}^q$.
(A3) $\boldsymbol{\Sigma}_{\mathcal{A}.\mathcal{A}}^{-1} \text{sign}(\boldsymbol{\beta}_{\mathcal{A}}^*)$ and $\hat{\boldsymbol{\Sigma}}_{\mathcal{A}.\mathcal{A}}^{-1} \text{sign}(\boldsymbol{\beta}_{\mathcal{A}}^*)$ have no zero entries,

where $\|A\|_{\max}$ and $\|A\|_{\min}$ represent the element-wise maximum and minimum absolute values of vector/matrix A, respectively. Condition (A1) is associated with the irrepresentability condition proposed by [52], indicating that the important variables in the true support are not highly correlated with the unimportant variables. Condition (A2) establishes an upper bound on the maximum norm of the precision matrix. Condition (A3) ensures that $\hat{\boldsymbol{\Sigma}}_{\mathcal{A}.\mathcal{A}}^{-1} \text{sign}(\boldsymbol{\beta}_{\mathcal{A}}^*)$ has probability one that it has no zero entries, and that all matrices $\boldsymbol{\Sigma}_{\mathcal{A}.\mathcal{A}}^{-1}$ and vectors $\boldsymbol{\beta}_{\mathcal{A}}^*$ satisfying that $\boldsymbol{\Sigma}_{\mathcal{A}.\mathcal{A}}^{-1} \text{sign}(\boldsymbol{\beta}_{\mathcal{A}}^*)$ has zero entries forms a zero measure set. These conditions are imposed in [51].

Theorem 4.2 *Under Assumptions A1, A2 and A3, and assume that* $\|\boldsymbol{\beta}_{\mathcal{A}}^*\|_{\min} > (a + 1)\lambda$, *then the LLA algorithm initialized by* $\hat{\boldsymbol{\beta}}^{LPO}$ *converges to* $\hat{\boldsymbol{\beta}}^{oracle}$ *after one iteration with probability* $1 - P_1 - P_2 - P_3$, *where* P_i *are the probabilities of events* $E_i, i = 1, 2, 3$ *and each* E_i *is defined as follows:*

(1) $E_1 = \{\|\hat{\boldsymbol{\beta}}^{LPO} - \boldsymbol{\beta}^*\|_{\max} > a_0\lambda\}$, *where* $a_0 = \min\{1, a_2\}$;
(2) $E_2 = \{\|\hat{\boldsymbol{\beta}}_{\mathcal{A}}^{oracle}\|_{\min} \leq a\lambda\}$;
(3) $E_3 = \{\|\hat{\boldsymbol{\Sigma}}_{\mathcal{A}^c}^T \hat{\boldsymbol{\beta}}^{oracle} - \hat{\boldsymbol{\mu}}_{\mathcal{A}^c}\|_{\max} > a_1\lambda\}$.

Furthermore, if $\lambda \gg (\log(pq)q^2/n)^{1/2}$, *then* $P_i \to 0, i = 1, 2, 3$ *as* $n \to \infty$.

Theorem 4.2 implies that the estimator of the LPO rule with folded concave penalty obtained by the LLA algorithm enjoys the oracle properties. The proof of Theorem 4.2 is given in the Appendix.

We adopt a two-step procedure for solving a weighted ℓ_1 LPO. In the first step, we solve ℓ_1-norm LPO to obtain an initial estimator $\hat{\boldsymbol{\beta}}^0$, and then compute weighted ℓ_1 LPO with $\alpha_j = \frac{1}{\lambda} P'_\lambda(|\hat{\beta}_j^0|)$ for $j = 1, \ldots, p$ to obtain the final estimator. The two-step procedure for weighted ℓ_1 LPO can be derived from Algorithms 1 and we summarize it in Algorithm 2.

Algorithm 2 Two-Step AS-ADMM for Weighted ℓ_1 LPO

Require: $\tilde{\boldsymbol{\beta}}^0, \lambda, \upsilon, \tilde{\mathbf{z}}^0, \tilde{\boldsymbol{\gamma}}^0, \tilde{\mathbf{m}\omega}_i^0$, and $\phi > 0, \theta = 1.618, t = 0$.
  Set $\alpha_j = 1$ for $j = 1, \ldots, p$
  while the stopping criterion is not satisfied, **do**
    Compute $\tilde{\boldsymbol{\beta}}^{t+1}$ by (4.9) and (4.10).
    Compute $\mathbf{m}\omega_i^{t+\frac{1}{2}}$, $\mathbf{z}^{t+1}$ and $\mathbf{m}\omega_i^{t+1}$ by (4.11), (4.13) and (4.12).
    Compute $\boldsymbol{\gamma}^{t+1}$ by (4.14).
  end while The solution is denoted as $\hat{\boldsymbol{\beta}}^{\ell_1}, \hat{\mathbf{z}}^{\ell_1}, \hat{\mathbf{m}\omega}^{\ell_1}$
Require: $\hat{\boldsymbol{\beta}}^0 = \hat{\boldsymbol{\beta}}^{\ell_1}, \hat{\mathbf{z}}^0 = \hat{\mathbf{z}}^{\ell_1}, \hat{\mathbf{m}\omega}^0 = \hat{\mathbf{m}\omega}^{\ell_1}$ and $\phi > 0, \theta = 1.618, t = 0$.
  Set $\alpha_j = \lambda^{-1} P'_\lambda(|\hat{\beta}_j^0|)$ for $j = 1, \ldots, p$.
  while the stopping criterion is not satisfied, **do**
    Compute $\tilde{\boldsymbol{\beta}}^{t+1}$ by (4.9) and (4.10).
    Compute $\mathbf{m}\omega_i^{t+\frac{1}{2}}$, $\mathbf{z}^{t+1}$ and $\mathbf{m}\omega_i^{t+1}$ by (4.11), (4.13) and (4.12).
    Compute $\boldsymbol{\gamma}^{t+1}$ by (4.14).
  end while

Based on Theorem 4.2, we can further derive Theorem 4.3, which demonstrates the estimation consistency of $\mathbf{w}_{\text{opt}}$.

Denote our proposed estimate for $\mathbf{w}_{\text{opt}}$ by

$$\hat{\mathbf{w}}_{\text{opt}} = \frac{\mu_0 \hat{\boldsymbol{\beta}}^{oracle}}{\hat{\boldsymbol{\mu}}^T \hat{\boldsymbol{\beta}}^{oracle}}. \tag{4.23}$$

Theorem 4.3 *Suppose the following conditions are valid*

C3.1 $\boldsymbol{\Theta} := \boldsymbol{\mu}^T \boldsymbol{\Sigma}^{-1} \boldsymbol{\mu} \geq \kappa$ *for some* $\kappa > 0$;
C3.2 $d = (\lambda + \|\hat{\boldsymbol{\mu}} - \boldsymbol{\mu}\|_{\max})\|\boldsymbol{\beta}^*\|_1 + \|\hat{\boldsymbol{\Sigma}} - \boldsymbol{\Sigma}\|_{\max}\|\boldsymbol{\beta}^*\|_1^2 = o_p(1)$,

then $\|\hat{\mathbf{w}}_{opt} - \mathbf{w}_{opt}\|_{\max} \to 0$ *in probability.*

Conditions C3.1 and C3.2 are used by [18]. Condition C3.1 asserts that the square of the Sharpe ratio maintains a lower bound and remains bounded away from zero. Condition C3.2 is likely to be fulfilled with high probability under the circumstance that the portfolio exhibits significant sparsity, indicated by $\|\boldsymbol{\beta}^*\|_1 = o(\sqrt{\log p/n})$.

4.3 Feature-Splitting Algorithm for PQR

Suppose that $\{\mathbf{x}_i, y_i\}$, $i = 1, \ldots, n$, is a random sample from linear regression model

$$y_i = \mathbf{x}_i^T \boldsymbol{\beta} + \varepsilon_i,$$

where $\boldsymbol{\beta}$ is p-dimensional vector of regression coefficients, and ε_i is a random error with $E(\varepsilon_i|\mathbf{x}_i) = 0$. In this chapter, we are interested in solving QR in ultrahigh-dimensional regime, in which $p \gg n$. Define $\mathbf{y} = (y_1, \ldots, y_n)^T$ as the response vector, and $\mathbf{X} = (\mathbf{x}_1, \ldots, \mathbf{x}_n)^T$ as the corresponding design matrix. For a given $\tau \in (0, 1)$, the quantile of interest, define the loss function $\rho_\tau(z) = z[\tau - I(z < 0)] = \tau(z)_+ + (1 - \tau)(z)_-$, where $I(\cdot)$ is the indicator function, $(z)_+ = \max\{0, z\}$, and $(z)_- = (-z)_+$. The QR is to minimize its objective function

$$L(\mathbf{y} - \mathbf{X}\boldsymbol{\beta}) = \frac{1}{n}\sum_{i=1}^{n} \rho_\tau(y_i - \mathbf{x}_i^T \boldsymbol{\beta}) \tag{4.24}$$

with respect to $\boldsymbol{\beta}$, and this leads to the QR estimate of $\boldsymbol{\beta}$. The minimization problem in QR can be reformulated as a linear programming problem. The Frisch-Newton algorithm can then be employed to solve this minimization problem, with computational complexity increasing as a cubic function of p when $p < n$.

4.3.1 Penalized Quantile Regression

In the context of ultrahigh-dimensional predictors, it is often assumed that $\boldsymbol{\beta}$ exhibits sparsity, indicating that only a small subset of its elements are nonzero. This implies that only a fraction of predictors are genuinely significant in the model. Consequently, identifying these significant predictors becomes pivotal in quantile regression within ultrahigh dimensions. Variable selection in quantile regression shares similarities with that in linear regression, for which penalized least-squares methods have been proposed. Hence, extending the Penalized Quantile Regression (PQR) method to quantile regression for variable selection is a natural progression. With PQR, the objective is to minimize the penalized quantile loss function

$$Q(\boldsymbol{\beta}) = L(\mathbf{y} - \mathbf{X}\boldsymbol{\beta}) + \sum_{j=1}^{p} p_{\lambda_j}(|\beta_j|), \tag{4.25}$$

where $p_{\lambda_j}(\cdot)$ is a penalty function with a regularization parameter λ_j that controls model complexity. While the algorithms to be developed allow for different regression coefficients to have distinct penalties, it's common practice to consider all $p_{\lambda_j}(\cdot)$ to be identical, denoted as $p_\lambda(\cdot)$. This chapter primarily focuses on two widely used penalties: the Lasso penalty (ℓ_1) defined as $p_\lambda(|\beta|) = \lambda|\beta|$, and the SCAD penalty. The SCAD penalty, with its first derivative defined as (4.16), is characterized by $p'_\lambda(0) := p'_\lambda(0+) = \lambda$ and $a = 3.7$, as suggested in [34]. Notably, the proposed algorithms are directly applicable to other folded concave penalties as well [53].

Minimizing the objective function of PQR in (4.25) poses a challenge due to the nonsmooth nature of both the loss function and the penalty function. This challenge is exacerbated when folded concave penalties like the SCAD penalty are employed in ultrahigh-dimensional PQR, introducing nonconvexity and ultrahigh dimensionality into the minimization problem. It's worth noting that PQR with the ℓ_1 penalty presents a convex minimization problem, and when $p \leq n$, it possesses a unique minimizer. However, for PQR with a folded concave penalty, achieving the minimization of the objective function in (4.25) may entail iteratively minimizing PQR with a reweighted ℓ_1 penalty, aided by the local linear approximation (LLA) to the penalty function. Specifically, given $\boldsymbol{\beta}^k = (\beta_1^k, \ldots, \beta_p^k)^T$ updated from the k-th step during iterations, we first approximate

$$p_\lambda(|\beta_j|) \approx q_\lambda(|\beta_j|; |\beta_j^k|) = p_\lambda(|\beta_j^k|) + p'_\lambda(|\beta_j^k|)(|\beta_j| - |\beta_j^k|), \tag{4.26}$$

which is referred to as the LLA. Then at the $(k+1)$-th step we minimize

$$Q^{k+1}(\boldsymbol{\beta}) = L(\mathbf{y} - \mathbf{X}\boldsymbol{\beta}) + \lambda \sum_{j=1}^{p} \lambda^{-1} p'_{\lambda_j}(|\beta_j^k|)|\beta_j| = L(\mathbf{y} - \mathbf{X}\boldsymbol{\beta}) + \lambda \sum_{j=1}^{p} \alpha_j |\beta_j|, \tag{4.27}$$

where $\alpha_j = \lambda^{-1} p'_{\lambda_j}(|\beta_j^k|) \geq 0$. The function in (4.27) is the objective function of PQR with reweighted ℓ_1 penalty and weights α_j's that are updated at every step.

The LLA was initially introduced in [44] for penalized likelihood models with finite-dimensional predictors. It was subsequently adapted in [45, 46] for penalized least squares in ultrahigh-dimensional linear regression models. Notably, if we initialize $\boldsymbol{\beta}^0 = \mathbf{0}$, then $\boldsymbol{\beta}^1$ corresponds to the PQR-Lasso estimator, defined as PQR with an ℓ_1 penalty. Subsequently, $\boldsymbol{\beta}^2$ can be viewed as the one-step sparse estimator, with the initial value being the PQR-Lasso estimator. With appropriately chosen tuning parameters, [45, 46] demonstrated that under certain regularity conditions on high-dimensional linear models, the resulting penalized least-squares estimator $\boldsymbol{\beta}^2$ exhibits the strong oracle property with a probability approaching one. This motivates our focus on developing a feature-splitting algorithm for PQR with weighted ℓ_1 penalty.

4.3.2 *Three-Block ADMM*

Define the PQR estimator with weighted ℓ_1-penalty to be

$$\hat{\boldsymbol{\beta}} = \underset{\boldsymbol{\beta}}{\text{argmin}}\, L(\mathbf{y} - \mathbf{X}\boldsymbol{\beta}) + \lambda \|\boldsymbol{\alpha} \circ \boldsymbol{\beta}\|_1, \tag{4.28}$$

where $\|\boldsymbol{\alpha} \circ \boldsymbol{\beta}\|_1 = \sum_{j=1}^{p} |\alpha_j \beta_j|$ with $\boldsymbol{\alpha}$ being the weight vector. The nonsmoothness of the objective function in (4.28) hampers the efficient application of gradient-based methods. To address this challenge and facilitate computation, we decompose problem (4.28) into the following constrained optimization problem,

$$\min_{\boldsymbol{\beta}, \mathbf{z}} L(\mathbf{z}) + \lambda \|\boldsymbol{\alpha} \circ \boldsymbol{\beta}\|_1, \qquad \text{s.t. } \mathbf{z} + \mathbf{X}\boldsymbol{\beta} = \mathbf{y}. \tag{4.29}$$

Problem (4.29) is a natural candidate of classical two-block ADMM algorithm. Define augmented Lagrangian function as

$$\mathcal{L}_{\phi}(\boldsymbol{\beta}, \mathbf{z}; \boldsymbol{\gamma}) = L(\mathbf{z}) + \lambda \|\boldsymbol{\alpha} \circ \boldsymbol{\beta}\|_1 + \langle \boldsymbol{\gamma}, \mathbf{z} + \mathbf{X}\boldsymbol{\beta} - \mathbf{y} \rangle + \frac{\phi}{2} \|\mathbf{z} + \mathbf{X}\boldsymbol{\beta} - \mathbf{y}\|_2^2, \tag{4.30}$$

where $\boldsymbol{\gamma} \in \mathbb{R}^n$ is the Lagrangian multiplier, and $\phi > 0$ is the parameter associated with the quadratic term. The classic iterative scheme at the iteration k for two-block ADMM is

$$\begin{aligned}
\boldsymbol{\beta}^{k+1} &= \underset{\boldsymbol{\beta}}{\text{argmin}}\, \mathcal{L}_{\phi}(\boldsymbol{\beta}, \mathbf{z}^k; \boldsymbol{\gamma}^k), \\
\mathbf{z}^{k+1} &= \underset{\mathbf{z}}{\text{argmin}}\, \mathcal{L}_{\phi}(\boldsymbol{\beta}^{k+1}, \mathbf{z}; \boldsymbol{\gamma}^k), \\
\boldsymbol{\gamma}^{k+1} &= \boldsymbol{\gamma}^k + \theta\phi(\mathbf{z}^{k+1} + \mathbf{X}\boldsymbol{\beta}^{k+1} - \mathbf{y}),
\end{aligned}$$

where θ is a tuning parameter controlling the step size. The impact of θ on algorithm convergence has been extensively discussed in the literature [54, 55], where convergence is established when θ is constrained within the interval $(0, (1 + \sqrt{5})/2)$. In our numerical experiments, we opt to set $\theta = 1.618$, slightly below $(1 + \sqrt{5})/2$, to expedite convergence. Although [39] proposed an efficient algorithm (*qradmm*) for solving PQR using the two-block ADMM algorithm, we observed in our numerical study that *qradmm* can encounter memory issues with larger values of p, despite performing well for moderate dimensions. This observation motivates us to partition the high-dimensional variable into smaller blocks and accelerate updates through parallelization.

We propose a new three-block semi-proximal ADMM framework designed to enable parallel updates of $\boldsymbol{\beta}$, effectively addressing the challenges posed by ultrahigh dimensionality. In the two-block ADMM approach for solving (4.29), the primary computational burden arises from the $\boldsymbol{\beta}$ update, which requires up to $O(np)$ operations and may hinder efficient algorithm execution in ultrahigh dimensions (p). This necessitates the development of a feature-splitting algorithm for PQR in ultrahigh dimensions. For a predetermined value of G, we partition $\mathbf{X}$ and $\boldsymbol{\beta}$ as follows:

$$\mathbf{X} = (\mathbf{X}_1, \ldots, \mathbf{X}_G), \;\; \boldsymbol{\beta} = (\boldsymbol{\beta}_1^T, \boldsymbol{\beta}_2^T, \ldots, \boldsymbol{\beta}_G^T)^T, \;\; \mathbf{X}\boldsymbol{\beta} = \sum_{g=1}^{G} \mathbf{X}_g \boldsymbol{\beta}_g.$$

Then problem (4.29) can be rewritten as a three-block optimization problem

$$\begin{aligned}
\min_{\boldsymbol{\beta},\mathbf{z},\mathbf{m}\omega} \quad & L(\mathbf{z}) + \sum_{g=1}^{G} \lambda \|\boldsymbol{\alpha}_g \circ \boldsymbol{\beta}_g\|_1, \\
\text{s.t.} \quad & \mathbf{X}_1\boldsymbol{\beta}_1 + \mathbf{z} + \mathbf{m}\omega_2 + \cdots + \mathbf{m}\omega_G = \mathbf{y}, \qquad \mathbf{X}_g\boldsymbol{\beta}_g = \mathbf{m}\omega_g, \quad g = 2, \ldots, G.
\end{aligned} \tag{4.31}$$

Intuitively, the slack variables $\mathbf{m}\omega_g, g = 2, \ldots, G$ store information regarding each local update $\boldsymbol{\beta}_g$. Each $\boldsymbol{\beta}_g$ is updated independently, and we treat $\boldsymbol{\beta} = (\boldsymbol{\beta}_1^T, \boldsymbol{\beta}_2^T, \ldots, \boldsymbol{\beta}_G^T)^T$ as a single variable block in the algorithm. Similarly, all $\boldsymbol{\omega}_g$ together constitute the third variable block. There may be multiple approaches to reformulating a problem into a form that ADMM can handle. For instance, in formulation (4.31), the role of $\mathbf{X}_1\boldsymbol{\beta}_1$ is not unique, and $\mathbf{X}_g\boldsymbol{\beta}_g, g = 1, \ldots, G$ are exchangeable. In this section, we adopt formulation (4.31) to elucidate the computational framework.

The augmented Lagrangian function for (4.31) is given by

$$\begin{aligned}
\mathcal{L}_\phi(\boldsymbol{\beta}, \mathbf{z}, \mathbf{m}\omega; \boldsymbol{\gamma}) =& \frac{1}{n}[\tau \mathbf{m}\mathbf{1}^T(\mathbf{z})_+ + (1-\tau)\mathbf{m}\mathbf{1}^T(\mathbf{z})_-] + \lambda \sum_{g=1}^{G} \|\boldsymbol{\alpha}_g \circ \boldsymbol{\beta}_g\|_1 \\
&+ \boldsymbol{\gamma}_1^T(\mathbf{X}_1\boldsymbol{\beta}_1 + \mathbf{z} + \mathbf{m}\omega_2 + \cdots + \mathbf{m}\omega_G - \mathbf{y}) \\
&+ \frac{\phi}{2}\|\mathbf{X}_1\boldsymbol{\beta}_1 + \mathbf{z} + \mathbf{m}\omega_2 + \cdots + \mathbf{m}\omega_G - \mathbf{y}\|_2^2 \\
&+ \sum_{g=2}^{G} \boldsymbol{\gamma}_g^T(\mathbf{X}_g\boldsymbol{\beta}_g - \mathbf{m}\omega_g) + \frac{\phi}{2}\sum_{g=2}^{G}\|\mathbf{X}_g\boldsymbol{\beta}_g - \mathbf{m}\omega_g\|_2^2.
\end{aligned} \tag{4.32}$$

As evident from (4.32), each $\boldsymbol{\beta}_g$ is decoupled in the quadratic term, facilitating natural parallelization for $\boldsymbol{\beta}$ updates. The two-block ADMM framework can be straightforwardly extended to solve (4.31), giving rise to the Gauss-Seidel multi-block ADMM algorithm. At the k-th iteration, this algorithm updates each variable with the following procedure

$$\begin{cases}
\boldsymbol{\beta}^{k+1} = \operatorname{argmin} \mathcal{L}_\phi(\boldsymbol{\beta}, \mathbf{z}^k, \mathbf{m}\omega^k; \boldsymbol{\gamma}^k) \\
\mathbf{z}^{k+1} = \operatorname{argmin} \mathcal{L}_\phi(\boldsymbol{\beta}^{k+1}, \mathbf{z}, \mathbf{m}\omega^k; \boldsymbol{\gamma}^k) \\
\mathbf{m}\omega^{k+1} = \operatorname{argmin} \mathcal{L}_\phi(\boldsymbol{\beta}^{k+1}, \mathbf{z}^{k+1}, \mathbf{m}\omega; \boldsymbol{\gamma}^k) \\
\boldsymbol{\gamma}_1^{k+1} = \boldsymbol{\gamma}_1^k + \theta\phi(\mathbf{X}_1\boldsymbol{\beta}_1^{k+1} + \mathbf{z}^{k+1} + \sum_{g=2}^{G} \mathbf{m}\omega_g^{k+1} - \mathbf{y}) \\
\boldsymbol{\gamma}_g^{k+1} = \boldsymbol{\gamma}_g^k + \theta\phi(\mathbf{X}_g\boldsymbol{\beta}_g^{k+1} - \mathbf{m}\omega_g^{k+1}), \quad g = 2, \ldots, G.
\end{cases} \tag{4.33}$$

The theoretical convergence of Procedure (4.33) has been unclear until the work of [56], who demonstrated that the Gauss-Seidel multi-block ADMM is not guaranteed to converge. In response to this convergence uncertainty, [43] introduced a symmetric

Gauss-Seidel-based semi-proximal ADMM (sGS-sPADMM) for convex programming problems. Motivated by the work of [43], we propose a three-block ADMM algorithm for solving PQR with a weighted ℓ_1 penalty, using a special iterative cycle ($\boldsymbol{\beta} \rightarrow \mathbf{m}\omega \rightarrow \mathbf{z} \rightarrow \mathbf{m}\omega$)

$$\begin{cases} \boldsymbol{\beta}^{k+1} & = \operatorname{argmin} \mathcal{L}_\phi(\boldsymbol{\beta}, \mathbf{z}^k, \mathbf{m}\omega^k; \boldsymbol{\gamma}^k) + \frac{\phi}{2}\|\boldsymbol{\beta} - \boldsymbol{\beta}^k\|^2_{\mathcal{T}_f} \\ \mathbf{m}\omega^{k+\frac{1}{2}} & = \operatorname{argmin} \mathcal{L}_\phi(\boldsymbol{\beta}^{k+1}, \mathbf{z}^k, \mathbf{m}\omega; \boldsymbol{\gamma}^k) \\ \mathbf{z}^{k+1} & = \operatorname{argmin} \mathcal{L}_\phi(\boldsymbol{\beta}^{k+1}, \mathbf{z}, \mathbf{m}\omega^{k+\frac{1}{2}}; \boldsymbol{\gamma}^k) + \frac{\phi}{2}\|\mathbf{z} - \mathbf{z}^k\|^2_{\mathcal{T}_g} \\ \mathbf{m}\omega^{k+1} & = \operatorname{argmin} \mathcal{L}_\phi(\boldsymbol{\beta}^{k+1}, \mathbf{z}^{k+1}, \mathbf{m}\omega; \boldsymbol{\gamma}^k) \\ \boldsymbol{\gamma}_1^{k+1} & = \boldsymbol{\gamma}_1^k + \theta\phi(\mathbf{X}_1\boldsymbol{\beta}_1^{k+1} + \mathbf{z}^{k+1} + \sum_{i=2}^K \mathbf{m}\omega_i^{k+1} - \mathbf{y}) \\ \boldsymbol{\gamma}_i^{k+1} & = \boldsymbol{\gamma}_i^k + \theta\phi(\mathbf{X}_i\boldsymbol{\beta}_i^{k+1} - \mathbf{m}\omega_i^{k+1}), \ i = 2, \ldots, K, \end{cases} \tag{4.34}$$

where $\mathcal{T}_f$ and $\mathcal{T}_g$ are some positive semidefinite matrices.

Given the augmented Lagrangian function defined in (4.32),

$$\boldsymbol{\beta}^{k+1} = \operatorname{argmin} \mathcal{L}_\phi(\boldsymbol{\beta}, \mathbf{z}^k, \mathbf{m}\omega^k; \boldsymbol{\gamma}^k)$$

becomes

$$\begin{aligned} \boldsymbol{\beta}_1^{k+1} &= \operatorname*{argmin}_{\boldsymbol{\beta}_1 \in \mathbb{R}^{p_1}} \lambda\|\boldsymbol{\alpha}_1 \circ \boldsymbol{\beta}_1\|_1 + \frac{\phi}{2}\|\mathbf{X}_1\boldsymbol{\beta}_1 + \sum_{g=2}^G \mathbf{m}\omega_g^k + \mathbf{z}^k - \mathbf{y} + \frac{\boldsymbol{\gamma}_1^k}{\phi}\|_2^2, \\ \boldsymbol{\beta}_g^{k+1} &= \operatorname*{argmin}_{\boldsymbol{\beta}_g \in \mathbb{R}^{p_g}} \lambda\|\boldsymbol{\alpha}_g \circ \boldsymbol{\beta}_g\|_1 + \frac{\phi}{2}\|\mathbf{X}_g\boldsymbol{\beta}_g - \mathbf{m}\omega_g^k + \frac{\boldsymbol{\gamma}_g^k}{\phi}\|_2^2, \quad g = 2, \ldots, G. \end{aligned} \tag{4.35}$$

The $\boldsymbol{\beta}$ subproblems constitute a series of weighted ℓ_1-penalized least-squares problems. However, when p_g is excessively large, the matrix $\mathbf{X}_g \in \mathbb{R}^{n \times p_g}$ may not be full column rank, potentially leading to ill-defined sequences. To mitigate this concern, an additional general position condition [57] can be employed, ensuring the existence of a unique solution for QR under rather general conditions. Standard quadratic solvers can then be applied to efficiently solve (4.35). In our numerical studies, we utilize the R solver "glmnet" to compute $\boldsymbol{\beta}$ via the coordinate descent (CD) algorithm.

$\mathbf{m}\omega_g$, $g = 2, \ldots, G$, and $\mathbf{z}$ are updated in the following cycle:

$$\begin{aligned} \mathbf{m}\omega_g^{k+\frac{1}{2}} &= \frac{1}{G}(\mathbf{y} - \mathbf{z}^k + G\mathbf{X}_g\boldsymbol{\beta}_g^{k+1} - \sum_{j=1}^G \mathbf{X}_j\boldsymbol{\beta}_j^{k+1}), \\ \mathbf{z}^{k+1} &= \Big(\mathbf{y} - \mathbf{X}_1\boldsymbol{\beta}_1^{k+1} - \sum_{g=2}^G \mathbf{m}\omega_g^{k+\frac{1}{2}} - \frac{\boldsymbol{\gamma}_1^k}{\phi} - \frac{\tau}{n\phi}\Big)_+ \\ &\quad - \Big(-\mathbf{y} + \mathbf{X}_1\boldsymbol{\beta}_1^{k+1} + \sum_{g=2}^G \mathbf{m}\omega_g^{k+\frac{1}{2}} + \frac{\boldsymbol{\gamma}_1^k}{\phi} + \frac{\tau - 1}{n\phi}\Big)_+, \\ \mathbf{m}\omega_g^{k+1} &= \frac{1}{G}(\mathbf{y} - \mathbf{z}^{k+1} + G\mathbf{X}_g\boldsymbol{\beta}_g^{k+1} - \sum_{j=1}^G \mathbf{X}_j\boldsymbol{\beta}_j^{k+1}), \end{aligned} \tag{4.36}$$

in which we perform an extra intermediate step to compute $\mathbf{m}\omega^{k+\frac{1}{2}}$ before computing $\mathbf{z}^{k+1}$. As seen from (4.36), the extra cost for update $\mathbf{m}\omega$ is negligible. The derivations of updates are given in 4.6.1.

Finally, we update $\boldsymbol{\gamma}_1$ and $\boldsymbol{\gamma}_g$ via gradient ascent,

$$\begin{aligned} \boldsymbol{\gamma}_1^{k+1} &= \boldsymbol{\gamma}_1^k + \theta\phi(\mathbf{X}_1\boldsymbol{\beta}_1^{k+1} + \mathbf{z}^{k+1} + \sum_{g=2}^{G} \mathbf{m}\omega_g^{k+1} - \mathbf{y}), \\ \boldsymbol{\gamma}_g^{k+1} &= \boldsymbol{\gamma}_g^k + \theta\phi(\mathbf{X}_g\boldsymbol{\beta}_g^{k+1} - \mathbf{m}\omega_g^{k+1}), \quad g = 2, \dots, G. \end{aligned} \tag{4.37}$$

We denote this algorithm as *FS-QRADMM-CD*, and provide a summary in Algorithm 3. Based on our numerical investigations, we find that FS-QRADMM-CD exhibits favorable practical performance.

Algorithm 3 FS-QRADMM-CD for weighted ℓ_1-penalized QR

Require: $\boldsymbol{\beta}^0, \mathbf{m}\omega^0, \mathbf{z}^0, \boldsymbol{\gamma}^0$, and $\phi > 0, \theta > 0$ are given.
  while the stopping criterion is not satisfied, **do**
    Compute $\boldsymbol{\beta}^{k+1}$ by (4.35) using CD algorithm.
    Compute $\mathbf{m}\omega^{k+\frac{1}{2}}$, $\mathbf{z}^{k+1}$ and $\mathbf{m}\omega^{k+1}$ by (4.36).
    Update $\boldsymbol{\gamma}^{k+1}$ by (4.37).
  end while

In addition to using the coordinate descent algorithm to update $\boldsymbol{\beta}$, we have an alternative solution for updating $\boldsymbol{\beta}$. To ensure that solutions from (4.35) are well defined, we introduce G self-adjoint positive semidefinite matrices, denoted as $\mathcal{T}_g, g = 1, \dots, G$, into (4.35). A guiding principle is to keep $\mathcal{T}_g$ as small as possible while keeping the optimization problems computationally tractable. Here, we include proximal terms $\frac{1}{2}\|\boldsymbol{\beta}_g - \boldsymbol{\beta}_g^k\|^2_{\mathbf{m}\mathcal{T}_g}$, $g = 1, \dots, G$, in each of the $\boldsymbol{\beta}$-subproblems, where the proximal operator $\mathcal{T}_g$ is positive definite. The positive definiteness of $\mathcal{T}_g$ makes $\{\boldsymbol{\beta}^k\}$ is automatically well defined. In this section, we choose $\mathbf{m}\mathcal{T}_g = \eta_g \mathbf{m} I_{p_g} - \phi\mathbf{X}_g^T\mathbf{X}_g$ with $\eta_g > \phi\lambda_{\max}(\mathbf{X}_g^T\mathbf{X}_g)$. Essentially, this is a linearization step of the $\boldsymbol{\beta}$ update, as it uses $\eta_g\mathbf{I}_{p_g}$ to approximate the Hessian matrix $\mathbf{X}_g^T\mathbf{X}_g$. The modified minimization problem admits a closed-form solution, which can be carried out component-wise.

$$\begin{aligned} \boldsymbol{\beta}_1^{k+1} &= \mathrm{S}\Big(\boldsymbol{\beta}_1^k - \frac{\phi}{\eta_1}\mathbf{X}_1^T(\mathbf{X}_1\boldsymbol{\beta}_1^k + \sum_{g=2}^{G}\mathbf{m}\omega_g^k + \mathbf{z}^k - \mathbf{y} + \frac{\boldsymbol{\gamma}_1^k}{\phi}), \frac{\alpha_1\lambda}{\eta_1}\Big), \\ \boldsymbol{\beta}_g^{k+1} &= \mathrm{S}\Big(\boldsymbol{\beta}_g^k - \frac{\phi}{\eta_g}\mathbf{X}_g^T(\mathbf{X}_g\boldsymbol{\beta}_g^k - \mathbf{m}\omega_g^k + \frac{\boldsymbol{\gamma}_g^k}{\phi}), \frac{\alpha_g\lambda}{\eta_g}\Big), \quad g = 2, \dots, G. \end{aligned} \tag{4.38}$$

Updates in (4.38) highlight one advantage of splitting the feature space into lower dimensions. The $\boldsymbol{\beta}$ update can be viewed as a one-step iteration of the proximal gradient. After feature-splitting, η_g's are relatively small compared to the "un-splitted" η, as η needs to be larger than $\phi\lambda_{\max}(\mathbf{X}^T\mathbf{X})$. Given that η increases significantly with

p for high-dimensional data, the step size for the update (i.e., $\frac{1}{\eta}$) can be relatively small, potentially slowing down the convergence of the algorithm. The updates for $\mathbf{m}\omega$, $\mathbf{z}$, and $\boldsymbol{\gamma}$ in this algorithm remain exactly the same as those in Algorithm 3. We denote this algorithm as *FS-QRADMM-prox* and summarize it in Algorithm 4. It's important to note that $\mathbf{m}\mathcal{T}_g \succ 0$ is also required in the proof of convergence for $\{\boldsymbol{\beta}^k\}$.

Algorithm 4 FS-QRADMM-prox for weighted ℓ_1-penalized QR

Require: $\boldsymbol{\beta}^0, \mathbf{m}\omega^0, \mathbf{z}^0, \boldsymbol{\gamma}^0$, and $\phi > 0, \theta > 0$ are given. $\mathbf{m}\mathcal{T}_g = \eta_g \mathbf{m} I_{p_g} - \phi \mathbf{X}_g^T \mathbf{X}_g$ with $\eta_g > \phi \lambda_{\max}(\mathbf{X}_g^T \mathbf{X}_g)$, $g = 1, \ldots, G$.
while the stopping criterion is not satisfied, **do**
  Compute $\boldsymbol{\beta}^{k+1}$ by (4.38).
  Compute $\mathbf{m}\omega^{k+\frac{1}{2}}$, $\mathbf{z}^{k+1}$ and $\mathbf{m}\omega^{k+1}$ by (4.36).
  Update $\boldsymbol{\gamma}^{k+1}$ by (4.37).
end while

Therefore, we compute $\boldsymbol{\beta}_2, \ldots, \boldsymbol{\beta}_G$ on separate processors/cores using parallel computing, and subsequently aggregate the updated information to compute other variables.

We establish the linear rate of convergence for Algorithm 4 in Theorem 4.4. The proximal term is essential for establishing the theory, and its proof is provided in Appendix 4.6.1.6.

Theorem 4.4 *For* $\theta \in \left(0, (1+\sqrt{5})/2\right)$, *the sequence* $(\boldsymbol{\beta}^k, \mathbf{z}^k, \mathbf{m}\boldsymbol{\omega}^k, \boldsymbol{\gamma}^k)$ *generated by Algorithm 4 converges to a limit point* $(\bar{\boldsymbol{\beta}}, \bar{\mathbf{z}}, \bar{\mathbf{m}\boldsymbol{\omega}}, \bar{\boldsymbol{\gamma}})$, *where* $(\bar{\boldsymbol{\beta}}, \bar{\mathbf{z}}, \bar{\mathbf{m}\boldsymbol{\omega}})$ *is the primal optimal and* $\bar{\boldsymbol{\gamma}}$ *is the dual optimal. Furthermore, there exists a constant* $\mu \in (0, 1)$ *such that* $Dist^{k+1} \le \mu Dist^k$, *where* $Dist^k$ *at the k-th iteration is defined as,*

$$\begin{aligned} Dist^k =& \|\mathbf{z}^k - \bar{\mathbf{z}}\|_2^2 + \frac{G-1}{G}\|\mathbf{z}^k - \mathbf{z}^{k-1}\|_2^2 + \sum_{g=1}^{G} \left\| \mathbf{X}_g(\boldsymbol{\beta}_g^k - \bar{\boldsymbol{\beta}}_g) - \frac{1}{G}(\mathbf{X}\boldsymbol{\beta}^k - \mathbf{X}\bar{\boldsymbol{\beta}}) \right\|_2^2 \\ &+ \frac{m_1}{G} \left\| \sum_{g=1}^{G} \mathbf{X}_g(\boldsymbol{\beta}_g^k - \bar{\boldsymbol{\beta}}_g) + (\mathbf{z}^k - \bar{\mathbf{z}}) \right\|_2^2 + \sum_{g=1}^{G} \|\boldsymbol{\beta}_g^k - \bar{\boldsymbol{\beta}}_g\|_{\mathbf{m}\mathcal{T}_g}^2, \end{aligned} \tag{4.39}$$

where $m_1 = 1 + d_1 - d_1\theta - (1-d_1)\min\{\theta, \frac{1}{\theta}\}$ *and* $d_1 \in (0, \frac{1}{2})$.

Remark. Indeed, the penalized quantile regression problem can be reformulated as a linear programming problem. Both the primal and dual problems are feasible, and strong duality holds, indicating that the optimal value of the dual problem equals the optimal value of the linear programming problem (primal problem). Consequently, the optimal values of both the primal and dual problems equal the optimal value of the penalized quantile regression problem.

The impact of G on convergence can be viewed from two perspectives. Firstly, increasing G diminishes the dimension of subproblems and reduces the value of η, thereby accelerating the computation of each sub-problem. However, on the other hand, a higher G results in an increased number of subproblems and may elevate the value of μ, thus potentially slowing down both practical and theoretical convergence. In our numerical experiments, we observed that selecting G from 5 to 10 tends to work well for values of p ranging from thousands to tens of thousands. This balance helps mitigate computational overhead while maintaining convergence efficiency.

4.3.3 PQR-Lasso and PQR-SCAD

In this section, PQR-Lasso refers to the Penalized Quantile Regression problem in (4.25) with the ℓ_1 penalty, i.e., $p_\lambda(|\beta|) = \lambda|\beta|$. Consequently, the PQR-Lasso can be directly solved using Algorithms 3 and 4 with all weights $\alpha_j = 1$ for $j = 1, \ldots, p$ in (4.28). The resulting solutions obtained from Algorithms 3 and 4 for the PQR-Lasso are denoted as FS-QRADMM-CD(Lasso) and FS-QRADMM-prox(Lasso) in Sect. 4.4.2, respectively.

Parallel to the PQR-Lasso, PQR-SCAD refers to the PQR in (4.25) with the SCAD penalty whose first-order derivative is defined in (4.16). Since the SCAD penalty is folded concave, the objective function of PQR-SCAD may have multiple local minimizers. To mitigate this issue, we recommend (a) using the proposed algorithm to obtain the PQR-Lasso estimate $\hat{\boldsymbol{\beta}}_L = (\hat{\beta}_{L,1}, \ldots, \hat{\beta}_{L,p})^T$, and then (b) further obtaining the PQR with a weighted ℓ_1 penalty, in which the weight α_j is $\lambda^{-1} p'_\lambda(|\hat{\beta}_{L,j}|)$ with $p'_\lambda(|\beta|)$ being the first-order derivative of the SCAD penalty. We refer to the resulting estimate as the two-step PQR-SCAD estimate. Note that both the ℓ_1 penalty and the SCAD-based weighted ℓ_1 penalty are convex. The two-step SCAD estimate is well defined when $L(\mathbf{y} - \mathbf{X}\boldsymbol{\beta})$ is strictly convex with respect to $\boldsymbol{\beta}$. Denote FS-QRADMM-CD(TS-SCAD) and FS-QRADMM-prox(TS-SCAD) to be the resulting solutions of Algorithms 3 and 4 for the two-step PQR-SCAD. The corresponding algorithms of FS-QRADMM-CD and FS-QRADMM-prox for the two-step PQR-SCAD are presented in Algorithms 5 and 6 in Appendix 4.6.2.

We denote FS-QRADMM-CD(TS-SCAD) and FS-QRADMM-prox(TS-SCAD) as the resulting solutions of Algorithms 3 and 4 for the two-step PQR-SCAD. The corresponding algorithms for FS-QRADMM-CD and FS-QRADMM-prox for the two-step PQR-SCAD are presented in Algorithms 5 and 6.

The two-step PQR-SCAD follows the same principle as the one-step sparse maximum likelihood estimation proposed in [44] for folded concave penalization problems. The second step in the two-step PQR-SCAD aims to address the bias inherent in the ℓ_1 penalty, which tends to over-penalize large coefficients and introduce bias into the resulting model. As demonstrated in Corollary 8 of [46], the two-step PQR-SCAD has the capability to identify the oracle estimator among multiple local minima with overwhelming probability, under certain regularity conditions. This theoretical result provides justification for the effectiveness of the two-step SCAD approach. In other words, the resulting solutions obtained from Algorithms 5 and 6 for the two-step PQR-SCAD exhibit a strong oracle property, as defined in [46].

Indeed, the two-step PQR-SCAD procedure can be generalized to the two-step PQR with a general folded concave penalty. The conditions that the folded concave penalty functions should satisfy are outlined in 4.2.3. As demonstrated in [46], the two-step PQR with a general folded concave penalty also exhibits the strong oracle property under certain regularity conditions.

It is desirable to have a data-driven method to select the regularization parameters in PQR-Lasso and PQR-SCAD. In our numerical study, we set the same penalty and tuning parameter for all coefficients, and λ is chosen by HBIC criterion proposed in [58],

$$\mathrm{HBIC}(\lambda) = \log\{\sum_{i=1}^{n} \rho_\tau(y_i - \mathbf{m}x_i^T\boldsymbol{\beta})\} + |\mathcal{A}|\frac{\log(\log n)\log(p)}{n}, \tag{4.40}$$

where $|\mathcal{A}|$ is the cardinality of active set. We select the λ that minimizes HBIC.

In the setting of penalized least squares, [45] suggests using different regularization parameters (λ_1 and λ) in the first and second steps to ensure that the resulting Lasso estimate satisfies a certain rate of convergence. Specifically, they recommend setting $\lambda_1 = \upsilon\lambda$, where $\upsilon > 0$ tends to 0 as $n \to \infty$. Following this recommendation, we choose $\lambda_1 = \upsilon\lambda$ in our numerical studies in 4.4.2, where $\upsilon = \lambda$ as suggested by [45].

4.4 Numerical Study

4.4.1 Numerical Study of LPO

In this section, we undertake a thorough simulation study to assess the convergence of the algorithm, computational efficiency, and the performance of AS-ADMM and two-step-AS-ADMM in portfolio allocation models. Throughout this section, we partition the effective parameter $\boldsymbol{\beta}$ into $K = 8$ groups.

Convergence Analysis

We examine an example to analyze the convergence of AS-ADMM. We generate the data from a multivariate normal distribution $N(\boldsymbol{\mu}, \boldsymbol{\Sigma})$, where $\mathbf{m}\mu = (1, \ldots, 1, 0, \ldots, 0)^T$ with first 10 elements being nonzero, and the covariance matrix $\boldsymbol{\Sigma} = (\rho_{ij})_{p\times p}$ with $\rho_{ij} = 0.25^{|i-j|}$ for $1 \le i, j \le p$. The sample size $n = 200$, and the dimension of the data $p = 500, 2000$. The number of groups $K = 8$. Figure 4.1 illustrates the plot of the prime residual versus iterations, demonstrating the convergence property of AS-ADMM. From Fig. 4.1, we observe that the

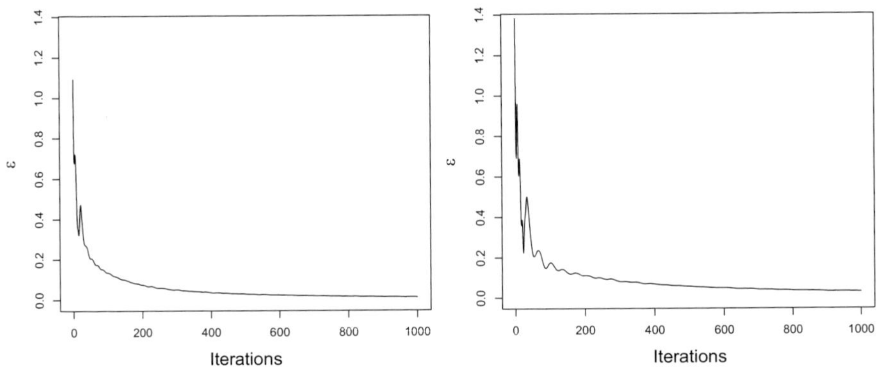

Fig. 4.1 AS-ADMM over 100 replications. Left panel is for $p = 500$ and right panel for $p = 2000$

prime residuals initially decline rapidly within the first 100 iterations, followed by a gradual decrease until convergence. When $p = 500$, the algorithm converges faster compared to the case when $p = 2000$. The algorithm is approximately converging within 800 iterations. Figure 4.1 confirms the convergence property of AS-ADMM.

Computing Time Comparison

In this section, we compare the computing time of AS-ADMM with *lpSolve* and classical ADMM algorithm. We generate the data from a multivariate normal distribution $N(\boldsymbol{\mu}, \boldsymbol{\Sigma})$, where $\mu_j \sim N(0, 1)$ for $j = 1, \ldots, p$ and the covariance matrix $\boldsymbol{\Sigma} = (\rho_{ij})_{p\times p}$ with $\rho_{ij} = 0.25^{|i-j|}$ for $1 \leq i, j \leq p$. We set the sample size $n = 500$ and the dimension of the data $p = 2000, 5000, 10000$. The number of groups $K = 8$.

The memory limit per replicate is 4GB. The results of the computing times are based on 100 replications, and we report the average computing time and its standard deviation of each replicate in Table 4.1.

Table 4.1 demonstrates that the computation speed of AS-ADMM is significantly faster than both *lpSolve* and ADMM. For instance, when $p = 2000$, the computation time of AS-ADMM is approximately $1/4$ of ADMM and $1/70$ of *lpSolve*. As we increase the value of p, *lpSolve* fails first when $p = 5000$. ADMM can complete the task when $p = 5000$, but its computation time is about 7 times that of AS-ADMM. When $p = 10000$, both *lpSolve* and ADMM fail to finish the job due to memory

Table 4.1 Computing time (in seconds) comparisons of *lpSolve*, ADMM and AS-ADMM

Algorithm	lpsolve	ADMM	AS-ADMM
$p = 2000$	260.32(151.88)	14.11(1.35)	3.75(0.51)
$p = 5000$	✗	89.24(8.01)	13.21 (1.05)
$p = 10000$	✗	✗	51.55 (21.07)

Table 4.2 Comparison of K on computing time (in seconds)

Algorithm	AS-ADMM(K = 4)	AS-ADMM(K = 8)	AS-ADMM(K = 16)
$p = 2000$	3.61(0.47)	3.75(0.51)	3.87(0.49)
$p = 5000$	14.21(1.32)	13.21(1.05)	13.49(1.22)
$p = 10000$	71.33(28.21)	51.55(21.07)	47.75(19.02)

limitations, while AS-ADMM can still complete the task in a relatively short time. These results highlight the benefits of the asset-splitting algorithm in terms of both computation speed and memory efficiency.

We further explore the impact of the number of groups K on the computational efficiency of the proposed algorithm. We consider two additional choices: $K = 4$ and $K = 16$. The results are summarized in Table 4.2.

When the dimension is moderately large, as observed in cases such as $K = 4$, $K = 8$, and $K = 16$, they exhibit similar computation times. It appears that employing a larger value for K might be advantageous as the dimension of the data increases. However, the selection of K should consider the computational power available on the equipment being used. We recommend using K values ranging from 4 to 8 for dimensions p ranging from thousands to tens of thousands. For the subsequent simulation study and real data example, we opt to use $K = 8$.

Portfolio Allocation Simulation

We denote R_i as the excess return over the risk-free interest rate of the i-th asset. The three-factor model for R_i is defined as follows:

$$R_i = b_{i1} f_1 + b_{i2} f_2 + b_{i3} f_3 + \epsilon_i, i = 1, \ldots, p, \tag{4.41}$$

where b_{ij}'s are the factor loadings of the i-th asset on the factor f_j, and ϵ_i is the noise. We assume the noise is independent of the three factors with $E(\epsilon_i) = 0$ and $var(\epsilon_i) = \sigma_i^2$. The matrix form of the three-factor model can be represented as

$$\mathbf{R} = \mathbf{Bm} f + \boldsymbol{\epsilon}. \tag{4.42}$$

The covariance of $\mathbf{R}$ is

$$\boldsymbol{\Sigma} = \text{Cov}(\mathbf{Bm} f) + \text{Cov}(\boldsymbol{\epsilon}) = \mathbf{B}\text{Cov}(\mathbf{m} f)\mathbf{B}^T + \text{diag}(\sigma_1^2, \ldots, \sigma_p^2) \tag{4.43}$$

We generate the factor loadings from the trivariate normal distribution $N(\boldsymbol{\mu}_b, \text{Cov}(\mathbf{b}))$, and keep it fixed throughout the simulations. The returns of the three factors over n periods are drawn from the trivariate normal distribution $N(\boldsymbol{\mu}_f, \text{Cov}(\mathbf{m} f))$. We consider three different distributions of the noises in the

Table 4.3 Parameters for factor loadings and factor returns used in the simulation

Parameters for factor loadings				Parameters for factor returns			
μ_b	Cov(b)			μ_f	Cov($\mathbf{m}f$)		
0.7828	0.02914	0.02387	0.01018	0.02355	1.2507	−0.0350	−0.2042
0.5180	0.02387	0.05395	−0.00696	0.01298	−0.0350	0.3156	−0.0023
0.4100	0.01018	−0.00696	0.08685	0.02071	−0.2042	−0.0023	0.1930

simulations: (1) $N(0, 0.02)$; (2) $0.02 \times (G(4, 2) - 2)$, where $G(a, b)$ is the Gamma distribution with a as the shape parameter and b as the rate parameter, and (3) $0.01\sqrt{2}t_4$, where t_4 is the t-distribution with degrees of freedom equal to 4. We use the parameters for factor loadings and factor returns in [17], which are given in Table 4.3.

The following criteria are used to evaluate the performance of different algorithms in the simulations.

1. Sharpe ratio: the ratio of the return to the risk as a measure of portfolio allocation, which represents the portfolio return per unit risk [59, 60];
2. Number of long positions (Long): the cardinality of the set $\{w_j | w_j > 0, j = 1, \ldots, p\}$;
3. Number of short positions (Short): the cardinality of the set $\{w_j | w_j < 0, j = 1, \ldots, p\}$.

In our simulation study, we consider two combinations of (n, p): $(n, p) = (480, 200)$ and (480, 2000). The algorithm parameters are selected based on cross-validation. We compare the performance of ADMM, *lpSolve*, AS-ADMM with ℓ_1 penalty (ℓ_1-AS-ADMM), and the two-step AS-ADMM (two-step-AS-ADMM) in the simulation. The results, based on 500 replications, are summarized in Tables 4.4 and 4.5. We have imposed a memory limit of 4GB and a time limit of 24 h for computation.

When the number of assets in the portfolio is not large (e.g., $p = 200$), all methods exhibit comparable performance, as evident from Table 4.4. As expected, ℓ_1-AS-ADMM performs similarly to ADMM and *lpSolve* since they are solving the same optimization problems. However, ℓ_1-AS-ADMM achieves a slightly better Sharpe ratio and selects fewer assets in the portfolio. The two-step method further improves portfolio performance by achieving better Sharpe ratios with fewer assets selected.

In scenarios with a relatively large number of assets, such as $p = 2000$, *lpSolve* fails to complete the task due to memory limitations. ℓ_1-AS-ADMM and ADMM demonstrate similar performance, but the portfolio selected by ℓ_1-AS-ADMM is sparser. Notably, when p significantly exceeds n, the two-step method achieves significantly better performance compared to the ℓ_1 method, with higher Sharpe ratios and fewer assets included in the portfolio. These results underscore the importance of implementing the two-step method when $p \gg n$.

Table 4.4 Numerical comparisons of ADMM, *lpSolve*, AS-ADMM with ℓ_1 penalty and two-step AS-ADMM when $n = 480$ and $p = 200$

Normal distribution

Algorithm	Sharpe ratio	Long	Short
ADMM	0.97(0.06)	114.97(6.64)	26.64(4.73)
lpSolve	0.97(0.07)	115.15(7.11)	22.79(4.51)
ℓ_1-AS-ADMM	0.99(0.06)	110.23(5.97)	22.42(4.37)
two-step-AS-ADMM	1.04(0.07)	99.71(6.95)	23.04(8.75)

Gamma distribution

Algorithm	Sharpe ratio	Long	Short
ADMM	0.97(0.06)	114.60(6.18)	26.61(4.65)
lpSolve	0.97(0.07)	115.09(6.56)	23.02(4.47)
ℓ_1-AS-ADMM	0.99(0.06)	110.00(5.69)	22.50(4.24)
two-step-AS-ADMM	1.03(0.07)	101.00(6.53)	23.76(7.03)

T distribution

Algorithm	Sharpe ratio	Long	Short
ADMM	0.97(0.07)	114.43(6.41)	26.27(4.78)
lpSolve	0.97(0.08)	115.21(7.19)	22.64(4.68)
ℓ_1-AS-ADMM	0.99(0.07)	109.94(6.08)	22.33(4.30)
two-step-AS-ADMM	1.02(0.07)	101.35(6.95)	23.07(4.75)

Table 4.5 Numerical comparisons of ADMM, *lpSolve*, AS-ADMM with ℓ_1 penalty and two-step AS-ADMM when $n = 480$ and $p = 2000$

Normal distribution

Algorithm	Sharpe ratio	Long	Short
ADMM	0.81(0.04)	445.97(102.26)	216.57(55.58)
lpSolve	✗	✗	✗
ℓ_1-AS-ADMM	0.81(0.05)	299.21(13.23)	122.59(9.47)
two-step-AS-ADMM	0.94(0.06)	142.12(37.20)	48.47(25.97)

Gamma distribution

Algorithm	Sharpe ratio	Long	Short
ADMM	0.81(0.04)	442.00(106.81)	215.39(57.95)
lpSolve	✗	✗	✗
ℓ_1-AS-ADMM	0.81(0.05)	298.05(13.19)	123.48(9.76)
two-step-AS-ADMM	0.94(0.07)	94.50(8.10)	19.04(4.90)

t-Distribution

Algorithm	Sharpe ratio	Long	Short
ADMM	0.80(0.04)	441.40(107.14)	215.66(57.94)
lpSolve	✗	✗	✗
ℓ_1-AS-ADMM	0.81(0.05)	297.00(13.56)	123.29(9.70)
two-step-AS-ADMM	0.94(0.08)	95.22(8.55)	19.01(5.17)

Table 4.6 Real data comparisons of ADMM, *lpSolve*, AS-ADMM with ℓ_1 penalty and two-step AS-ADMM

Senario S1			
Algorithm	Sharpe ratio	Long	Short
ADMM	1.018	17	1
lpSolve	0.972	11	3
ℓ_1-AS-ADMM	1.198	5	0
Two-step-AS-ADMM	1.567	5	2
Senario S2			
Algorithm	Sharpe ratio	Long	Short
ADMM	1.683	16	0
lpSolve	1.611	11	2
ℓ_1-AS-ADMM	1.729	7	0
Two-step-AS-ADMM	2.033	3	1

Real Data Example

In this section, we illustrate the proposed procedure through an empirical analysis of a real-world dataset. The dataset consists of daily returns for stocks in the S&P 500 index with complete records from January 2013 to June 2014. There are a total of 440 stocks with complete records, with 252 records for each stock in the year 2013 and 126 records for each stock in the year 2014. We partition the data into training and testing datasets. The training dataset consists of data from the year 2013, while the testing dataset consists of data from the year 2014. In this section, we compare the performance of the proposed algorithms with existing ones in the following two scenarios.

S1. Obtain the weights of the stocks using the data from **July 2013 to December 2013** and test the of the portfolio on the data in 2014.

S2. Obtain the weights of the stocks using the data from **January 2013 to December 2013** and test the performance of the portfolio on the data in 2014.

We use the criteria in Sect. 4.4.1 to evaluate the performances of different methods. The results are summarized in Table 4.6, from which it can be seen that the performances of all the methods under scenario S2 are better than S1, which is mainly because they have a larger training data set under S2. The algorithms with ℓ_1 penalty have similar performances, and the proposed ℓ_1-AS-ADMM have a slightly better Sharpe ratio and sparser portfolio. The two-step method can further improve the performance in terms of both the Sharpe ratio and the sparsity of the portfolio.

4.4.2 *Numerical Study of PQR*

In this section, we evaluate the performance of the proposed algorithms in PQR through simulation studies and demonstrate the application of the newly proposed procedure via empirical analysis. For all ADMM-based methods, we implement the warm-start technique introduced in [61, 62], which uses the solution from the previous λ to initialize computation at the current λ. The way of splitting the features has no influences on the convergence property of the algorithm. We equally distribute the features into K groups without adjusting the order in our numerical studies. The stopping criterion of ADMM-based algorithms is provided in the Appendix.

Simulation Study

In this simulation, we compare the performance of Algorithms 3 and 4 with R packages *rqPen* [36], *qradmm* [39], *hqReg* [63] and *Conquer* [64]. Since *qradmm* package is boosted by FORTRAN, we re-implement its core algorithm, i.e., a two-block proximal ADMM, using R code for a relatively fair comparison. The R package *rqPen* implements an iterative coordinate descent algorithm (QICD) proposed in [38] to solve sparse quantile regression. QICD applies a convex majorization function on the concave penalty term and solves the majorized objective function by coordinate descent. The R-package *qradmm* implements a two-block proximal ADMM for PQR with ℓ_1 penalty proposed in [39]. We use the R packages *hqreg* and *conquer* to implement the methods proposed by [39, 64]. The regularization parameter λ in all algorithms to be compared is selected by the HBIC criterion defined in (4.40).

We take the simulation setting similar to that of [38]. We generate $\widetilde{\mathbf{z}} = (Z_1, Z_2, \ldots, Z_p)^T$ from $\mathbf{m}N_p(\mathbf{m}0, \mathbf{\Sigma})$, where $\mathbf{\Sigma} = (\sigma_{ij})$ with $\sigma_{ij} = 0.5^{|i-j|}$. Then set $X_1 = \Phi(Z_1)$ and $X_j = Z_j$ for $j = 2, \ldots, p$, where $\Phi(\cdot)$ is the cumulative distribution function of $N(0, 1)$. The response variable Y is generated from the following heteroscedastic regression model,

$$Y = X_6 + X_{100} + X_{500} + X_{1000} + 0.7X_1\varepsilon, \tag{4.44}$$

where $\varepsilon \sim N(0, 1)$. We consider three different quantiles $\tau = 0.3, 0.5$ and 0.7. Note that X_1 does not affect the center of the conditional distribution Y given $\mathbf{x}$, but affects the conditional distribution when $\tau = 0.3$ or 0.7. In our simulation, we set $n = 400$ and $p = 1000$ and 50000. For each case, we conduct 500 replications.

The following criteria are used to compare the performance of different algorithms.

1. Average absolute error: the average and standard deviation of $\|\hat{\boldsymbol{\beta}} - \boldsymbol{\beta}\|_1 = \sum_{j=1}^{p} |\hat{\beta}_j - \beta_j|$ over 500 replications.
2. Size: the average number of nonzero $\hat{\beta}_j$'s over 500 replications.
3. P_1: the proportion of models that select all active features except for X_1 over 500 replications
4. P_2: the proportion of models that select X_1 over 500 replications.

The proportion P_2 is expected to be close to 0 when $\tau = 0.5$, and be close to 1 when $\tau = 0.3$ and 0.7.

Table 4.7 Comparison of algorithms for PQR when $p = 1000$ and $n = 400$

$n = 400$, $p = 1000$	τ	$\|\hat{\beta} - \beta\|_1$	P1 (%)	P2 (%)	Size
FS-QRADMM-CD (Lasso)	0.3	0.295 (0.003)	100	100	5.56 (0.03)
	0.5	0.210 (0.003)	100	5.4	4.36 (0.03)
	0.7	0.281 (0.003)	100	100	5.56 (0.03)
FS-QRADMM-prox (Lasso)	0.3	0.295 (0.003)	100	100	5.62 (0.03)
	0.5	0.198 (0.003)	100	4.6	4.34 (0.02)
	0.7	0.301 (0.003)	100	100	5.56 (0.03)
qradmm (Lasso)	0.3	0.310 (0.003)	100	100	5.68 (0.03)
	0.5	0.230 (0.003)	100	9	5.32 (0.06)
	0.7	0.327 (0.005)	100	100	6.73 (0.08)
rqPen (Lasso)	0.3	0.598 (0.004)	100	61.2	5.10 (0.04)
	0.5	0.267 (0.003)	100	0	4.23 (0.02)
	0.7	0.601 (0.004)	100	56.6	5.04 (0.04)
hqReg (Lasso)	0.3	0.593 (0.006)	100	50	4.95 (0.04)
	0.5	0.235 (0.003)	100	0	4.31 (0.03)
	0.7	0.589 (0.006)	100	51.6	4.97 (0.04)
Conquer (Lasso)	0.3	0.590 (0.005)	100	45	4.73 (0.03)
	0.5	0.231 (0.002)	100	0	4.27 (0.02)
	0.7	0.586 (0.005)	100	45	4.72 (0.03)
FS-QRADMM-CD (TS-SCAD)	0.3	0.119 (0.002)	100	100	5.00 (0.00)
	0.5	0.035 (0.001)	100	0.2	4.00 (0.00)
	0.7	0.125 (0.002)	100	100	5.00 (0.00)
FS-QRADMM-prox (TS-SCAD)	0.3	0.115 (0.002)	100	100	5.00 (0.00)
	0.5	0.040 (0.001)	100	0.2	4.00 (0.00)
	0.7	0.123 (0.001)	100	100	5.00 (0.00)
qradmm (TS-SCAD)	0.3	0.122 (0.002)	100	100	5.00 (0.00)
	0.5	0.038 (0.001)	100	0.4	4.00 (0.00)
	0.7	0.129 (0.002)	100	100	5.00 (0.00)
rqPen (TS-SCAD)	The algorithm runs out of memory for $\tau = 0.3, 0.5, 0.7$				
Conquer(SCAD)	0.3	0.339 (0.004)	100	63.4	4.67 (0.02)
	0.5	0.049 (0.001)	100	0	4.07 (0.01)
	0.7	0.350 (0.004)	100	57	4.60 (0.02)

The simulation results over 500 replications are summarized in Tables 4.7 and 4.8. Compared to the PQR-Lasso, the two-step PQR-SCAD produces models with significantly smaller absolute error and better selection accuracy in general. FS-QRADMM-CD(TS-SCAD) and FS-QRADMM-prox (TS-SCAD) exhibit the best performances with respect to estimation and variable selection accuracy. When $p =$

Table 4.8 Comparison of algorithms for PQR when $p = 50000$ and $n = 400$

$n = 400$, $p = 50000$	τ	$\|\hat{\boldsymbol{\beta}} - \boldsymbol{\beta}\|_1$	P1 (%)	P2 (%)	Size
FS-QRADMM-CD (Lasso)	0.3	0.320 (0.003)	100	98.2	5.34 (0.02)
	0.5	0.250 (0.003)	100	2	4.25 (0.03)
	0.7	0.349 (0.003)	100	100	5.15 (0.03)
FS-QRADMM-prox (Lasso)	0.3	0.326 (0.003)	100	92.4	4.93 (0.01)
	0.5	0.121 (0.001)	100	0	4.01 (0.00)
	0.7	0.394 (0.002)	100	95.6	5.01 (0.11)
qradmm (Lasso)	The algorithm runs out of memory for $\tau = 0.3, 0.5, 0.7$				
rqPen (Lasso)	The algorithm runs out of memory for $\tau = 0.3, 0.5, 0.7$				
hqReg (Lasso)	0.3	0.812 (0.004)	100	2.2	4.23 (0.02)
	0.5	0.365 (0.003)	100	0	4.20 (0.02)
	0.7	0.808 (0.004)	100	4.4	4.26 (0.02)
Conquer (Lasso)	0.3	0.717 (0.003)	100	18	5.88 (0.07)
	0.5	0.303 (0.002)	100	0	8.63 (0.11)
	0.7	0.705 (0.003)	100	26.8	6.13 (0.07)
FS-QRADMM-CD (TS-SCAD)	0.3	0.180 (0.003)	100	98.8	4.99 (0.00)
	0.5	0.047 (0.001)	100	0	4.00 (0.00)
	0.7	0.172 (0.003)	100	99.6	5.00 (0.03)
FS-QRADMM-prox (TS-SCAD)	0.3	0.158 (0.002)	100	100	5.00 (0.00)
	0.5	0.069 (0.005)	100	2.2	7.31 (0.47)
	0.7	0.244 (0.007)	100	99.2	6.64 (0.14)
qradmm (TS-SCAD)	The algorithm runs out of memory for $\tau = 0.3, 0.5, 0.7$				
rqPen (TS-SCAD)	The algorithm runs out of memory for $\tau = 0.3, 0.5, 0.7$				
Conquer (SCAD)	0.3	0.396 (0.002)	100	48.4	5.04 (0.04)
	0.5	0.058 (0.001)	100	0	5.78 (0.07)
	0.7	0.390 (0.003)	100	49.6	5.12 (0.05)

1000, three ADMM-based methods perform comparably well and outperform *rqPen*, *hqReg*, and *Conquer* by a significant margin. *rqPen*, *hqReg*, and *Conquer* obtain relatively larger estimation errors and are more likely to miss X_1 when $\tau = 0.3$ and 0.7. The current version of *rqPen* runs out of memory when solving the two-step PQR-SCAD, as noted in the table. Nonetheless, when $p = 50000$, both *Qradmm*

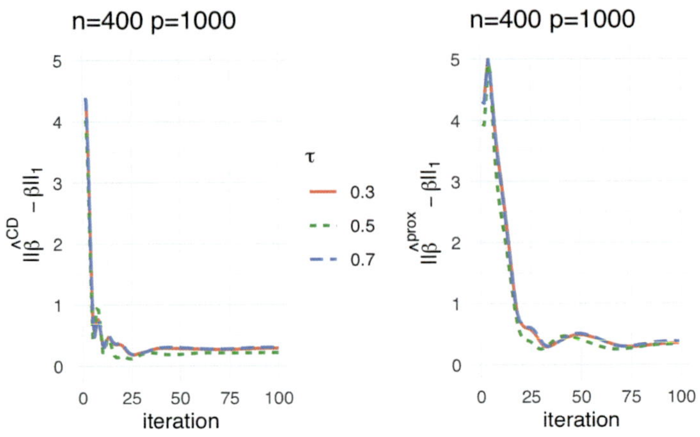

Fig. 4.2 Convergence curves of $\|\hat{\boldsymbol{\beta}} - \boldsymbol{\beta}^*\|_1$ of FS-QRADMM-CD(Lasso) (left panel) and FS-QRADMM-prox(Lasso) (right panel) over 500 replications

and *rqPen* fail due to their demanding memory usage. In fact, we notice that the efficiency of *Qradmm* deteriorates sharply as p increases. *hqReg* and *Conquer* are able to finish the job when $p = 50000$, but the proposed methods still outperform *hqReg* and *Conquer*.

Figure 4.2 plots the curves of $\|\hat{\boldsymbol{\beta}} - \boldsymbol{\beta}^*\|_1$ with respect to the iteration steps averaged over 500 replications when $n = 400$, $p = 1000$, and $\tau = (0.3, 0.5, 0.7)$. We can see that Algorithms 3 and 4 converge to true $\boldsymbol{\beta}$ within approximately 20 iterations.

We also examine the implementation of the proposed methods when $n = 30000$ and $p = 1000$ to investigate their performance when the sample size is large. The results are summarized in Table 4.10 in Sect. 4.6.2 in the Appendix.

A Real Data Example

QR models are widely utilized in the analysis of consumer markets due to their robustness against outliers. In this section, we apply the proposed algorithm for an empirical analysis of a supermarket dataset studied in [65] and compare it with other existing algorithms. This dataset comprises the daily number of customers and the daily sales volumes of 6398 products from a supermarket in China over 464 days. Following [65], we set the response variable to be the daily number of customers, and the predictors to be the daily sale volumes of products. Since the sample size $n = 464$ is much smaller than the dimension $p = 6398$, it is reasonable to assume that only a small proportion of predictors have significant effects on the response. The distribution of the number of customers is highly skewed, motivating us to consider PQR with the proposed algorithm in this example. We standardize both the response and predictors for our analysis.

Table 4.9 Performances of ADMM and *lpSolve* of sparse quantile regression on the Chinese Supermarket Data

	τ	$\frac{1}{n}\sum_{i=1}^{n}\rho_\tau(y_i - \hat{y}_i)$	Size
FS-QRADMM-CD (Lasso)	0.3	0.118 (0.001)	97.35 (0.62)
	0.5	0.116 (0.001)	100.56 (0.53)
	0.7	0.127 (0.001)	97.37 (0.71)
FS-QRADMM-prox (Lasso)	0.3	0.117 (0.001)	103.23 (1.13)
	0.5	0.113 (0.001)	118.61 (0.93)
	0.7	0.127 (0.001)	96.02 (1.06)
qradmm (Lasso)	0.3	0.115 (0.000)	119.01 (0.49)
	0.5	0.116 (0.001)	121.26 (0.37)
	0.7	0.130 (0.000)	127.35 (0.58)
rqPen (Lasso)	0.3	0.117 (0.001)	113.62 (0.73)
	0.5	0.115 (0.001)	117.17 (0.56)
	0.7	0.128 (0.001)	120.11 (0.62)
hqReg (Lasso)	0.3	0.117 (0.001)	49.31 (0.46)
	0.5	0.116 (0.001)	90.85 (0.65)
	0.7	0.127 (0.001)	43.42 (0.42)
Conquer (Lasso)	0.3	0.118 (0.001)	80.8 (1.47)
	0.5	0.114 (0.001)	42.9(0.58)
	0.7	0.125 (0.001)	39.6 (0.49)
FS-QRADMM-CD (TS-SCAD)	0.3	0.112 (0.000)	63.86 (0.39)
	0.5	0.111 (0.001)	69.77 (0.47)
	0.7	0.116 (0.001)	72.71 (0.61)
FS-QRADMM-prox (TS-SCAD)	0.3	0.116 (0.000)	97.03 (0.96)
	0.5	0.110 (0.000)	100.33 (1.02)
	0.7	0.113 (0.000)	95.66 (0.79)
qradmm (TS-SCAD)	0.3	0.113 (0.001)	469.88 (2.56)
	0.5	0.114 (0.001)	477.72 (2.33)
	0.7	0.120 (0.001)	521.33 (1.99)
rqPen(TS-SCAD)	The algorithm runs out of memory for the three τs		
Conquer(SCAD)	0.3	0.115 (0.001)	65.7 (1.95)
	0.5	0.113 (0.001)	65.7(0.82)
	0.7	0.120 (0.001)	36.0 (0.48)

We randomly split observations into training and testing datasets of sizes 300 and 164, respectively, and fit PQR-Lasso and two-step QR-SCAD on the training data with $\tau = 0.3, 0.5$ and 0.7. The regularization parameter λ is chosen by HBIC criterion. We report the averaged predictive error and its standard deviation on the testing data over 100 replications in Table 4.9. The predictive error is measured by the loss function $\frac{1}{n}\sum_{i=1}^{n}\rho_\tau(y_i - \hat{y}_i)$. We also report the average model sizes and their corresponding standard deviation to evaluate the interpretability of models selected from different methods. For PQR-Lasso, we observe that ADMM-based algorithms have similar performances with those of *rqPen*, *hqReg* in terms of prediction error. The average values and standard deviations of the loss function are very close among those methods. In general, all methods perform the best when $\tau = 0.5$. The proposed method selects fewer products than *qradmm*, *rqPen*,*hqReg* do in most scenarios, which indicates a better model interpretability.

Similar results are also observed with the two-step PQR-SCAD. However, the *rqPen* for the two-step SCAD fails in this example due to the limitation of computing memory. The proposed algorithms have a similar prediction error to that of *qradmm*, but the model sizes are much smaller. When $\tau = 0.7$, the proposed methods outperform *qradmm*, with fewer products included in the QR model. Conquer with SCAD penalty has similar performances to the proposed method under this scenario. We can also notice that PQR-SCAD gives better loss than PQR-Lasso does, and the two-step PQR-SCAD procedures select fewer products when the proposed algorithms are implemented.

4.5 Conclusion and Discussion

In this chapter, we apply the three-block ADMM algorithm based on feature splitting to different domains. In the field of portfolio allocation, we introduce the AS-ADMM algorithm, which addresses high-dimensional portfolio selection problems through asset splitting. We investigate its convergence and demonstrate the advantages of the proposed algorithm in terms of computational efficiency and finite sample performance through simulation results. Additionally, we propose a novel regularization method with folded concave penalty, recasting it into a weighted ℓ_1 model using the LLA algorithm and establishing its oracle property. Numerical results indicate that the newly proposed procedure improves the performance of the ℓ_1 model, particularly when the portfolio comprises a vast number of assets.

On the other hand, we apply the same algorithmic concept to penalized quantile regression (PQR). By introducing a parallelizable algorithm based on the three-block ADMM with feature-splitting, we propose an efficient method for PQR and establish its convergence. Numerical experiments show that the proposed algorithm outperforms existing ones for PQR, particularly in scenarios with huge data dimensions where existing algorithms struggle due to memory constraints. Furthermore, we highlight the potential extension of the proposed algorithms to other statistical models, such as support vector machines, which share a similar loss function with QR. This presents an interesting avenue for future research.

4.6 Appendix

4.6.1 Proofs

This section consists of some technical lemmas and their technical proofs as well as proofs of Theorems 1.1–1.4.

Rate of Convergence

To prove Theorem 4.1, we first present a set of general theoretical results. Define functions f, g, h and linear operators F, G, H as follows:

$$f(\boldsymbol{\beta}) = \sum_{i=1}^{K} \alpha_i \|\boldsymbol{\beta}_i\|_1, \quad h(\mathbf{m}\omega) = \mathbf{m}0, \quad g(\mathbf{z}) = \delta_{\mathcal{Z}_0}(\mathbf{z}), \tag{4.45}$$

$$\begin{aligned} F &= Diag(\mathbf{A}_1, \ldots \mathbf{A}_K), \\ G &= (\mathbf{m}I_n, 0, \ldots, 0)^T, \\ H &= \begin{pmatrix} \mathbf{m}I_n & \mathbf{m}I_n & \cdots & \mathbf{m}I_n \\ -\mathbf{m}I_n & 0 & \cdots & 0 \\ \vdots & \ddots & & \vdots \\ 0 & \cdots & -\mathbf{m}I_n & 0 \\ 0 & 0 & \cdots & -\mathbf{m}I_n \end{pmatrix}. \end{aligned} \tag{4.46}$$

Next, we discuss the convergence rate for solving a general three-block constrained optimization problem, and Theorem 4.1 is a direct application to the nonconvex Dantzig selector.

By the definition of f, g, h, F, G, H, problem (4.4) has the following general formulation,

$$\min_{\boldsymbol{\beta}, \mathbf{z}, \mathbf{m}\omega} \{f(\boldsymbol{\beta}) + g(\mathbf{z}) + h(\mathbf{m}\omega) \mid F\boldsymbol{\beta} + G\mathbf{z} + H\mathbf{m}\omega = \mathbf{c}\}, \tag{4.47}$$

where

$$F\boldsymbol{\beta} = \begin{pmatrix} \mathbf{A}_1\boldsymbol{\beta}_1 \\ \mathbf{A}_2\boldsymbol{\beta}_2 \\ \vdots \\ \mathbf{A}_K\boldsymbol{\beta}_K \end{pmatrix}, \quad G\mathbf{z} = \begin{pmatrix} \mathbf{z} \\ 0 \\ \vdots \\ 0 \end{pmatrix}, \quad H\mathbf{m}\omega = \begin{pmatrix} \mathbf{m}\omega_2 + \cdots + \mathbf{m}\omega_K \\ -\mathbf{m}\omega_2 \\ \vdots \\ -\mathbf{m}\omega_K \end{pmatrix}. \tag{4.48}$$

By the definition of H, $H^T H$ is positive definite. (4.47) can be rewritten into a 2-block optimization problem,

$$\min_{\boldsymbol{\beta}, \mathbf{z}} \{f(\boldsymbol{\beta}) + g(\mathbf{z}) \mid Q(F\boldsymbol{\beta} + G\mathbf{z} - \mathbf{c}) = 0\}. \tag{4.49}$$

Throughout the analysis, we assume that there exists $(\hat{\boldsymbol{\beta}}, \hat{\mathbf{z}}, \hat{\mathbf{m}\omega}) \in \mathbb{R}^{p_1} \times \mathbb{R}^{p_2} \times \mathbb{R}^{p_3}$, such that $F\hat{\boldsymbol{\beta}} + G\hat{\mathbf{z}} + H\hat{\mathbf{m}\omega} = \mathbf{c}$. Sun et al. [43] discovered a critical connection between the 3-block and 2-block formulation. They proved that the iterative scheme (4.6) is equivalent to the following 2-block semi-proximal algorithm,

$$\begin{cases} \boldsymbol{\beta}^{t+1} & = \operatorname{argmin} \mathcal{L}_\phi(\boldsymbol{\beta}, \mathbf{z}^t; \boldsymbol{\gamma}^t) + \frac{\phi}{2}\|\boldsymbol{\beta} - \boldsymbol{\beta}^t\|^2_{F^T\mathcal{P}F+\mathcal{T}_f}, \\ \mathbf{z}^{t+1} & = \operatorname{argmin} \mathcal{L}_\phi(\boldsymbol{\beta}^{t+1}, \mathbf{z}; \boldsymbol{\gamma}^t) + \frac{\phi}{2}\|\mathbf{z} - \mathbf{z}^t\|^2_{G^T\mathcal{P}G+\mathcal{T}_g}, \\ \boldsymbol{\gamma}^{t+1} & = \boldsymbol{\gamma}^t + \theta\phi Q(F\boldsymbol{\beta}^{t+1} + G\mathbf{z}^{t+1} - \mathbf{c}). \end{cases} \tag{4.50}$$

This connection allows us to leverage some convergence theories for proximal 2-block ADMM. The augmented Lagrangian function for (4.49) is given by

$$\mathcal{L}_\phi(\mathbf{z}, \boldsymbol{\beta}; \boldsymbol{\gamma}) = f(\boldsymbol{\beta}) + g(\mathbf{z}) + \langle \boldsymbol{\gamma}, Q(F\boldsymbol{\beta} + G\mathbf{z} - \mathbf{c})\rangle + \frac{\phi}{2}\|Q(F\boldsymbol{\beta} + G\mathbf{z} - \mathbf{c})\|_2^2.$$

The KKT optimality condition of (4.49) is

$$0 \in (QF)^T\boldsymbol{\gamma} + \partial f(\boldsymbol{\beta}), \quad 0 \in (QG)^T\boldsymbol{\gamma} + \partial g(\mathbf{z}), \qquad Q(\mathbf{c} - F\boldsymbol{\beta} - G\mathbf{z}) = 0. \tag{4.51}$$

Denote the solution set to (4.51) as $\bar{\Omega}$. Let $\mathbf{m}\bar{u} = (\bar{\boldsymbol{\beta}}, \bar{\mathbf{z}}, \bar{\boldsymbol{\gamma}})$ be an optimal solution to (4.49). We have the following lemma on the convergence of the proposed algorithm by utilizing its equivalence to the procedure (4.50).

Lemma 4.1 *Suppose that $\mathcal{T}_f$ and $\mathcal{T}_g$ are chosen such that $\mathcal{T}_f + F^TF$ and $\mathcal{T}_g + G^TG$ are positive definite. Then under the condition $\theta \in (0, (1+\sqrt{5})/2)$, the sequence $(\boldsymbol{\beta}^t, \mathbf{z}^t, \mathbf{m}\boldsymbol{\omega}^t, \boldsymbol{\gamma}^t)$ generated by algorithm (1) converges to a limit point $(\bar{\boldsymbol{\beta}}, \bar{\mathbf{z}}, \mathbf{m}\bar{\boldsymbol{\omega}}, \bar{\boldsymbol{\gamma}})$ with $(\bar{\boldsymbol{\beta}}, \bar{\mathbf{z}}, \mathbf{z}\bar{\boldsymbol{\omega}})$ solving (4.47) and $\bar{\boldsymbol{\gamma}}$ is the dual optimal.*

Lemma 4.4 can be easily derived from Theorem 3.2 in [66]. To derive the convergence rate, we impose the following assumption that is imposed by [66].

Assumption 1 Suppose $\mathbf{u}^t$ converges to $\bar{\mathbf{u}} \in \bar{\Omega}$. There exists a positive constant q such that

$$\begin{aligned} \|\mathbf{u}^t - \bar{\mathbf{u}}\|_2^2 \le & q^2\Big(\|\boldsymbol{\beta}^t - \operatorname{prox}_f(\boldsymbol{\beta}^t - (QF)^T\boldsymbol{\gamma}^t)\|_2^2 + \|\mathbf{z}^t - \operatorname{prox}_g(\mathbf{z}^t - (QG)^T\boldsymbol{\gamma}^t)\|_2^2 \\ & + \|Q(\mathbf{c} - F\boldsymbol{\beta}^t - G\mathbf{z}^t)\|_2^2\Big), \end{aligned} \tag{4.52}$$

for sufficiently large t.

For any convex function P, $\operatorname{prox}_P(\cdot)$ denotes the proximal mapping associated with P; i.e.,

$$\operatorname{prox}_P(x) = \operatorname*{argmin}_y \left\{\frac{1}{2}\|x - y\|_2^2 + P(y)\right\}. \tag{4.53}$$

Denote

$$\mathcal{H} = C \times \operatorname{Diag}(F^T\mathcal{P}F + \mathcal{T}_f, G^T\mathcal{P}G + \mathcal{T}_g, (\theta\phi)^{-2}I),$$

where

$$C = \max\big\{3\phi^2\|F^T\mathcal{P}F+\mathcal{T}_f\|_2,\ 3\phi\lambda_{\max}(FF^T),$$
$$2\phi^2\|G^T\mathcal{P}G+\mathcal{T}_g\|_2,$$
$$3(1-\theta)^2\phi\lambda_{\max}(QFF^TQ)) + 2(1-\theta)^2\phi\lambda_{\max}(QGG^TQ) + \frac{1}{\phi}\big\},$$

then we have the following relationship.

Lemma 4.2 *Suppose the sequence* $\mathbf{u}^t = (\boldsymbol{\beta}^t, \mathbf{z}^t, \boldsymbol{\gamma}^t)$ *is generated by algorithm* (1) *and Assumption 1 holds, then* $\forall t \geq 0$,

$$\|\mathbf{u}^{t+1} - \bar{\mathbf{u}}\|_2^2 \leq q^2\|\mathbf{u}^{t+1} - \mathbf{u}^t\|_{\mathcal{H}}^2. \tag{4.54}$$

Lemma 4.3 *Suppose that Assumption 1 holds, and assume that both* $F^TF+\mathcal{T}_f$ *and* $G^TG+\mathcal{T}_g$ *are positive definite. Then for all k sufficiently large and* $\theta \in (0, \frac{1+\sqrt{5}}{2})$, *there exists* $\mu \in (0,1)$ *such that*

$$\|\mathbf{u}^{t+1} - \bar{\mathbf{u}}\|_{\mathcal{H}_1} + \|\mathbf{z}^{t+1} - \mathbf{z}^t\|_{G^T\mathcal{P}G+\mathcal{T}_g}^2 \leq \mu\Big(\|\mathbf{u}^t - \bar{\mathbf{u}}\|_{\mathcal{H}_1} + \|\mathbf{z}^t - \mathbf{z}^{t-1}\|_{G^T\mathcal{P}G+\mathcal{T}_g}^2\Big), \tag{4.55}$$

where

$$\mathcal{H}_1 = \begin{pmatrix} F^T(m_1Q+\mathcal{P})F+\mathcal{T}_f & m_1F^TQG & \cdots & \mathbf{m}0 \\ m_1G^TQF & G^T(\mathcal{P}+(m_1+1)Q)G & \cdots & \mathbf{m}0 \\ \mathbf{m}0 & \mathbf{m}0 & \cdots & \theta^{-1}\phi^{-2}\mathbf{m}I \end{pmatrix} \text{with} m_1 \in (0,1). \tag{4.56}$$

Proof From Theorem 3.1 in [66], we can derive the following results.

$$\begin{aligned}
&\Big\{\|\boldsymbol{\beta}^t - \bar{\boldsymbol{\beta}}\|_{F^T\mathcal{P}F+\mathcal{T}_f}^2 + \|\mathbf{z}^t - \bar{\mathbf{z}}\|_{G^T\mathcal{P}G+\mathcal{T}_g}^2 + \|\mathbf{z}^t - \mathbf{z}^{t-1}\|_{G^T\mathcal{P}G+\mathcal{T}_g}^2 \\
&+ (1-\min\{\theta, \frac{1}{\theta}\})\|Q(F\boldsymbol{\beta}^t + G\mathbf{z}^t - \mathbf{c})\|_2^2 + \theta^{-1}\phi^{-2}\|\boldsymbol{\gamma}^t - \bar{\boldsymbol{\gamma}}\|_2^2\Big\} \\
&- \Big\{\|\boldsymbol{\beta}^{t+1} - \bar{\boldsymbol{\beta}}\|_{F^T\mathcal{P}F+\mathcal{T}_f}^2 + \|\mathbf{z}^{t+1} - \bar{\mathbf{z}}\|_{G^T\mathcal{P}G+\mathcal{T}_g}^2 + \|\mathbf{z}^{t+1} - \mathbf{z}^t\|_{G^T\mathcal{P}G+\mathcal{T}_g}^2 \\
&+ (1-\min\{\theta, \frac{1}{\theta}\})\|Q(F\boldsymbol{\beta}^{t+1} + G\mathbf{z}^{t+1} - \mathbf{c})\|_2^2 + \theta^{-1}\phi^{-2}\|\boldsymbol{\gamma}^{t+1} - \bar{\boldsymbol{\gamma}}\|_2^2\Big\} \\
\geq& \|\mathbf{z}^{t+1} - \mathbf{z}^t\|_{G^T\mathcal{P}G+\mathcal{T}_g+(\theta-\theta^2+\min(\theta^2,1))GQ^TG}^2 + \|\boldsymbol{\beta}^{t+1} - \boldsymbol{\beta}^t\|_{F^T\mathcal{P}F+\mathcal{T}_f}^2 \\
&+ (1-\theta+\min\{\theta,\theta^{-1}\})\|Q(F\boldsymbol{\beta}^{t+1} + G\mathbf{z}^{t+1} - \mathbf{c})\|_2^2
\end{aligned} \tag{4.57}$$

When $\theta \in (0, \frac{1+\sqrt{5}}{2})$, it is ensured that $(1-\theta+\phi\min\{\theta,\theta^{-1}\}) > 0$. Let $d_1 \in (0, \frac{1}{2})$, then we have

$$
\begin{aligned}
&\Big\{\|\boldsymbol{\beta}^t-\bar{\boldsymbol{\beta}}\|^2_{F^T\mathcal{P}F+\mathcal{T}_f}+\|\mathbf{z}^t-\bar{\mathbf{z}}\|^2_{G^TG+\mathcal{T}_g}+\|\mathbf{z}^t-\mathbf{z}^{t-1}\|^2_{G^T\mathcal{P}G+\mathcal{T}_g}\\
&\quad+(1+d_1-d_1\theta-(1-d_1)\min\{\theta,\frac{1}{\theta}\})\|Q(F\boldsymbol{\beta}^t+G\mathbf{z}^t-\mathbf{c})\|_2^2\\
&\quad+\theta^{-1}\phi^{-2}\|\boldsymbol{\gamma}^t-\bar{\boldsymbol{\gamma}}\|_2^2\Big\}\\
&\quad-\Big\{\|\boldsymbol{\beta}^{t+1}-\bar{\boldsymbol{\beta}}\|^2_{F^T\mathcal{P}F+\mathcal{T}_f}+\|\mathbf{z}^{t+1}-\bar{\mathbf{z}}\|^2_{G^TG+\mathcal{T}_g}+\|\mathbf{z}^{t+1}-\mathbf{z}^t\|^2_{G^T\mathcal{P}G+\mathcal{T}_g}\\
&\quad+(1+d_1-d_1\theta-(1-d_1)\min\{\theta,\frac{1}{\theta}\})\|Q(F\boldsymbol{\beta}^{t+1}+G\mathbf{z}^{t+1}-\mathbf{c})\|_2^2\\
&\quad+\theta^{-1}\phi^{-2}\|\boldsymbol{\gamma}^{t+1}-\bar{\boldsymbol{\gamma}}\|_2^2\Big\}\\
&\geq\|\mathbf{z}^{t+1}-\mathbf{z}^t\|^2_{G^T\mathcal{P}G+\mathcal{T}_g+(\theta-\theta^2+\min(\theta^2,1))G^TQG}+\|\boldsymbol{\beta}^{t+1}-\boldsymbol{\beta}^t\|^2_{F^T\mathcal{P}F+\mathcal{T}_f}\\
&\quad+(1-d_1)(1-\theta+\min\{\theta,\theta^{-1}\})\|Q(F\boldsymbol{\beta}^{t+1}+G\mathbf{z}^{t+1}-\mathbf{c})\|_2^2\\
&\quad+d_1(1-\theta+\min\{\theta,\theta^{-1}\})\|Q(F\boldsymbol{\beta}^t+G\mathbf{z}^t-\mathbf{c})\|_2^2\\
&=\|\mathbf{z}^{t+1}-\mathbf{z}^t\|^2_{G^T\mathcal{P}G+\mathcal{T}_g+(\theta-\theta^2+\min(\theta^2,1))G^TQG}+\|\boldsymbol{\beta}^{t+1}-\boldsymbol{\beta}^t\|^2_{F^T\mathcal{P}F+\mathcal{T}_f}\\
&\quad+(1-2d_1))(1-\theta+\min\{\theta,\theta^{-1}\})\theta^{-2}\phi^{-2}\|\boldsymbol{\gamma}^{t+1}-\boldsymbol{\gamma}^t\|_2^2\\
&\quad+d_1(1-\theta+\min\{\theta,\theta^{-1}\})(\|Q(F\boldsymbol{\beta}^t+G\mathbf{z}^t-\mathbf{c})\|_2^2\\
&\quad+\|Q(F\boldsymbol{\beta}^{t+1}+G\mathbf{z}^{t+1}-\mathbf{c})\|_2^2)\\
&\geq\|\mathbf{z}^{t+1}-\mathbf{z}^t\|^2_{G^T\mathcal{P}G+\mathcal{T}_g+(\theta-\theta^2+\min(\theta^2,1))G^TQG}+\|\boldsymbol{\beta}^{t+1}-\boldsymbol{\beta}^t\|^2_{F^T\mathcal{P}F+\mathcal{T}_f}\\
&\quad+(1-2d_1)(1-\theta+\min\{\theta,\theta^{-1}\})\theta^{-2}\phi^{-2}\|\boldsymbol{\gamma}^{t+1}-\boldsymbol{\gamma}^t\|_2^2\\
&\quad+\frac{1}{2}d_1(1-\theta+\min\{\theta,\theta^{-1}\})\|QF(\boldsymbol{\beta}^{t+1}-\boldsymbol{\beta}^t)+QG(\mathbf{z}^{t+1}-\mathbf{z}^t)\|_2^2
\end{aligned}
\tag{4.58}
$$

Note that $Q(F\boldsymbol{\beta}^{t+1}+G\mathbf{z}^{t+1}-\mathbf{c})=QF(\boldsymbol{\beta}^{t+1}-\bar{\boldsymbol{\beta}})+QG(\mathbf{z}^{t+1}-\bar{\mathbf{z}})$, and we have

$$
\begin{aligned}
&\Big\{\|\boldsymbol{\beta}^t-\bar{\boldsymbol{\beta}}\|^2_{F^T\mathcal{P}F+\mathcal{T}_f}+\|\mathbf{z}^t-\bar{\mathbf{z}}\|^2_{G^TG+\mathcal{T}_g}+\|\mathbf{z}^t-\mathbf{z}^{t-1}\|^2_{G^T\mathcal{P}G+\mathcal{T}_g}\\
&\quad+(1+d_1-d_1\theta-(1-d_1)\min\{\theta,\frac{1}{\theta}\})\|QF(\boldsymbol{\beta}^t-\bar{\boldsymbol{\beta}})+QG(\mathbf{z}^t-\bar{\mathbf{z}})\|_2^2\\
&\quad+\theta^{-1}\phi^{-2}\|\boldsymbol{\gamma}^t-\bar{\boldsymbol{\gamma}}\|_2^2\Big\}\\
&\quad-\Big\{\|\boldsymbol{\beta}^{t+1}-\bar{\boldsymbol{\beta}}\|^2_{F^T\mathcal{P}F+\mathcal{T}_f}+\|\mathbf{z}^{t+1}-\bar{\mathbf{z}}\|^2_{G^TG+\mathcal{T}_g}+\|\mathbf{z}^{t+1}-\mathbf{z}^t\|^2_{G^T\mathcal{P}G+\mathcal{T}_g}\\
&\quad+(1+d_1-d_1\theta-(1-d_1)\min\{\theta,\frac{1}{\theta}\})\|QF(\boldsymbol{\beta}^{t+1}-\bar{\boldsymbol{\beta}})+QG(\mathbf{z}^{t+1}-\bar{\mathbf{z}}))\|_2^2\\
&\quad+\theta^{-1}\phi^{-2}\|\boldsymbol{\gamma}^{t+1}-\bar{\boldsymbol{\gamma}}\|_2^2\Big\}\\
&\geq\|\mathbf{z}^{t+1}-\mathbf{z}^t\|^2_{G^T\mathcal{P}G+\mathcal{T}_g+(\theta-\theta^2+\min(\theta^2,1))G^TQG}+\|\boldsymbol{\beta}^{t+1}-\boldsymbol{\beta}^t\|^2_{F^T\mathcal{P}F+\mathcal{T}_f}\\
&\quad+(1-2d_1)(1-\theta+\min\{\theta,\theta^{-1}\})\theta^{-2}\phi^{-2}\|\boldsymbol{\gamma}^{t+1}-\boldsymbol{\gamma}^t\|_2^2\\
&\quad+\frac{1}{2}d_1(1-\theta+\min\{\theta,\theta^{-1}\})\|QF(\boldsymbol{\beta}^{t+1}-\boldsymbol{\beta}^t)+QG(\mathbf{z}^{t+1}-\mathbf{z}^t)\|_2^2\\
&\geq(1-\theta+\min\{\theta,\theta^{-1}\})\min\{\frac{1}{2}d_1,1-2d_1\}\Big(\|\boldsymbol{\beta}^{t+1}-\boldsymbol{\beta}^t\|^2_{F^T\mathcal{P}F+\mathcal{T}_f}\\
&\quad+\|\mathbf{z}^{t+1}-\mathbf{z}^t\|^2_{G^TG+\mathcal{T}_g}+\theta^{-2}\phi^{-2}\|\boldsymbol{\gamma}^{t+1}-\boldsymbol{\gamma}^t\|_2^2\Big)
\end{aligned}
\tag{4.59}
$$

Let $m_1=1+d_1-d_1\theta-(1-d_1)\min\{\theta,\frac{1}{\theta}\}$ in $\mathcal{H}_1$ defined in (4.56), and $m_2=(1-\theta+\min\{\theta,\theta^{-1}\})\min\{\frac{1}{2}d_1,1-2d_1\}$. Note that when $\theta\in(0,\frac{1+\sqrt{5}}{2})$, the following relationship holds.

$$F^T F + T_f \succ 0 \text{ and } G^T G + T_g \succ 0 \Longleftrightarrow H_1 \succ 0.$$

Combining with Lemma 4.2, we have

$$\begin{aligned}
&\|\mathbf{u}^t - \bar{\mathbf{u}}\|_{\mathcal{H}_1} + \|\mathbf{z}^t - \mathbf{z}^{t-1}\|^2_{G^T\mathcal{P}G+\mathcal{T}_g} - (\|\mathbf{u}^{t+1} - \bar{\mathbf{u}}\|_{\mathcal{H}_1} + \|\mathbf{z}^{t+1} - \mathbf{z}^t\|^2_{G^T\mathcal{P}G+\mathcal{T}_g})\\
&\geq \frac{m_2}{C}\Big(C \times (\|\boldsymbol{\beta}^{t+1} - \boldsymbol{\beta}^t\|^2_{F^T\mathcal{P}F+\mathcal{T}_f} + \|\mathbf{z}^{t+1} - \mathbf{z}^t\|^2_{G^T\mathcal{P}G+\mathcal{T}_g}\\
&+ \theta^{-2}\phi^{-2}\|\boldsymbol{\gamma}^{t+1} - \boldsymbol{\gamma}^t\|_2^2\Big)\\
&= \frac{m_2}{C}\|\mathbf{m}u^{t+1} - \mathbf{m}u^t\|^2_{\mathcal{H}} \geq \frac{m_2 d_2}{Cq^2}\|\mathbf{u}^{t+1} - \bar{\mathbf{u}}\|_2^2 + \frac{m_2(1-d_2)}{Cq^2}\|\mathbf{z}^{t+1} - \mathbf{z}^t\|^2_{G^T\mathcal{P}G+\mathcal{T}_g}\\
&\geq \frac{m_2 d_2}{Cq^2\lambda_{\max}(\mathcal{H}_1)}\|\mathbf{u}^{t+1} - \bar{\mathbf{u}}\|^2_{\mathcal{H}_1} + \frac{m_2(1-d_2)}{Cq^2}\|\mathbf{z}^{t+1} - \mathbf{z}^t\|^2_{G^T\mathcal{P}G+\mathcal{T}_g}.
\end{aligned} \tag{4.60}$$

Take $d_2 = \frac{\lambda_{\max}(\mathcal{H}_1)}{1+\lambda_{\max}(\mathcal{H}_1)}$, then we can obtain (4.55) with $\mu = \left[1 + \frac{m_2}{Cq^2(1+\lambda_{\max}(\mathcal{H}_1))}\right]^{-1}$. □

Proof of Theorem 4.1

The proof of Theorem 4.1 is a direct application of Lemma 4.1, 4.2 and 4.3.

Proof (4.4) is a special case of (4.47) by setting

$$f(\boldsymbol{\beta}) = \|\boldsymbol{\alpha} \circ \boldsymbol{\beta}\|_1,\ g(\mathbf{z}) = \delta_{\mathcal{Z}_0}(\mathbf{z}),\ \mathbf{c} = (\hat{\boldsymbol{\mu}}, 0, \ldots, 0)^T,\ h(\mathbf{m}\omega) \equiv 0, \tag{4.61}$$

and let F, G and H: $\mathbb{R}^p \mapsto \mathbb{R}^{Kp}$ be linear mappings such that

$$\begin{aligned}
F(\boldsymbol{\beta}_1, \ldots, \boldsymbol{\beta}_K) &= \begin{pmatrix} \hat{\boldsymbol{\Sigma}}_1\boldsymbol{\beta}_1 \\ \hat{\boldsymbol{\Sigma}}_2\boldsymbol{\beta}_2 \\ \vdots \\ \hat{\boldsymbol{\Sigma}}_K\boldsymbol{\beta}_K \end{pmatrix}, \qquad G(\mathbf{z}) = \begin{pmatrix} -\mathbf{z} \\ 0 \\ \vdots \\ 0 \end{pmatrix},\\
H(\mathbf{m}\omega_2, \ldots, \mathbf{m}\omega_K) &= \begin{pmatrix} \mathbf{m}\omega_2 + \cdots + \mathbf{m}\omega_K \\ -\mathbf{m}\omega_2 \\ \vdots \\ -\mathbf{m}\omega_K \end{pmatrix}.
\end{aligned} \tag{4.62}$$

The augmented Lagrangian function for (4.4) is

$$
\begin{aligned}
\mathcal{L}\phi(\boldsymbol{\beta}, \mathbf{z}, \mathbf{m}\omega; \boldsymbol{\gamma}) = &\sum_{i=1}^{K} \|\boldsymbol{\alpha}_i \circ \boldsymbol{\beta}_i\|_1 + \delta_{\mathcal{Z}_0}(\mathbf{z}) \\
&+ \boldsymbol{\gamma}_1^T (\hat{\boldsymbol{\Sigma}}_1 \boldsymbol{\beta}_1 + \sum_{i=2}^{K} \mathbf{m}\omega_i - \mathbf{z} - \hat{\boldsymbol{\mu}}) \\
&+ \sum_{i=2}^{K} \boldsymbol{\gamma}_i^T (\hat{\boldsymbol{\Sigma}}_i \boldsymbol{\beta}_i - \mathbf{m}\omega_i) \\
&+ \frac{\phi}{2} \|\hat{\boldsymbol{\Sigma}}_1 \boldsymbol{\beta}_1 + \sum_{i=2}^{K} \mathbf{m}\omega_i - \mathbf{z} - \hat{\boldsymbol{\mu}}\|_2^2 \\
&+ \frac{\phi}{2} \sum_{i=2}^{K} \|\hat{\boldsymbol{\Sigma}}_i \boldsymbol{\beta}_i - \mathbf{m}\omega_i\|_2^2.
\end{aligned}
$$

$f = \|\cdot\|_1$ and $g = \delta_{\mathcal{Z}_0}(\mathbf{z})$ are piecewise linear-quadratic functions thus $\text{prox}_f(\cdot)$ and $\text{prox}_g(\cdot)$ are piecewise polyhedral [67], which implies Assumption 1 holds [68]. By Lemma 4.1, under the condition $\theta \in (0, (1+\sqrt{5})/2)$, the sequence $(\boldsymbol{\beta}^t, \mathbf{z}^t, \mathbf{m}\omega^t, \boldsymbol{\gamma}^t)$ generated by algorithm (1) converges to a limit point $(\bar{\boldsymbol{\beta}}, \bar{\mathbf{z}}, \bar{\mathbf{m}\omega}, \bar{\boldsymbol{\gamma}})$ with $(\bar{\boldsymbol{\beta}}, \bar{\mathbf{z}}, \bar{\mathbf{m}\omega})$ solving (4.4) and $\bar{\boldsymbol{\gamma}}$ is the dual optimal.

To derive the rate of convergence,

$$
\mathcal{P} = H(H^T H)^{-1} H^T = \begin{pmatrix} \frac{K-1}{K} I_p & \cdots & -\frac{1}{K} I_p & -\frac{1}{K} I_p \\ \vdots & \ddots & & \vdots \\ \vdots & & \ddots & \vdots \\ -\frac{1}{K} I_p & \cdots & \cdots & \frac{K-1}{K} I_p \end{pmatrix},
$$

then it follows that

$$
\begin{aligned}
&\|\boldsymbol{\beta}^{t+1} - \bar{\boldsymbol{\beta}}\|_{F^T \mathcal{P} F}^2 = \sum_{i=1}^{K} \|\hat{\boldsymbol{\Sigma}}_i (\boldsymbol{\beta}_i^{t+1} - \bar{\boldsymbol{\beta}}_i)\|_2^2 - \frac{1}{K+1} \|\sum_{i=0}^{K} \hat{\boldsymbol{\Sigma}}_i (\boldsymbol{\beta}_i^{t+1} - \bar{\boldsymbol{\beta}}_i)\|_2^2, \\
&m_1 \|QF(\boldsymbol{\beta}^{t+1} - \bar{\boldsymbol{\beta}}) + QG(\mathbf{z}^{t+1} - \bar{\mathbf{z}})\|_2^2 \\
&\quad = \frac{m_1}{K+1} \|\sum_{i=1}^{K} \hat{\boldsymbol{\Sigma}}_i (\boldsymbol{\beta}_i^{t+1} - \bar{\boldsymbol{\beta}}_i) + (\mathbf{z}^{t+1} - \bar{\mathbf{z}})\|_2^2, \\
&\|\mathbf{z}^{t+1} - \bar{\mathbf{z}}\|_{G^T G + \mathcal{T}_g}^2 = \|\mathbf{z}^{t+1} - \bar{\mathbf{z}}\|_2^2, \\
&\|\mathbf{z}^{t+1} - \mathbf{z}^t\|_{G^T \mathcal{P} G + \mathcal{T}_g}^2 = \frac{K}{K+1} \|\mathbf{z}^t - \mathbf{z}^{t-1}\|_2^2.
\end{aligned}
\tag{4.63}
$$

Since $\mathbf{u}^t = (\boldsymbol{\beta}^t, \mathbf{z}^t, \boldsymbol{\gamma}^t)$, and in (4.63) we derive the explicit forms of $\|\boldsymbol{\beta}^{t+1} - \bar{\boldsymbol{\beta}}\|_{F^T \mathcal{P} F}^2$, $\|\mathbf{z}^{t+1} - \bar{\mathbf{z}}\|_{G^T G + \mathcal{T}_g}^2$, $\|\mathbf{z}^{t+1} - \mathbf{z}^t\|_{G^T \mathcal{P} G + \mathcal{T}_g}^2$ and $m_1 \|QF(\boldsymbol{\beta}^{t+1} - \bar{\boldsymbol{\beta}}) + QG(\mathbf{z}^{t+1} - \bar{\mathbf{z}})\|_2^2$ for LPO problem, then plug them back to (4.60) will point to the conclusion of Theorem 4.1. □

Proof of Theorem 4.2

Proof Assume that none of the E_1, E_2, and E_3 is true, which is of the probability $1 - P_1 - P_2 - P_3$ at least. Then since $\|\boldsymbol{\beta}^*_{\mathcal{A}}\|_{\min} > (a+1)\lambda$, we have that

$$
\begin{aligned}
|\hat{\beta}_j^{\text{LPO}}| &\le \|\hat{\boldsymbol{\beta}}^{\text{LPO}} - \boldsymbol{\beta}^*\|_{\max} \le a_0\lambda \le a_2\lambda, \quad \text{for } j \in \mathcal{A}^c, \\
|\hat{\beta}_j^{\text{LPO}}| &\ge \|\boldsymbol{\beta}^*_{\mathcal{A}}\|_{\min} - \|\hat{\boldsymbol{\beta}}^{\text{LPO}} - \boldsymbol{\beta}^*\|_{\max} > a\lambda, \quad \text{for } j \in \mathcal{A}.
\end{aligned} \tag{4.64}
$$

Consider the properties of the folded concave penalty, we have that

$$
\begin{aligned}
&\rho'(|\hat{\beta}_j^{\text{LPO}}|) \ge a_1, \quad \text{for } j \in \mathcal{A}^c, \\
&\rho'(|\hat{\beta}_j^{\text{LPO}}|) = 0 \ \text{ for } j \in \mathcal{A}.
\end{aligned} \tag{4.65}
$$

We derive $L(\boldsymbol{\beta})$ for $j \in \mathcal{A}$,

$$
\begin{aligned}
\nabla_j L(\boldsymbol{\beta}) &= [\boldsymbol{\beta}^T\hat{\boldsymbol{\Sigma}}\boldsymbol{\beta}]^{-2}\big\{-[\boldsymbol{\beta}^T\hat{\boldsymbol{\Sigma}}\boldsymbol{\beta}]\frac{d[\boldsymbol{\beta}^T S_B\boldsymbol{\beta}]}{d\boldsymbol{\beta}_j} + [\boldsymbol{\beta}^T S_B\boldsymbol{\beta}]\frac{d[\boldsymbol{\beta}^T\hat{\boldsymbol{\Sigma}}\boldsymbol{\beta}]}{d\boldsymbol{\beta}_j}\big\} \\
&= [\boldsymbol{\beta}^T\hat{\boldsymbol{\Sigma}}\boldsymbol{\beta}]^{-2}\big\{-[\boldsymbol{\beta}^T\hat{\boldsymbol{\Sigma}}\boldsymbol{\beta}]2\hat{\mu}_j\hat{\boldsymbol{\mu}}^T\boldsymbol{\beta} + [\boldsymbol{\beta}^T S_B\boldsymbol{\beta}]2\hat{\boldsymbol{\Sigma}}_j^T\boldsymbol{\beta}\big\}.
\end{aligned} \tag{4.66}
$$

Then we equate $\nabla_j L(\boldsymbol{\beta})$ to zero and obtain

$$
\begin{aligned}
&\nabla_j L(\boldsymbol{\beta}) = 0 \\
\iff &[\boldsymbol{\beta}^T\hat{\boldsymbol{\Sigma}}\boldsymbol{\beta}]\hat{\mu}_j\hat{\boldsymbol{\mu}}^T\boldsymbol{\beta} - [\boldsymbol{\beta}^T S_B\boldsymbol{\beta}]\hat{\boldsymbol{\Sigma}}_j^T\boldsymbol{\beta} = 0 \\
\iff &\hat{\mu}_j\hat{\boldsymbol{\delta}}^T\boldsymbol{\beta} - \frac{[\boldsymbol{\beta}^T S_B\boldsymbol{\beta}]}{[\boldsymbol{\beta}^T\hat{\boldsymbol{\Sigma}}\boldsymbol{\beta}]}\hat{\boldsymbol{\Sigma}}_j^T\boldsymbol{\beta} = 0 \\
\iff &\hat{\mu}_j\hat{\boldsymbol{\mu}}^T\boldsymbol{\beta} + L(\boldsymbol{\beta})\hat{\boldsymbol{\Sigma}}_j^T\boldsymbol{\beta} = 0.
\end{aligned} \tag{4.67}
$$

Therefore $\hat{\boldsymbol{\beta}}^{\text{oracle}} = (\hat{\boldsymbol{\beta}}_{\mathcal{A}}^{\text{oracle}}, \mathbf{m}0)$ is the solution to

$$
\begin{aligned}
&\min_{\boldsymbol{\beta}:\boldsymbol{\beta}_{\mathcal{A}^c}=\mathbf{m}0} \ L(\boldsymbol{\beta}) \\
&\text{s.t.} \ \ \hat{\boldsymbol{\mu}}_{\mathcal{A}}\hat{\boldsymbol{\mu}}^T\boldsymbol{\beta} + L(\boldsymbol{\beta})\hat{\boldsymbol{\Sigma}}_{\mathcal{A}}^T\boldsymbol{\beta} = \mathbf{m}0,
\end{aligned} \tag{4.68}
$$

where $\hat{\boldsymbol{\Sigma}}_{\mathcal{A}}$ denotes the sub-matrix of $\hat{\boldsymbol{\Sigma}}$ whose column indices are in $\mathcal{A}$. This is a generalized eigenvalue problem, and the solution is given by

$$
\hat{\boldsymbol{\beta}}_{\mathcal{A}} = \hat{\boldsymbol{\Sigma}}_{\mathcal{A},\mathcal{A}}^{-1}\hat{\boldsymbol{\mu}}_{\mathcal{A}}, \quad \hat{\boldsymbol{\beta}}_{\mathcal{A}^c} = \mathbf{m}0, \tag{4.69}
$$

where $\hat{\boldsymbol{\Sigma}}_{\mathcal{A},\mathcal{A}} = (\hat{\Sigma}_{i,j})_{i\in\mathcal{A},j\in\mathcal{A}}$. Consider the LLA problem,

$$\begin{aligned} &\min_{\boldsymbol{\beta}\in R^p} \quad \sum_j \rho'(|\hat{\beta}_j^0|)|\beta_j| \\ &\text{s.t} \quad |\hat{\boldsymbol{\Sigma}}_j^T \boldsymbol{\beta} - \hat{\mu}_j| \le \lambda \rho'(|\hat{\beta}_j^0|), \quad j = 1, \ldots, p. \end{aligned} \tag{4.70}$$

Given that (4.65) holds, it remains to prove that $\hat{\boldsymbol{\beta}}^{\text{oracle}} = (\hat{\boldsymbol{\beta}}_{\mathcal{A}}, \mathbf{m}0)$ satisfies the following conditions:

(B1') $\hat{\boldsymbol{\Sigma}}_{\mathcal{A}}^T \boldsymbol{\beta} - \hat{\boldsymbol{\mu}}_{\mathcal{A}} = 0$;
(B2') $|\hat{\boldsymbol{\Sigma}}_j^T \boldsymbol{\beta} - \hat{\mu}_j| \le \lambda \rho'(|\hat{\beta}_j^{(0)}|)$, for $j \in \mathcal{A}^c$.

For $j \in \mathcal{A}^c$, since $\rho'(|\hat{\beta}_j^{(0)}|) \ge a_1$, we have

$$\begin{aligned} &\left\{|\hat{\boldsymbol{\Sigma}}^T \boldsymbol{\beta} - \hat{\mu}_j| \le a_1 \lambda\right\} \\ \subseteq &\left\{|\hat{\boldsymbol{\Sigma}}_j^T \boldsymbol{\beta} - \hat{\mu}_j| \le \lambda \rho'(|\hat{\beta}_j^{(0)}|)\right\}. \end{aligned} \tag{4.71}$$

Therefore when E_3 is not true, $\hat{\boldsymbol{\beta}}^{\text{oracle}}$ satisfies (B2'). When $\hat{\boldsymbol{\beta}}_{\mathcal{A}^c} = \mathbf{m}0$,

$$\hat{\boldsymbol{\Sigma}}_{\mathcal{A}}^T \boldsymbol{\beta} - \hat{\boldsymbol{\mu}}_{\mathcal{A}} = 0 \Longleftrightarrow \hat{\boldsymbol{\Sigma}}_{\mathcal{A},\mathcal{A}} \boldsymbol{\beta}_{\mathcal{A}} - \hat{\boldsymbol{\mu}}_{\mathcal{A}} = \mathbf{m}0. \tag{4.72}$$

By (4.69), we have $\hat{\boldsymbol{\beta}}_{\mathcal{A}}^{\text{oracle}} = \hat{\boldsymbol{\Sigma}}_{\mathcal{A},\mathcal{A}}^{-1} \hat{\boldsymbol{\mu}}_{\mathcal{A}}$. Therefore $\hat{\boldsymbol{\beta}}^{\text{oracle}}$ satisfies (B1').

Denote the solution of the next iteration of the LLA algorithm as $\hat{\boldsymbol{\beta}}^{(2)}$. Since $\hat{\boldsymbol{\beta}}_{\mathcal{A}^c}^{\text{oracle}} = \mathbf{m}0$, we have $\rho'(|\hat{\beta}_j^{\text{oracle}}|) = \rho'(0) > a_1$ for $j \in \mathcal{A}^c$. $\rho'(|\hat{\beta}_j^{\text{oracle}}|) = 0$ for $j \in \mathcal{A}$ under the event $\{\|\hat{\boldsymbol{\beta}}_{\mathcal{A}}^{\text{oracle}}\|_{\min} > a\lambda\}$. Therefore, we have $\hat{\boldsymbol{\beta}}^{(2)} = \hat{\boldsymbol{\beta}}^{(oracle)}$ is the unique minimizer of the following problem:

$$\begin{aligned} &\min_{\boldsymbol{\beta}\in R^p} \quad \sum_j \rho'(|\hat{\beta}_j^{\text{oracle}}|)|\beta_j| \\ &\text{s.t} \quad |\hat{\boldsymbol{\Sigma}}_j^T \boldsymbol{\beta} - \hat{\mu}_j| \le \lambda \rho'(|\hat{\beta}_j^{\text{oracle}}|), \quad j = 1, \ldots, p. \end{aligned} \tag{4.73}$$

Hence, the LLA algorithm finds the oracle estimator again and stays there. Next we show that $P_i \to 0, i = 1, 2, 3$ as $n \to \infty$.

According to the sign consistency result from Theorem 2 in [51], it is straightforward to derive that when $\lambda \gg (\log(pq)q^2/n)^{1/2}$, $P_1 \to 0$ as $n \to \infty$.

Since $|\hat{\beta}_j^{\text{oracle}}| \ge \|\boldsymbol{\beta}_{\mathcal{A}}^*\|_{\min} - \|\hat{\boldsymbol{\beta}}^{\text{oracle}} - \boldsymbol{\beta}^*\|_{\max}$ for $j \in \mathcal{A}$,

$$\begin{aligned} P_2 &= P(\|\hat{\boldsymbol{\beta}}_{\mathcal{A}}^{\text{oracle}}\|_{\min} \le a\lambda) \\ &\le P(\|\boldsymbol{\beta}_{\mathcal{A}}^*\|_{\min} - \|\hat{\boldsymbol{\beta}}^{\text{oracle}} - \boldsymbol{\beta}^*\|_{\max} \le a\lambda) \\ &\le P(\|\hat{\boldsymbol{\beta}}^{\text{oracle}} - \boldsymbol{\beta}^*\|_{\max} \ge \lambda). \end{aligned} \tag{4.74}$$

From the Theorem 2 of [49], we have that $\|\hat{\boldsymbol{\beta}}^{\text{oracle}} - \boldsymbol{\beta}^*\|_{\max} \leq O(\lambda_n)$ with probability going to 1 if $\lambda_n \ll \|\boldsymbol{\beta}^*_{\mathcal{A}}\|_{\min}$ and $\lambda_n \gg O\big((\log(pq)q^2/n)^{1/2}\big)$. Thus by the assumption that $\lambda \gg (\log(pq)q^2/n)^{1/2}$ and $\|\boldsymbol{\beta}^*_{\mathcal{A}}\|_{\min} > (a+1)\lambda$, we have that $P_2 \to 0$ as $n \to \infty$.

An intermediate result of Theorem 2 in [51] implies that $\|\hat{\boldsymbol{\Sigma}}^T_{\mathcal{A}^c}\hat{\boldsymbol{\beta}}^{\text{oracle}} - \hat{\boldsymbol{\mu}}_{\mathcal{A}^c}\|_{\max} = O\Big(\sqrt{\frac{\log\big((p-q)\log n\big)}{n}}\Big)$ with probability at least $1 - O(\log^{-1}(n))$, therefore we also have

$$P_3 = P(\|\hat{\boldsymbol{\Sigma}}^T_{\mathcal{A}^c}\hat{\boldsymbol{\beta}}^{\text{oracle}} - \hat{\boldsymbol{\mu}}_{\mathcal{A}^c}\|_{\max} > a_1\lambda) \to 0, \tag{4.75}$$

as $n \to 0$. □

Proof of Theorem 4.3

Let $\boldsymbol{\beta} = \boldsymbol{\Sigma}^{-1}\boldsymbol{\mu}$, we have

$$\mathbf{w}_{opt} = \mu_0 \frac{\boldsymbol{\beta}}{\boldsymbol{\mu}^T\boldsymbol{\beta}}, \tag{4.76}$$

Then

$$\begin{aligned} \hat{\mathbf{w}}_{opt} - \mathbf{w}_{opt} &= \mu_0\Big[\frac{\hat{\boldsymbol{\beta}}^{oracle} - \boldsymbol{\beta}^*}{\hat{\boldsymbol{\mu}}^T\hat{\boldsymbol{\beta}}^{oracle}} + \frac{\boldsymbol{\beta}^*(\boldsymbol{\mu}^T\boldsymbol{\beta}^* - \hat{\boldsymbol{\mu}}^T\hat{\boldsymbol{\beta}}^{oracle})}{\hat{\boldsymbol{\mu}}^T\hat{\boldsymbol{\beta}}^{oracle}\ \boldsymbol{\mu}^T\boldsymbol{\beta}^*}\Big] \\ &= \mu_0\Big[\frac{\hat{\boldsymbol{\beta}}^{oracle} - \boldsymbol{\beta}^*}{\hat{\boldsymbol{\mu}}^T\hat{\boldsymbol{\beta}}^{oracle}} + \frac{\boldsymbol{\beta}^*(\boldsymbol{\Theta} - \hat{\boldsymbol{\mu}}^T\hat{\boldsymbol{\beta}}^{oracle})}{\hat{\boldsymbol{\mu}}^T\hat{\boldsymbol{\beta}}^{oracle}\ \boldsymbol{\Theta}}\Big] \end{aligned} \tag{4.77}$$

$$\|(\boldsymbol{\beta}^*)^T\hat{\Sigma}\hat{\boldsymbol{\beta}}^{oracle} - (\boldsymbol{\beta}^*)^T\boldsymbol{\mu}\|_{\max} = \|(\boldsymbol{\beta}^*)^T\hat{\Sigma}\hat{\boldsymbol{\beta}}^{oracle} - (\boldsymbol{\beta}^*)^T\hat{\boldsymbol{\mu}} + (\boldsymbol{\beta}^*)^T\hat{\boldsymbol{\mu}} - (\boldsymbol{\beta}^*)^T\boldsymbol{\mu}\|_{\max}$$

then

$$\|(\boldsymbol{\beta}^*)^T\hat{\Sigma}\hat{\boldsymbol{\beta}}^{oracle} - (\boldsymbol{\beta}^*)^T\hat{\boldsymbol{\mu}} + (\boldsymbol{\beta}^*)^T\hat{\boldsymbol{\mu}} - (\boldsymbol{\beta}^*)^T\boldsymbol{\mu}\|_{\max} \leq \|\boldsymbol{\beta}^*\|_1(\|\hat{\boldsymbol{\mu}} - \boldsymbol{\mu}\|_{\max} + \|\hat{\Sigma}\hat{\boldsymbol{\beta}}^{oracle} - \hat{\boldsymbol{\mu}}\|_{\max}).$$

By the definition of $\hat{\boldsymbol{\beta}}^{oracle}$, we have

$$\|(\boldsymbol{\beta}^*)^T\hat{\Sigma}\hat{\boldsymbol{\beta}}^{oracle} - (\boldsymbol{\beta}^*)^T\boldsymbol{\mu}\|_{\max} \leq (\lambda + \|\hat{\boldsymbol{\mu}} - \boldsymbol{\mu}\|_{\max})\|\boldsymbol{\beta}^*\|_1, \tag{4.78}$$

and similarly we can have

$$\|(\boldsymbol{\beta}^*)^T\hat{\Sigma}\hat{\boldsymbol{\beta}}^{oracle} - \boldsymbol{\mu}^T\hat{\boldsymbol{\beta}}^{oracle}\|_{\max} \leq (\lambda + \|\hat{\boldsymbol{\mu}} - \boldsymbol{\mu}\|_{\max})\|\boldsymbol{\beta}^*\|_1. \tag{4.79}$$

Based on the above two inequalities and condition C3.1 and C3.2, we can have

$$\|\boldsymbol{\mu}^T \hat{\boldsymbol{\beta}}^{oracle} - \boldsymbol{\Theta}\|_{\max} \le 2 * (\lambda + \|\hat{\boldsymbol{\mu}} - \boldsymbol{\mu}\|_{\max})\|\boldsymbol{\beta}^*\|_1 \le o_p(1), \tag{4.80}$$

$$\|\hat{\boldsymbol{\mu}}^T \hat{\boldsymbol{\beta}}^{oracle} - \boldsymbol{\Theta}\|_{\max} \le (2\lambda + 3\|\hat{\boldsymbol{\mu}} - \boldsymbol{\mu}\|_{\max})\|\boldsymbol{\beta}^*\|_1 \le o_p(1). \tag{4.81}$$

We also have

$$\kappa \le \Theta \le \hat{\boldsymbol{\mu}}^T \hat{\boldsymbol{\beta}}^{oracle} + \|\hat{\boldsymbol{\mu}}^T \hat{\boldsymbol{\beta}}^{oracle} - \boldsymbol{\Theta}\|_{\max}, \tag{4.82}$$

so

$$\hat{\boldsymbol{\mu}}^T \hat{\boldsymbol{\beta}}^{oracle} \ge \kappa - o_p(1). \tag{4.83}$$

Based on (4.81), (4.83), condition C3.1 and C3.2, and $\|\hat{\boldsymbol{\beta}}^{\text{oracle}} - \boldsymbol{\beta}^*\|_{\max} = O(\sqrt{\frac{\log(q\log n)}{n}})$ with probability at least $1 - O(\log^{-1}(n))$ from the proof of Theorem 4.2, we can conclude $\|\hat{\mathbf{w}}_{opt} - \mathbf{w}_{opt}\|_{\max} \to 0$ in probability.

Sub-Problems in Algorithm 4

In this subsection, we derive the updates for $\boldsymbol{\beta}$, $\mathbf{z}$ and $\mathbf{m}\omega$ in Algorithm 4. For ease of notation, define a set of functions f, g, h.

$$f(\boldsymbol{\beta}) = n\lambda \sum_{g=1}^{G} \|\boldsymbol{\alpha}_g \circ \boldsymbol{\beta}_g\|_1, \;\; h(\mathbf{m}\omega) = \mathbf{m}0, \;\; g(\mathbf{z}) = \tau \mathbf{m}1^T(\mathbf{z})_+ + (1-\tau)\mathbf{m}1^T(\mathbf{z})_-, \tag{4.84}$$

Thus, f, g, h are closed proper convex functions. Further define matrices $\mathbf{m}F$, G, H

$$\mathbf{m}F = Diag(\mathbf{X}_1, \mathbf{X}_2, \ldots, \mathbf{X}_G), \quad \mathbf{m}G = (I_n, 0, \ldots, 0)^T,$$
$$\mathbf{m}H = \begin{pmatrix} I_n & I_n & \cdots & I_n \\ -I_n & 0 & \cdots & 0 \\ \vdots & \ddots & & \vdots \\ 0 & \cdots & -I_n & 0 \\ 0 & 0 & \cdots & -I_n \end{pmatrix}. \tag{4.85}$$

Then Problem (4.31) can be expressed as a general three-block constrained optimization problem,

$$\min_{\boldsymbol{\beta}, \mathbf{z}, \mathbf{m}\omega} \{f(\boldsymbol{\beta}) + g(\mathbf{z}) + h(\mathbf{m}\omega) \mid \mathbf{m}F\boldsymbol{\beta} + \mathbf{m}G\mathbf{z} + \mathbf{m}H\mathbf{m}\omega = \mathbf{c}\}, \tag{4.86}$$

where by definitions of $\mathbf{m}F$, $\mathbf{m}G$ and $\mathbf{m}H$,

$$\mathbf{m}F\boldsymbol{\beta} = \begin{pmatrix} \mathbf{X}_1\boldsymbol{\beta}_1 \\ \mathbf{X}_2\boldsymbol{\beta}_2 \\ \vdots \\ \mathbf{X}_G\boldsymbol{\beta}_G \end{pmatrix}, \quad \mathbf{m}G\mathbf{z} = \begin{pmatrix} \mathbf{z} \\ 0 \\ \vdots \\ 0 \end{pmatrix}, \quad \mathbf{m}H\mathbf{m}\omega = \begin{pmatrix} \mathbf{m}\omega_2 + \cdots + \mathbf{m}\omega_G \\ -\mathbf{m}\omega_2 \\ \vdots \\ -\mathbf{m}\omega_G \end{pmatrix}. \tag{4.87}$$

As in sGS-sPADMM proposed by [43], we update the three-block variables using a special cycle,

$$\begin{cases} \boldsymbol{\beta}^{k+1} & = \operatorname{argmin} \mathcal{L}_\phi(\boldsymbol{\beta}, \mathbf{z}^k, \mathbf{m}\omega^k; \boldsymbol{\gamma}^k) + (\frac{\phi}{2}\|\boldsymbol{\beta} - \boldsymbol{\beta}^k\|^2_{\mathcal{T}_f}), \\ \mathbf{m}\omega^{k+\frac{1}{2}} & = (\mathbf{m}H^T\mathbf{m}H)^{-1}\mathbf{m}H(\mathbf{c} - \mathbf{m}F\boldsymbol{\beta}^{k+1} - \mathbf{m}G\mathbf{z}^k), \\ \mathbf{z}^{k+1} & = \operatorname{argmin} \mathcal{L}_\phi(\boldsymbol{\beta}^{k+1}, \mathbf{z}, \mathbf{m}\omega^{k+\frac{1}{2}}; \boldsymbol{\gamma}^k) + (\frac{\phi}{2}\|\mathbf{z} - \mathbf{z}^k\|^2_{\mathcal{T}_h}), \\ \mathbf{m}\omega^{k+1} & = (\mathbf{m}H^T\mathbf{m}H)^{-1}\mathbf{m}H(\mathbf{c} - \mathbf{m}F\boldsymbol{\beta}^{k+1} - \mathbf{m}G\mathbf{z}^{k+1}), \\ \boldsymbol{\gamma}^{k+1} & = \boldsymbol{\gamma}^k + \theta\phi(\mathbf{m}F\boldsymbol{\beta}^{k+1} + \mathbf{m}G\mathbf{z}^{k+1} + \mathbf{m}H\mathbf{m}\omega^{k+1} - \mathbf{c}), \end{cases} \tag{4.88}$$

where $\mathcal{T}_f$ and $\mathcal{T}_h$ are optionally added self-adjoint positive semidefinite operators. To update $\mathbf{m}\omega$, we need to compute $(\mathbf{m}H^T\mathbf{m}H)^{-1}$. Since

$$\mathbf{m}H^T\mathbf{m}H = \begin{pmatrix} I_n & & \\ & \ddots & \\ & & I_n \end{pmatrix} + \begin{pmatrix} I_n \\ \vdots \\ I_n \end{pmatrix}\begin{pmatrix} I_n \\ \vdots \\ I_n \end{pmatrix}^T,$$

we apply the Sherman-Morrison-Woodebury formula to compute $(\mathbf{m}H^T\mathbf{m}H)^{-1}$ and it follows that

$$\begin{aligned} \mathbf{m}\omega_i^{k+\frac{1}{2}} &= (\mathbf{m}H^T\mathbf{m}H)^{-1}\mathbf{m}H^T(\mathbf{c} - \mathbf{m}F\boldsymbol{\beta}^{k+1} - \mathbf{m}G\mathbf{z}^k) \\ &= \frac{1}{G}(\mathbf{y} - \mathbf{z}^k + G\mathbf{X}_i\boldsymbol{\beta}_i^{k+1} - \sum_{j=1}^{G}\mathbf{X}_j\boldsymbol{\beta}_j^{k+1}), \quad i = 2, \ldots, G, \end{aligned} \tag{4.89}$$

In the $\mathbf{z}$-subproblem, we set $\mathcal{T}_h = 0$ and then we have

$$\begin{aligned} \mathbf{z}^{k+1} &= \operatorname*{argmin}_{\mathbf{z}} \mathcal{L}_\phi(\boldsymbol{\beta}^{k+1}, \mathbf{z}, \mathbf{m}\omega^{k+\frac{1}{2}}; \boldsymbol{\gamma}^k) \\ &= \operatorname*{argmin}_{\mathbf{z}} \left\{ \frac{1}{n}\sum_{i=1}^{n}\rho_\tau(z_i) + \boldsymbol{\gamma}_1^T(\mathbf{X}_1\boldsymbol{\beta}_1^{k+1} + \mathbf{z} + \sum_{i=2}^{G}\mathbf{m}\omega_i^{k+\frac{1}{2}} - \mathbf{y}) \right. \\ &\qquad \left. + \frac{\phi}{2}\|\mathbf{X}_1\boldsymbol{\beta}_1^{k+1} + \mathbf{z} + \sum_{i=2}^{G}\mathbf{m}\omega_i^{k+\frac{1}{2}} - \mathbf{y}\|_2^2 \right\} \\ &= \operatorname*{argmin}_{\mathbf{z}} \frac{1}{n}\sum_{i=1}^{n}\rho_\tau(z_i) + \frac{\phi}{2}\|\mathbf{X}_1\boldsymbol{\beta}_1^{k+1} + \mathbf{z} + \sum_{i=2}^{G}\mathbf{m}\omega_i^{k+\frac{1}{2}} - \mathbf{y} + \frac{\boldsymbol{\gamma}_1^k}{\phi}\|_2^2. \end{aligned} \tag{4.90}$$

The closed-form solution of the **z**-subproblem can be easily derived as

$$\mathbf{z}^{k+1} = \max\Big(\mathbf{y} - \mathbf{X}_1\boldsymbol{\beta}_1^{k+1} - \sum_{i=2}^{G}\mathbf{m}\omega_i^{k+\frac{1}{2}} - \frac{\boldsymbol{\gamma}_1^k}{\phi} - \frac{\tau}{n\phi}, 0\Big) \\ - \max\Big(-(\mathbf{y} - \mathbf{X}_1\boldsymbol{\beta}_1^{k+1} - \sum_{i=2}^{G}\mathbf{m}\omega_i^{k+\frac{1}{2}} - \frac{\boldsymbol{\gamma}_1^k}{\phi} - \frac{\tau}{n\phi} + \frac{1}{n\phi}), 0\Big). \tag{4.91}$$

Proof of Theorem 4.4

We first show Lemmas 4.4, 4.5 and 4.6, which are used in the proof of Theorem 1. From (4.85), we have Fact 1 below.

Fact 4.1 $\mathbf{m}H^T\mathbf{m}H$ *is positive definite.*

Assumptions 2 are imposed to obtain theoretical guarantee on feasibility and convergence of the sequence $(\boldsymbol{\beta}^k, \mathbf{z}^k, \mathbf{m}\omega^k, \boldsymbol{\gamma}^k)$.

Assumption 2 There exists $(\hat{\boldsymbol{\beta}}, \hat{\mathbf{z}}, \hat{\mathbf{m}\boldsymbol{\omega}}) \in \mathbb{R}^{p_1} \times \mathbb{R}^{p_2} \times \mathbb{R}^{p_3}$, such that $\mathbf{m}F\hat{\boldsymbol{\beta}} + \mathbf{m}G\hat{\mathbf{z}} + \mathbf{m}H\hat{\mathbf{z}}\boldsymbol{\omega} = \mathbf{c}$.

For algorithm (4.88), the projection matrix $\mathcal{P} = \mathbf{m}H(\mathbf{m}H^T\mathbf{m}H)^{-1}\mathbf{m}H^T$ plays an important role in the convergence analysis. Let $Q = I - \mathcal{P}$. Since $\mathbf{m}\omega$ can be expressed as $\mathbf{m}\omega(\boldsymbol{\beta}, \mathbf{z}) = (\mathbf{m}H^T\mathbf{m}H)^{-1}\mathbf{m}H^T(\mathbf{c} - \mathbf{m}F\boldsymbol{\beta} - \mathbf{m}G\mathbf{z})$, it follows that $\mathbf{m}H\mathbf{m}\omega = \mathcal{P}(\mathbf{c} - \mathbf{m}F\boldsymbol{\beta} - \mathbf{m}G\mathbf{z})$. Given that $h(\mathbf{m}\omega) = \mathbf{m}0$ in our case, we can now rewrite (4.86) as

$$\min_{\boldsymbol{\beta},\mathbf{z}}\{f(\boldsymbol{\beta}) + g(\mathbf{z}) \mid Q(\mathbf{m}F\boldsymbol{\beta} + \mathbf{m}G\mathbf{z} - \mathbf{c}) = 0\}. \tag{4.92}$$

Stopping Criterion. In the implementation of Algorithm 4, we use the same stopping criterion as the one introduced in [40]. The primal and dual residuals are often used in characterizing the convergence stage. Define $\mathbf{m}r^{k+1} = (\|\mathbf{X}_1\boldsymbol{\beta}_1^{k+1} + \mathbf{z}^{k+1} + \mathbf{m}\omega_2^{k+1} + \cdots + \mathbf{m}\omega_G^{k+1} - \mathbf{y}\|_2^2 + \|\sum_{g=2}^{G}(\mathbf{X}_g\boldsymbol{\beta}_g^{k+1} - \mathbf{m}\omega_g^{k+1})\|_2^2)^{0.5}$ as the primal residual and $\mathbf{m}s^{k+1} = \|\phi/G(\mathbf{X}_1^T, \ldots, \mathbf{X}_G^T)^T(\mathbf{z}^{k+1} - \mathbf{z}^k)\|_2$ as the dual residual at the $(k+1)$th iteration. The termination criterion is

$$\|\mathbf{m}r^k\|_2 \le \epsilon^{\text{pri}} \text{ and } \|\mathbf{m}s^k\|_2 \le \epsilon^{\text{dual}}, \tag{4.93}$$

where $\epsilon^{\text{pri}} > 0$ and $\epsilon^{\text{dual}} > 0$ are feasibility tolerances chosen as

$$\epsilon^{\text{pri}} = \sqrt{n}\epsilon^{\text{abs}} + \frac{\epsilon^{\text{rel}}}{\sqrt{G}}\max(\|\mathbf{X}\boldsymbol{\beta}^k\|_2, \|\mathbf{z}^k\|_2, \|\mathbf{c}\|_2), \text{ and } \epsilon^{\text{dual}} = \sqrt{p}\epsilon^{\text{abs}} + \frac{\epsilon^{\text{rel}}}{G}\|\mathbf{m}F^TQ\boldsymbol{\gamma}^k\|_2.$$

A common choice could be $\epsilon^{\text{abs}} = 0.001$ and $\epsilon^{\text{rel}} = 0.001$.

The augmented Lagrangian function for (4.92) is given by

$$\mathcal{L}_\phi(\mathbf{z}, \boldsymbol{\beta}; \boldsymbol{\gamma}) = f(\boldsymbol{\beta}) + g(\mathbf{z}) + \langle \boldsymbol{\gamma}, Q(\mathbf{m}F\boldsymbol{\beta} + \mathbf{m}G\mathbf{z} - \mathbf{c})\rangle + \frac{\phi}{2}\|Q(\mathbf{m}F\boldsymbol{\beta} + \mathbf{m}G\mathbf{z} - \mathbf{c})\|_2^2.$$

Using similar arguments in [43], it follows that applying the updates in 4.88 to problem (4.86) is equivalent to applying the following 2-block semi-proximal ADMM to (4.92),

$$\begin{cases} \boldsymbol{\beta}^{k+1} &= \text{argmin}\, \mathcal{L}_\phi(\boldsymbol{\beta}, \mathbf{z}^k; \boldsymbol{\gamma}^k) + \frac{\phi}{2}\|\boldsymbol{\beta} - \boldsymbol{\beta}^k\|^2_{\mathbf{m}F^T\mathcal{P}\mathbf{m}F+\mathcal{T}_f}, \\ \mathbf{z}^{k+1} &= \text{argmin}\, \mathcal{L}_\phi(\boldsymbol{\beta}^{k+1}, \mathbf{z}; \boldsymbol{\gamma}^k) + \frac{\phi}{2}\|\mathbf{z} - \mathbf{z}^k\|^2_{\mathbf{m}G^T\mathcal{P}\mathbf{m}G+\mathcal{T}_h}, \\ \boldsymbol{\gamma}^{k+1} &= \boldsymbol{\gamma}^k + \theta\phi Q(\mathbf{m}F\boldsymbol{\beta}^{k+1} + \mathbf{m}G\mathbf{z}^{k+1} - \mathbf{c}). \end{cases} \tag{4.94}$$

The Karush-Kuhn-Tucker (KKT) optimality condition of (4.92) is

$$0 \in (Q\mathbf{m}F)^T\boldsymbol{\gamma} + \partial f(\boldsymbol{\beta}), \quad 0 \in (Q\mathbf{m}G)^T\boldsymbol{\gamma} + \partial g(\mathbf{z}), \quad Q(\mathbf{c} - \mathbf{m}F\boldsymbol{\beta} - \mathbf{m}G\mathbf{z}) = 0. \tag{4.95}$$

Denote the solution set to (4.95) as $\bar{\Omega}$, then we can replace Assumption 2 by assuming that $\bar{\Omega}$ is non-empty. Let $\bar{\mathbf{m}u} = (\bar{\boldsymbol{\beta}}, \bar{\mathbf{z}}, \bar{\boldsymbol{\gamma}})$ be an optimal solution to (4.92). We have the following lemma on the convergence of the proposed algorithm by utilizing its equivalence to the updates in (4.94).

Lemma 4.4 *Suppose Assumption 2 holds. $\mathcal{T}_f$ and $\mathcal{T}_h$ are chosen such that $\mathcal{T}_f + \mathbf{m}F^T\mathbf{m}F$ and $\mathcal{T}_h + \mathbf{m}G^T\mathbf{m}G$ are positive definite. Then under the condition $\theta \in (0, (1+\sqrt{5})/2)$, the sequence $(\boldsymbol{\beta}^k, \mathbf{z}^k, \mathbf{m}\boldsymbol{\omega}^k, \mathbf{m}\boldsymbol{\gamma}^k)$ generated by (4.88) converges to a limit point $(\bar{\boldsymbol{\beta}}, \bar{\mathbf{z}}, \bar{\mathbf{m}\boldsymbol{\omega}}, \bar{\boldsymbol{\gamma}})$ with $(\bar{\boldsymbol{\beta}}, \bar{\mathbf{z}}, \bar{\mathbf{m}\boldsymbol{\omega}})$ solving (4.86) and $\bar{\boldsymbol{\gamma}}$ is the dual optimal.*

Lemma 4.4 follows by a direct application of Theorem 3.2 in [66]. Based on (4.85), we have the following fact.

Fact 4.2 *Suppose $\mathbf{u}^k$ converges to $\bar{\mathbf{u}} \in \bar{\Omega}$. There exists a positive constant q such that*

$$\|\mathbf{u}^k - \bar{\mathbf{u}}\|_2^2 \le q^2\Big(\|\boldsymbol{\beta}^k - \text{prox}_f(\boldsymbol{\beta}^k - (Q\mathbf{m}F)^T\boldsymbol{\gamma}^k)\|_2^2 + \|\mathbf{z}^k - \text{prox}_h(\mathbf{z}^k - (Q\mathbf{m}G)^T\boldsymbol{\gamma}^k)\|_2^2 + \|Q(\mathbf{c} - \mathbf{m}F\boldsymbol{\beta}^k - \mathbf{m}G\mathbf{z}^k)\|_2^2\Big), \tag{4.96}$$

for a sufficiently large k.

For any convex function P, $\text{prox}_P(\cdot)$ denotes the proximal mapping associated with P. That is,

$$\text{prox}_P(x) = \underset{y}{\text{argmin}}\left\{\frac{1}{2}\|x - y\|_2^2 + P(y)\right\}. \tag{4.97}$$

Denote $\mathcal{H} = C \times \text{Diag}(\mathbf{m}F^T\mathcal{P}\mathbf{m}F + \mathcal{T}_f, \mathbf{m}G^T\mathbf{m}G + \mathcal{T}_h, (\theta^{-2}\phi^{-1})I)$, where $C = \max\{3\phi^2\|\mathbf{m}F^T\mathcal{P}\mathbf{m}F + \mathcal{T}_f\|_2, 3\phi^2\lambda_{\max}(\mathbf{m}F\mathbf{m}F^T), 2\phi^2\|\mathbf{m}G^T\mathcal{P}\mathbf{m}G + \mathcal{T}_h\|_2,$

$3(1-\theta)^2\phi\lambda_{\max}(Q\mathbf{m}F\mathbf{m}F^TQ)) + 2(1-\theta)^2\phi\lambda_{\max}(QG\mathbf{m}G^TQ) + \frac{1}{\phi}\}$, then we have the following relationship.

Lemma 4.5 *Suppose the sequence* $\mathbf{u}^k = (\boldsymbol{\beta}^k, \mathbf{z}^k, \boldsymbol{\gamma}^k)$ *is generated by algorithm* (4.88), *then for any* $k \geq 0$,

$$\|\mathbf{u}^{k+1} - \bar{\mathbf{u}}\|_2^2 \leq q^2\|\mathbf{u}^{k+1} - \mathbf{u}^k\|_{\mathcal{H}}^2. \tag{4.98}$$

Proof Consider the optimality conditions of subproblems in (4.94), we have

$$\begin{aligned} 0 \in \quad & \partial f(\boldsymbol{\beta}^{k+1}) + (Q\mathbf{m}F)^T\boldsymbol{\gamma}^k + \phi(Q\mathbf{m}F)^TQ(\mathbf{m}F\boldsymbol{\beta}^{k+1} + \mathbf{m}G\mathbf{z}^k - \mathbf{c}) \\ & + \phi(\mathbf{m}F^T\mathcal{P}\mathbf{m}F + \mathcal{T}_f)(\boldsymbol{\beta}^{k+1} - \boldsymbol{\beta}^k), \\ 0 \in \quad & \partial g(\mathbf{z}^{k+1}) + (Q\mathbf{m}G)^T\boldsymbol{\gamma}^k + \phi(Q\mathbf{m}G)^TQ(\mathbf{m}F\boldsymbol{\beta}^{k+1} + \mathbf{m}G\mathbf{z}^{k+1} - \mathbf{c}) \\ & + \phi(\mathbf{m}G^T\mathcal{P}\mathbf{m}G + \mathcal{T}_h)(\mathbf{z}^{k+1} - \mathbf{z}^k), \\ 0 = \quad & (\theta\phi)^{-1}(\boldsymbol{\gamma}^{k+1} - \boldsymbol{\gamma}^k) - Q(\mathbf{m}F\boldsymbol{\beta}^{k+1} + \mathbf{m}G\mathbf{z}^{k+1} - \mathbf{c}). \end{aligned} \tag{4.99}$$

Then we have $Q(\mathbf{m}F\boldsymbol{\beta}^{k+1} + \mathbf{m}G\mathbf{z}^k - \mathbf{c}) = (\theta\phi)^{-1}(\boldsymbol{\gamma}^{k+1} - \boldsymbol{\gamma}^k) - Q\mathbf{m}G(\mathbf{z}^{k+1} - \mathbf{z}^k)$ and it follows that

$$\begin{aligned} \boldsymbol{\beta}^{k+1} =& \text{prox}_f\Big(\boldsymbol{\beta}^{k+1} - (Q\mathbf{m}F)^T(\boldsymbol{\gamma}^k + \theta^{-1}(\boldsymbol{\gamma}^{k+1} - \boldsymbol{\gamma}^k) - \phi Q\mathbf{m}G(\mathbf{z}^{k+1} - \mathbf{z}^k)) \\ & + \phi(\mathbf{m}F^T\mathcal{P}\mathbf{m}F + \mathcal{T}_f)(\boldsymbol{\beta}^{k+1} - \boldsymbol{\beta}^k)\Big), \\ \mathbf{z}^{k+1} =& \text{prox}_g\Big(\mathbf{z}^{k+1} - (Q\mathbf{m}G)^T(\boldsymbol{\gamma}^k + \theta^{-1}(\boldsymbol{\gamma}^{k+1} - \boldsymbol{\gamma}^k)) + \phi(\mathbf{m}G^T\mathcal{P}\mathbf{m}G + \mathcal{T}_h)(\mathbf{z}^{k+1} - \mathbf{z}^k)\Big), \\ \boldsymbol{\gamma}^{k+1} =& \boldsymbol{\gamma}^k + \theta\phi Q(\mathbf{m}F\boldsymbol{\beta}^{k+1} + \mathbf{m}G\mathbf{z}^{k+1} - \mathbf{c}). \end{aligned}$$

and we have

$$\begin{aligned} & \|\boldsymbol{\beta}^{k+1} - \text{prox}_f(\boldsymbol{\beta}^{k+1} - (Q\mathbf{m}F)^T\boldsymbol{\gamma}^{k+1})\|_2^2 + \|\mathbf{z}^{k+1} - \text{prox}_g(\mathbf{z}^{k+1} - (Q\mathbf{m}G)^T\boldsymbol{\gamma}^{k+1})\|_2^2 \\ & + \|Q(\mathbf{c} - \mathbf{m}F\boldsymbol{\beta}^{k+1} - \mathbf{m}G\mathbf{z}^{k+1})\|_2^2 \\ \leq \; & \|\mathbf{u}^{k+1} - \mathbf{u}^k\|_{\mathcal{H}}^2. \end{aligned} \tag{4.100}$$

We first bound the term $\|\boldsymbol{\beta}^{k+1} - \text{prox}_f(\boldsymbol{\beta}^{k+1} - (Q\mathbf{m}F)^T\boldsymbol{\gamma}^{k+1})\|_2^2$. By the fact that the proximal mapping is Lipschitz continuous with constant 1, i.e., $\|\text{prox}_h(x) - \text{prox}_h(y)\|_2 \leq \|x - y\|_2$ for any mapping h,

$$\begin{aligned}
&\|\boldsymbol{\beta}^{k+1} - \text{prox}_f(\boldsymbol{\beta}^{k+1} - (Q\mathbf{m}F)^T\boldsymbol{\gamma}^{k+1})\|_2^2 \\
\leq &\|\boldsymbol{\beta}^{k+1} - (Q\mathbf{m}F)^T\big(\boldsymbol{\gamma}^k + \theta^{-1}(\boldsymbol{\gamma}^{k+1} - \boldsymbol{\gamma}^k) - \phi Q\mathbf{m}G(\mathbf{z}^{k+1} - \mathbf{z}^k)\big) \\
&+ \phi(\mathbf{m}F^T\mathcal{P}\mathbf{m}F + \mathcal{T}_f)(\boldsymbol{\beta}^{k+1} - \boldsymbol{\beta}^k) - \boldsymbol{\beta}^{k+1} + (Q\mathbf{m}F)^T\boldsymbol{\gamma}^{k+1}\|_2^2 \\
= &\|\phi(\mathbf{m}F^T\mathcal{P}\mathbf{m}F + \mathcal{T}_f)(\boldsymbol{\beta}^{k+1} - \boldsymbol{\beta}^k) + \phi\mathbf{m}F^TQ\mathbf{m}G(\mathbf{z}^{k+1} - \mathbf{z}^k) \\
&+ \left(1 - \frac{1}{\theta}\right)(Q\mathbf{m}F)^T(\boldsymbol{\gamma}^{k+1} - \boldsymbol{\gamma}^k)\|_2^2 \\
= &\|\phi(\mathbf{m}F^T\mathcal{P}\mathbf{m}F + \mathcal{T}_f)(\boldsymbol{\beta}^{k+1} - \boldsymbol{\beta}^k)\|_2^2 + \|\phi\mathbf{m}F^TQ\mathbf{m}G(\mathbf{z}^{k+1} - \mathbf{z}^k)\|_2^2 \\
&+ \left(1 - \frac{1}{\theta}\right)^2\|(Q\mathbf{m}F)^T(\boldsymbol{\gamma}^{k+1} - \boldsymbol{\gamma}^k)\|_2^2 \\
&+ 2\phi^2(\boldsymbol{\beta}^{k+1} - \boldsymbol{\beta}^k)(\mathbf{m}F^T\mathcal{P}\mathbf{m}F + \mathcal{T}_f)^T\mathbf{m}F^TQ\mathbf{m}G(\mathbf{z}^{k+1} - \mathbf{z}^k) \\
&+ 2\left(1 - \frac{1}{\theta}\right)\phi(\boldsymbol{\beta}^{k+1} - \boldsymbol{\beta}^k)(\mathbf{m}F^T\mathcal{P}\mathbf{m}F + \mathcal{T}_f)^T(Q\mathbf{m}F)^T(\boldsymbol{\gamma}^{k+1} - \boldsymbol{\gamma}^k) \\
&+ 2\left(1 - \frac{1}{\theta}\right)\phi(\mathbf{z}^{k+1} - \mathbf{z}^k)^T\mathbf{m}G^TQ\mathbf{m}F(Q\mathbf{m}F)^T(\boldsymbol{\gamma}^{k+1} - \boldsymbol{\gamma}^k)
\end{aligned} \tag{4.101}$$

By taking into account the fact that

$$\begin{aligned}
&2\phi^2(\boldsymbol{\beta}^{k+1} - \boldsymbol{\beta}^k)(\mathbf{m}F^T\mathcal{P}\mathbf{m}F + \mathcal{T}_f)^T\mathbf{m}F^TQ\mathbf{m}G(\mathbf{z}^{k+1} - \mathbf{z}^k) \\
\leq &\|\phi(\mathbf{m}F^T\mathcal{P}\mathbf{m}F + \mathcal{T}_f)(\boldsymbol{\beta}^{k+1} - \boldsymbol{\beta}^k)\|_2^2 + \|\phi\mathbf{m}F^TQ\mathbf{m}G(\mathbf{z}^{k+1} - \mathbf{z}^k)\|_2^2, \\
&2(1 - \frac{1}{\theta})\phi(\boldsymbol{\beta}^{k+1} - \boldsymbol{\beta}^k)(\mathbf{m}F^T\mathcal{P}\mathbf{m}F + \mathcal{T}_f)^T(Q\mathbf{m}F)^T(\boldsymbol{\gamma}^{k+1} - \boldsymbol{\gamma}^k) \\
\leq &\|\phi(\mathbf{m}F^T\mathcal{P}\mathbf{m}F + \mathcal{T}_f)(\boldsymbol{\beta}^{k+1} - \boldsymbol{\beta}^k)\|_2^2 + (1 - \frac{1}{\theta})^2\|(Q\mathbf{m}F)^T(\boldsymbol{\gamma}^{k+1} - \boldsymbol{\gamma}^k)\|_2^2,
\end{aligned}$$

and

$$\begin{aligned}
&2\left(1 - \frac{1}{\theta}\right)\phi(\mathbf{z}^{k+1} - \mathbf{z}^k)^T\mathbf{m}G^TQ\mathbf{m}F(Q\mathbf{m}F)^T(\boldsymbol{\gamma}^{k+1} - \boldsymbol{\gamma}^k) \\
\leq &\left(1 - \frac{1}{\theta}\right)^2\|(Q\mathbf{m}F)^T(\boldsymbol{\gamma}^{k+1} - \boldsymbol{\gamma}^k)\|_2^2 + \|\phi\mathbf{m}F^TQ\mathbf{m}G(\mathbf{z}^{k+1} - \mathbf{z}^k)\|_2^2,
\end{aligned}$$

and the inequality that

$$\begin{aligned}
\|\mathbf{m}F^TQ\mathbf{m}G(\mathbf{z}^{k+1} - \mathbf{z}^k)\|_2^2 &= (Q\mathbf{m}G(\mathbf{z}^{k+1} - \mathbf{z}^k))^T\mathbf{m}F\mathbf{m}F^T(Q\mathbf{m}G(\mathbf{z}^{k+1} - \mathbf{z}^k))^T \\
&\leq \lambda_{\max}(\mathbf{m}F\mathbf{m}F^T)\|Q\mathbf{m}G(\mathbf{z}^{k+1} - \mathbf{z}^k)\|_2^2,
\end{aligned}$$

where $\lambda_{\max}(\mathbf{m}F\mathbf{m}F^T)$ is the largest eigenvalue of $\mathbf{m}F\mathbf{m}F^T$, (4.101) can be reduced to

$$
\begin{aligned}
&\|\boldsymbol{\beta}^{k+1} - \mathrm{prox}_f(\boldsymbol{\beta}^{k+1} - (Q\mathbf{m}F)^T\boldsymbol{\gamma}^{k+1})\|_2^2 \\
\le& 3\phi^2\|\mathbf{m}F^T\mathcal{P}\mathbf{m}F + \mathcal{T}_f\|_2\|\boldsymbol{\beta}^{k+1} - \boldsymbol{\beta}^k\|^2_{\mathbf{m}F^T\mathcal{P}\mathbf{m}F+\mathcal{T}_f} + 3\phi^2\lambda_{\max}(\mathbf{m}F\mathbf{m}F^T)\|\mathbf{z}^{k+1} - \mathbf{z}^k\|^2_{\mathbf{m}G^TQ\mathbf{m}G} \\
&+ 3\left(1 - \frac{1}{\theta}\right)^2 \|(Q\mathbf{m}F)^T(\boldsymbol{\gamma}^{k+1} - \boldsymbol{\gamma}^k)\|_2^2.
\end{aligned} \tag{4.102}
$$

Similarly we can bound the term $\|\mathbf{z}^{k+1} - \mathrm{prox}_g(\mathbf{z}^{k+1} - (Q\mathbf{m}G)^T\boldsymbol{\gamma}^{k+1})\|_2^2$,

$$
\begin{aligned}
&\|\mathbf{z}^{k+1} - \mathrm{prox}_h(\mathbf{z}^{k+1} - (Q\mathbf{m}G)^T\boldsymbol{\gamma}^{k+1})\|_2^2 \\
\le& 2\phi^2\|\mathbf{m}G^T\mathcal{P}\mathbf{m}G + \mathcal{T}_h\|_2\|\mathbf{z}^{k+1} - \mathbf{z}^k\|^2_{\mathbf{m}G^T\mathcal{P}\mathbf{m}G+\mathcal{T}_h} + 2\left(1 - \frac{1}{\theta}\right)^2 \|(Q\mathbf{m}G)^T(\boldsymbol{\gamma}^{k+1} - \boldsymbol{\gamma}^k)\|_2^2.
\end{aligned} \tag{4.103}
$$

From the update of $\boldsymbol{\gamma}$, we have

$$
\|Q(\mathbf{c} - \mathbf{m}F\boldsymbol{\beta}^{k+1} - \mathbf{m}G\mathbf{z}^{k+1})\|_2^2 = (\theta\phi)^{-2}\|\boldsymbol{\gamma}^{k+1} - \boldsymbol{\gamma}^k\|_2^2. \tag{4.104}
$$

Combining (4.102), (4.103) and (4.104), we can obtain that

$$
\begin{aligned}
&\|\boldsymbol{\beta}^{k+1} - \mathrm{prox}_f(\boldsymbol{\beta}^{k+1} - (Q\mathbf{m}F)^T\boldsymbol{\gamma}^{k+1})\|_2^2 + \|\mathbf{z}^{k+1} - \mathrm{prox}_h(\mathbf{z}^{k+1} - \mathbf{m}G\boldsymbol{\gamma}^{k+1})\|_2^2 \\
&+ \|Q(\mathbf{c} - \mathbf{m}F\boldsymbol{\beta}^{k+1} - \mathbf{m}G\mathbf{z}^{k+1})\|_2^2 \\
\le& 3\phi^2\|\mathbf{m}F^T\mathcal{P}\mathbf{m}F + \mathcal{T}_f\|_2\|\boldsymbol{\beta}^{k+1} - \boldsymbol{\beta}^k\|^2_{\mathbf{m}F^T\mathcal{P}\mathbf{m}F+\mathcal{T}_f} + 3\phi^2\lambda_{\max}(\mathbf{m}F\mathbf{m}F^T)\|\mathbf{z}^{k+1} - \mathbf{z}^k\|^2_{\mathbf{m}G^TQ\mathbf{m}G} \\
&+ (\theta\phi)^{-2}\|\boldsymbol{\gamma}^{k+1} - \boldsymbol{\gamma}^k\|_2^2 \\
&+ 3\left(1 - \frac{1}{\theta}\right)^2 \|(Q\mathbf{m}F)^T(\boldsymbol{\gamma}^{k+1} - \boldsymbol{\gamma}^k)\|_2^2 + 2\phi^2\|\mathbf{m}G^T\mathcal{P}\mathbf{m}G + \mathcal{T}_h\|_2\|\mathbf{z}^{k+1} - \mathbf{z}^k\|^2_{\mathbf{m}G^T\mathcal{P}\mathbf{m}G+\mathcal{T}_h} \\
&+ 2\left(1 - \frac{1}{\theta}\right)^2 \|(Q\mathbf{m}G)^T(\boldsymbol{\gamma}^{k+1} - \boldsymbol{\gamma}^k)\|_2^2 \\
\le& C \times \left(\|\boldsymbol{\beta}^{k+1} - \boldsymbol{\beta}^k\|^2_{\mathbf{m}F^T\mathcal{P}\mathbf{m}F+\mathcal{T}_f} + \|\mathbf{z}^{k+1} - \mathbf{z}^k\|^2_{\mathbf{m}G^T\mathbf{m}G+\mathcal{T}_h} + \theta^{-2}\phi^{-1}\|\boldsymbol{\gamma}^{k+1} - \boldsymbol{\gamma}^k\|_2^2\right)
\end{aligned} \tag{4.105}
$$

□

Lemma 4.6 *Suppose that Assumptions 2 holds, and assume that both* $\mathbf{m}F^T\mathbf{m}F + \mathcal{T}_f$ *and* $\mathbf{m}G^T\mathbf{m}G + \mathcal{T}_h$ *are positive definite. Then for all k sufficiently large and* $\theta \in (0, \frac{1+\sqrt{5}}{2})$, *there exists* $\mu \in (0, 1)$ *such that*

$$
\|\mathbf{u}^{k+1} - \bar{\mathbf{u}}\|_{\mathcal{H}_1} + \|\mathbf{z}^{k+1} - \mathbf{z}^k\|^2_{\mathbf{m}G^T\mathcal{P}\mathbf{m}G+\mathcal{T}_h} \le \mu\left(\|\mathbf{u}^k - \bar{\mathbf{u}}\|_{\mathcal{H}_1} + \|\mathbf{z}^k - \mathbf{z}^{k-1}\|^2_{\mathbf{m}G^T\mathcal{P}\mathbf{m}G+\mathcal{T}_h}\right), \tag{4.106}
$$

where

$$
\mathcal{H}_1 = \begin{pmatrix} \mathbf{m}F^T(m_1Q + \mathcal{P})\mathbf{m}F + \mathcal{T}_f & m_1\mathbf{m}F^TQ\mathbf{m}G & \cdots & \mathbf{m}0 \\ m_1\mathbf{m}G^TQ\mathbf{m}F & \mathbf{m}G^T(\mathcal{P} + (m_1+1)Q)\mathbf{m}G & \cdots & \mathbf{m}0 \\ \mathbf{m}0 & \mathbf{m}0 & \cdots & \theta^{-1}\phi^{-2}I \end{pmatrix} \tag{4.107}
$$

with $m_1 \in (0, 1)$.

Proof From Theorem 1 in [66], we can derive the following results.

$$\begin{aligned}
&\Big\{\|\boldsymbol{\beta}^k-\bar{\boldsymbol{\beta}}\|^2_{\mathbf{m}F^T\mathcal{P}\mathbf{m}F+\mathcal{T}_f}+\|\mathbf{z}^k-\bar{\mathbf{z}}\|^2_{\mathbf{m}G^T\mathbf{m}G+\mathcal{T}_h}+\|\mathbf{z}^k-\mathbf{z}^{k-1}\|^2_{\mathbf{m}G^T\mathcal{P}\mathbf{m}G+\mathcal{T}_h}\\
&+(1-\min\{\theta,\frac{1}{\theta}\})\|Q(\mathbf{m}F\boldsymbol{\beta}^k+\mathbf{m}G\mathbf{z}^k-\mathbf{c})\|_2^2+\theta^{-1}\phi^{-2}\|\boldsymbol{\gamma}^k-\bar{\boldsymbol{\gamma}}\|_2^2\Big\}\\
&-\Big\{\|\boldsymbol{\beta}^{k+1}-\bar{\boldsymbol{\beta}}\|^2_{\mathbf{m}F^T\mathcal{P}\mathbf{m}F+\mathcal{T}_f}+\|\mathbf{z}^{k+1}-\bar{\mathbf{z}}\|^2_{\mathbf{m}G^T\mathbf{m}G+\mathcal{T}_h}+\|\mathbf{z}^{k+1}-\mathbf{z}^{k}\|^2_{\mathbf{m}G^T\mathcal{P}\mathbf{m}G+\mathcal{T}_h}\\
&+(1-\min\{\theta,\frac{1}{\theta}\})\|Q(\mathbf{m}F\boldsymbol{\beta}^{k+1}+\mathbf{m}G\mathbf{z}^{k+1}-\mathbf{c})\|_2^2+\theta^{-1}\phi^{-2}\|\boldsymbol{\gamma}^{k+1}-\bar{\boldsymbol{\gamma}}\|_2^2\Big\}\\
\geq&\|\mathbf{z}^{k+1}-\mathbf{z}^k\|^2_{\mathbf{m}G^T\mathcal{P}\mathbf{m}G+\mathcal{T}_h+(\theta-\theta^2+\min(\theta^2,1))\mathbf{m}GQ^T\mathbf{m}G}+\|\boldsymbol{\beta}^{k+1}-\boldsymbol{\beta}^k\|^2_{\mathbf{m}F^T\mathcal{P}\mathbf{m}F+\mathcal{T}_f}\\
&+(1-\theta+\min\{\theta,\theta^{-1}\})\|Q(\mathbf{m}F\boldsymbol{\beta}^{k+1}+\mathbf{m}G\mathbf{z}^{k+1}-\mathbf{c})\|_2^2
\end{aligned}\tag{4.108}$$

When $\theta\in(0,\frac{1+\sqrt{5}}{2})$, it is ensured that $(1-\theta+\phi\min\{\theta,\theta^{-1}\})>0$. Let $d_1\in(0,\frac{1}{2})$, then we have

$$\begin{aligned}
&\Big\{\|\boldsymbol{\beta}^k-\bar{\boldsymbol{\beta}}\|^2_{\mathbf{m}F^T\mathcal{P}\mathbf{m}F+\mathcal{T}_f}+\|\mathbf{z}^k-\bar{\mathbf{z}}\|^2_{\mathbf{m}G^T\mathbf{m}G+\mathcal{T}_h}+\|\mathbf{z}^k-\mathbf{z}^{k-1}\|^2_{\mathbf{m}G^T\mathcal{P}\mathbf{m}G+\mathcal{T}_h}\\
&+(1+d_1-d_1\theta-(1-d_1)\min\{\theta,\frac{1}{\theta}\})\|Q(\mathbf{m}F\boldsymbol{\beta}^k+\mathbf{m}G\mathbf{z}^k-\mathbf{c})\|_2^2\\
&+\theta^{-1}\phi^{-2}\|\boldsymbol{\gamma}^k-\bar{\boldsymbol{\gamma}}\|_2^2\Big\}\\
&-\Big\{\|\boldsymbol{\beta}^{k+1}-\bar{\boldsymbol{\beta}}\|^2_{\mathbf{m}F^T\mathcal{P}\mathbf{m}F+\mathcal{T}_f}+\|\mathbf{z}^{k+1}-\bar{\mathbf{z}}\|^2_{\mathbf{m}G^T\mathbf{m}G+\mathcal{T}_h}+\|\mathbf{z}^{k+1}-\mathbf{z}^{k}\|^2_{\mathbf{m}G^T\mathcal{P}\mathbf{m}G+\mathcal{T}_h}\\
&+(1+d_1-d_1\theta-(1-d_1)\min\{\theta,\frac{1}{\theta}\})\|Q(\mathbf{m}F\boldsymbol{\beta}^{k+1}+\mathbf{m}G\mathbf{z}^{k+1}-\mathbf{c})\|_2^2\\
&+\theta^{-1}\phi^{-2}\|\boldsymbol{\gamma}^{k+1}-\bar{\boldsymbol{\gamma}}\|_2^2\Big\}
\end{aligned}\tag{4.109}$$

$$\begin{aligned}
\geq\ &\|\mathbf{z}^{k+1}-\mathbf{z}^k\|^2_{\mathbf{m}G^T\mathcal{P}\mathbf{m}G+\mathcal{T}_h+(\theta-\theta^2+\min(\theta^2,1))\mathbf{m}G^TQ\mathbf{m}G}+\|\boldsymbol{\beta}^{k+1}-\boldsymbol{\beta}^k\|^2_{\mathbf{m}F^T\mathcal{P}\mathbf{m}F+\mathcal{T}_f}\\
&+(1-d_1)(1-\theta+\min\{\theta,\theta^{-1}\})\|Q(\mathbf{m}F\boldsymbol{\beta}^{k+1}+\mathbf{m}G\mathbf{z}^{k+1}-\mathbf{c})\|_2^2\\
&+d_1(1-\theta+\min\{\theta,\theta^{-1}\})\|Q(\mathbf{m}F\boldsymbol{\beta}^{k}+\mathbf{m}G\mathbf{z}^{k}-\mathbf{c})\|_2^2\\
=\ &\|\mathbf{z}^{k+1}-\mathbf{z}^k\|^2_{\mathbf{m}G^T\mathcal{P}\mathbf{m}G+\mathcal{T}_h+(\theta-\theta^2+\min(\theta^2,1))\mathbf{m}G^TQ\mathbf{m}G}+\|\boldsymbol{\beta}^{k+1}-\boldsymbol{\beta}^k\|^2_{\mathbf{m}F^T\mathcal{P}\mathbf{m}F+\mathcal{T}_f}\\
&+(1-2d_1))(1-\theta+\min\{\theta,\theta^{-1}\})\theta^{-2}\phi^{-2}\|\boldsymbol{\gamma}^{k+1}-\boldsymbol{\gamma}^k\|_2^2\\
&+d_1(1-\theta+\min\{\theta,\theta^{-1}\})(\|Q(\mathbf{m}F\boldsymbol{\beta}^{k}+\mathbf{m}G\mathbf{z}^{k}-\mathbf{c})\|_2^2\\
&+\|Q(\mathbf{m}F\boldsymbol{\beta}^{k+1}+\mathbf{m}G\mathbf{z}^{k+1}-\mathbf{c})\|_2^2)\\
\geq\ &\|\mathbf{z}^{k+1}-\mathbf{z}^k\|^2_{\mathbf{m}G^T\mathcal{P}\mathbf{m}G+\mathcal{T}_h+(\theta-\theta^2+\min(\theta^2,1))\mathbf{m}G^TQ\mathbf{m}G}+\|\boldsymbol{\beta}^{k+1}-\boldsymbol{\beta}^k\|^2_{\mathbf{m}F^T\mathcal{P}\mathbf{m}F+\mathcal{T}_f}\\
&+(1-2d_1)(1-\theta+\min\{\theta,\theta^{-1}\})\theta^{-2}\phi^{-2}\|\boldsymbol{\gamma}^{k+1}-\boldsymbol{\gamma}^k\|_2^2\\
&+\frac{1}{2}d_1(1-\theta+\min\{\theta,\theta^{-1}\})\|Q\mathbf{m}F(\boldsymbol{\beta}^{k+1}-\boldsymbol{\beta}^k)+Q\mathbf{m}G(\mathbf{z}^{k+1}-\mathbf{z}^k)\|_2^2
\end{aligned}\tag{4.110}$$

Note that $Q(\mathbf{m}F\boldsymbol{\beta}^{k+1}+\mathbf{m}G\mathbf{z}^{k+1}-\mathbf{c})=Q\mathbf{m}F(\boldsymbol{\beta}^{k+1}-\bar{\boldsymbol{\beta}})+Q\mathbf{m}G(\mathbf{z}^{k+1}-\bar{\mathbf{z}})$, and we have

$$
\begin{aligned}
&\Big\{\|\boldsymbol{\beta}^k-\bar{\boldsymbol{\beta}}\|^2_{\mathbf{m}F^T\mathcal{P}\mathbf{m}F+\mathcal{T}_f}+\|\mathbf{z}^k-\bar{\mathbf{z}}\|^2_{\mathbf{m}G^T\mathbf{m}G+\mathcal{T}_h}+\|\mathbf{z}^k-\mathbf{z}^{k-1}\|^2_{\mathbf{m}G^T\mathcal{P}\mathbf{m}G+\mathcal{T}_h}\\
&+(1+d_1-d_1\theta-(1-d_1)\min\{\theta,\frac{1}{\theta}\})\|Q\mathbf{m}F(\boldsymbol{\beta}^k-\bar{\boldsymbol{\beta}})+Q\mathbf{m}G(\mathbf{z}^k-\bar{\mathbf{z}})\|_2^2\\
&+\theta^{-1}\phi^{-2}\|\boldsymbol{\gamma}^k-\bar{\boldsymbol{\gamma}}\|_2^2\Big\}\\
&-\Big\{\|\boldsymbol{\beta}^{k+1}-\bar{\boldsymbol{\beta}}\|^2_{\mathbf{m}F^T\mathcal{P}\mathbf{m}F+\mathcal{T}_f}+\|\mathbf{z}^{k+1}-\bar{\mathbf{z}}\|^2_{\mathbf{m}G^T\mathbf{m}G+\mathcal{T}_h}+\|\mathbf{z}^{k+1}-\mathbf{z}^{k}\|^2_{\mathbf{m}G^T\mathcal{P}\mathbf{m}G+\mathcal{T}_h}\\
&+(1+d_1-d_1\theta-(1-d_1)\min\{\theta,\frac{1}{\theta}\})\|Q\mathbf{m}F(\boldsymbol{\beta}^{k+1}-\bar{\boldsymbol{\beta}})+Q\mathbf{m}G(\mathbf{z}^{k+1}-\bar{\mathbf{z}}))\|_2^2\\
&+\theta^{-1}\phi^{-2}\|\boldsymbol{\gamma}^{k+1}-\bar{\boldsymbol{\gamma}}\|_2^2\Big\}\\
\geq\;&\|\mathbf{z}^{k+1}-\mathbf{z}^k\|^2_{\mathbf{m}G^T\mathcal{P}\mathbf{m}G+\mathcal{T}_h+(\theta-\theta^2+\min(\theta^2,1))\mathbf{m}G^TQ\mathbf{m}G}+\|\boldsymbol{\beta}^{k+1}-\boldsymbol{\beta}^k\|^2_{\mathbf{m}F^T\mathcal{P}\mathbf{m}F+\mathcal{T}_f}\\
&+(1-2d_1)(1-\theta+\min\{\theta,\theta^{-1}\})\theta^{-2}\phi^{-2}\|\boldsymbol{\gamma}^{k+1}-\boldsymbol{\gamma}^k\|_2^2\\
&+\frac{1}{2}d_1(1-\theta+\min\{\theta,\theta^{-1}\})\|Q\mathbf{m}F(\boldsymbol{\beta}^{k+1}-\boldsymbol{\beta}^k)+Q\mathbf{m}G(\mathbf{z}^{k+1}-\mathbf{z}^k)\|_2^2\\
\geq\;&(1-\theta+\min\{\theta,\theta^{-1}\})\min\{\frac{1}{2}d_1,1-2d_1,\theta\}\Big(\|\boldsymbol{\beta}^{k+1}-\boldsymbol{\beta}^k\|^2_{\mathbf{m}F^T\mathcal{P}\mathbf{m}F+\mathcal{T}_f}\\
&+\|\mathbf{z}^{k+1}-\mathbf{z}^k\|^2_{\mathbf{m}G^T\mathbf{m}G+\mathcal{T}_h}+\theta^{-2}\phi^{-2}\|\boldsymbol{\gamma}^{k+1}-\boldsymbol{\gamma}^k\|_2^2\Big)
\end{aligned}\tag{4.111}
$$

Let $m_1=1+d_1-d_1\theta-(1-d_1)\min\{\theta,\frac{1}{\theta}\}$ in $\mathcal{H}_1$ defined in (4.107), and $m_2=(1-\theta+\min\{\theta,\theta^{-1}\})\min\{\frac{1}{2}d_1,1-2d_1,\theta\}$. Note that when $\theta\in(0,\frac{1+\sqrt{5}}{2})$, the following relationship holds.

$$
\mathbf{m}F^T\mathbf{m}F+\mathcal{T}_f\succ 0 \text{ and } \mathbf{m}G^T\mathbf{m}G+\mathcal{T}_h\succ 0\Longleftrightarrow \mathcal{H}_1\succ 0.
$$

Combining with Lemma 4.5, we have

$$
\begin{aligned}
&\|\mathbf{u}^k-\bar{\mathbf{u}}\|^2_{\mathcal{H}_1}+\|\mathbf{z}^k-\mathbf{z}^{k-1}\|^2_{\mathbf{m}G^T\mathbf{m}G+\mathcal{T}_h}-(\|\mathbf{u}^{k+1}-\bar{\mathbf{u}}\|_{\mathcal{H}_1}+\|\mathbf{z}^{k+1}-\mathbf{z}^k\|^2_{\mathbf{m}G^T\mathbf{m}G+\mathcal{T}_h})\\
\geq\;&\frac{m_2}{C}\Big(C\times(\|\boldsymbol{\beta}^{k+1}-\boldsymbol{\beta}^k\|^2_{\mathbf{m}F^T\mathcal{P}\mathbf{m}F+\mathcal{T}_f}+\|\mathbf{z}^{k+1}-\mathbf{z}^k\|^2_{\mathbf{m}G^T\mathbf{m}G+\mathcal{T}_h}\\
&+\theta^{-2}\phi^{-2}\|\boldsymbol{\gamma}^{k+1}-\boldsymbol{\gamma}^k\|_2^2)\Big)\\
=\;&\frac{m_2}{C}\|\mathbf{m}u^{k+1}-\mathbf{m}u^k\|^2_{\mathcal{H}}\geq\frac{m_2d_2}{Cq^2}\|\mathbf{u}^{k+1}-\bar{\mathbf{u}}\|_2^2+\frac{m_2(1-d_2)}{Cq^2}\|\mathbf{z}^{k+1}-\mathbf{z}^k\|^2_{\mathbf{m}G^T\mathbf{m}G+\mathcal{T}_h}\\
\geq\;&\frac{m_2d_2}{Cq^2\lambda_{\max}(\mathcal{H}_1)}\|\mathbf{u}^{k+1}-\bar{\mathbf{u}}\|^2_{\mathcal{H}_1}+\frac{m_2(1-d_2)}{Cq^2}\|\mathbf{z}^{k+1}-\mathbf{z}^k\|^2_{\mathbf{m}G^T\mathbf{m}G+\mathcal{T}_h}.
\end{aligned}\tag{4.112}
$$

Take $d_2=\frac{\lambda_{\max}(\mathcal{H}_1)}{1+\lambda_{\max}(\mathcal{H}_1)}$, then we can obtain (4.106) with $\mu=\left[1+\frac{m_2}{Cq^2(1+\lambda_{\max}(\mathcal{H}_1))}\right]^{-1}$. □

Proof of Theorem 4.4. Since $f=\|\cdot\|_1$ and $g=\frac{1}{n}\mathbf{m}1_n^T(\cdot)_+$ are piecewise linear-quadratic functions, thus both $\text{prox}_f(\cdot)$ and $\text{prox}_g(\cdot)$ are piecewise polyhedral [67] which implies Fact 2 [66]. Since we take $\mathbf{m}\mathcal{T}_g=\eta_i I_{p_g}-\mathbf{X}_g^T\mathbf{X}_g$, $g=1,\ldots,G$, then $\mathcal{T}_g+\mathbf{m}F^T\mathbf{m}F=\text{Diag}(\eta_1 I_{p_1},\ldots,\eta_K I_{p_K})$ is positive definite, and this together with the fact that $\mathbf{m}G^T\mathbf{m}G=I_n\succ 0$ imply that the sequence $(\boldsymbol{\beta}^k,\mathbf{z}^k,\mathbf{m}\omega^k,\boldsymbol{\gamma}^k)$ is automatically well defined. By Lemma 4.4, under the condition $\theta\in(0,(1+\sqrt{5})/2)$, the sequence $(\boldsymbol{\beta}^k, \mathbf{z}^k, \mathbf{m}\omega^k, \boldsymbol{\gamma}^k)$ generated by algorithm (4.88) converges to a limit point $(\bar{\boldsymbol{\beta}}, \bar{\mathbf{z}}, \bar{\mathbf{m}\omega}, \bar{\boldsymbol{\gamma}})$ with $(\bar{\boldsymbol{\beta}}, \bar{\mathbf{z}}, \bar{\mathbf{m}\omega})$ solving (4.31) and $\bar{\boldsymbol{\gamma}}$ is the dual optimal.

To derive the rate of convergence, we first compute $\mathcal{H}_1$. By definition,

$$\mathcal{P} = \mathbf{m}H(\mathbf{m}H^T\mathbf{m}H)^{-1}\mathbf{m}H^T = \frac{1}{G}\begin{pmatrix} (G-1)I & -I & \cdots & -I \\ -I & (G-1)I & \cdots & -I \\ \vdots & \ddots & & \vdots \\ \vdots & & \ddots & \vdots \\ -I & -I & \cdots & (G-1)I \end{pmatrix}.$$

It follows that

$$\begin{aligned}
\|\boldsymbol{\beta}^{k+1} - \bar{\boldsymbol{\beta}}\|^2_{\mathbf{m}F^T\mathcal{P}\mathbf{m}F} &= \sum_{i=1}^{G}\|\mathbf{X}_i(\boldsymbol{\beta}_i^{k+1} - \bar{\boldsymbol{\beta}}_i)\|_2^2 - \frac{1}{G}\|\sum_{i=1}^{G}\mathbf{X}_i^T(\boldsymbol{\beta}_i^{k+1} - \bar{\boldsymbol{\beta}}_i)\|_2^2, \\
m_1\|Q\mathbf{m}F(\boldsymbol{\beta}^{k+1} - \bar{\boldsymbol{\beta}}) + Q\mathbf{m}G(\mathbf{z}^{k+1} - \bar{\mathbf{z}})\|_2^2 &= \frac{m_1}{G}\|\sum_{i=1}^{G}\mathbf{X}_i(\boldsymbol{\beta}_i^{k+1} - \bar{\boldsymbol{\beta}}_i) + (\mathbf{z}^{k+1} - \bar{\mathbf{z}})\|_2^2, \\
\|\mathbf{z}^{k+1} - \bar{\mathbf{z}}\|^2_{\mathbf{m}G^T\mathbf{m}G+\mathcal{T}_h} &= \|\mathbf{z}^{k+1} - \bar{\mathbf{z}}\|_2^2, \\
\|\mathbf{z}^{k+1} - \mathbf{z}^k\|^2_{\mathbf{m}G^T\mathcal{P}\mathbf{m}G+\mathcal{T}_h} &= \frac{G-1}{G}\|\mathbf{z}^{k+1} - \mathbf{z}^k\|_2^2.
\end{aligned} \tag{4.113}$$

Plugging equations (4.113) back into (4.112), we derive the results in Theorem 4.4 easily.

4.6.2 *Linear Programming and ADMM*

To be self-contained, we provide some basics of linear programming and ADMM that repeatedly used in the main text.

Linear Programming

The minimization problem (4.3) is a convex minimization problem, and it can be directly recast as a linear programming problem:

Algorithm 5 FS-QRADMM-CD for Two-Step PQR-SCAD

Require: $\tilde{\boldsymbol{\beta}}^0, \lambda, \upsilon, \tilde{\mathbf{z}}^0, \tilde{\boldsymbol{\gamma}}^0, \tilde{\mathbf{m}\omega}_i^0$, and $\phi > 0, \theta = 1.618, k = 0$.
while the stopping criterion is not satisfied, **do**
Update $\tilde{\boldsymbol{\beta}}^{k+1}$ by

$$\tilde{\boldsymbol{\beta}}_1^{k+1} = \operatorname*{argmin}_{\boldsymbol{\beta}_1 \in \mathbb{R}^{p_1}} n\upsilon\lambda \|\boldsymbol{\beta}_1\|_1 + \frac{\phi}{2} \|\mathbf{X}_1\boldsymbol{\beta}_1 + \sum_{g=2}^{G} \tilde{\mathbf{m}\omega}_g^k + \tilde{\mathbf{z}}^k - \mathbf{y} + \frac{\tilde{\boldsymbol{\gamma}}_1^k}{\phi}\|_2^2,$$

$$\tilde{\boldsymbol{\beta}}_g^{k+1} = \operatorname*{argmin}_{\boldsymbol{\beta}_g \in \mathbb{R}^{p_g}} n\upsilon\lambda \|\boldsymbol{\beta}_g\|_1 + \frac{\phi}{2} \|\mathbf{X}_g\boldsymbol{\beta}_g - \tilde{\mathbf{m}\omega}_g^k + \frac{\tilde{\boldsymbol{\gamma}}_g^k}{\phi}\|_2^2, \quad g = 2, \ldots, G$$

Compute $\tilde{\mathbf{m}\omega}^{k+\frac{1}{2}}, \tilde{\mathbf{z}}^{k+1}$ and $\tilde{\mathbf{m}\omega}^{k+1}$ by (4.36).
Update $\tilde{\boldsymbol{\gamma}}^{k+1}$ by (4.37).
end while The solution is denoted as $\hat{\boldsymbol{\beta}}^{\ell_1}, \hat{\mathbf{z}}^{\ell_1}, \hat{\mathbf{m}\omega}^{\ell_1}$.
Require: $\hat{\boldsymbol{\beta}}^0 = \hat{\boldsymbol{\beta}}^{\ell_1}, \hat{\mathbf{z}}^0 = \hat{\mathbf{z}}^{\ell_1}, \hat{\mathbf{m}\omega}^0 = \hat{\mathbf{m}\omega}^{\ell_1}$ and $\phi > 0, \theta = 1.618, k = 0$. Compute $\alpha_j = \lambda^{-1} p_\lambda(|\hat{\beta}_j^0|)$ for $j = 1, \ldots, p$.
while the stopping criterion is not satisfied, **do**
Update $\hat{\boldsymbol{\beta}}^{k+1}$ by (4.35).
Compute $\hat{\mathbf{m}\omega}^{k+\frac{1}{2}}, \hat{\mathbf{z}}^{k+1}$ and $\hat{\mathbf{m}\omega}^{k+1}$ by (4.36).
Update $\hat{\boldsymbol{\gamma}}^{k+1}$ by (4.37).
end while

Algorithm 6 FS-QRADMM-prox for Two-Step PQR-SCAD

Require: $\tilde{\boldsymbol{\beta}}^0, \lambda, \upsilon, \tilde{\mathbf{z}}^0, \tilde{\boldsymbol{\gamma}}^0, \tilde{\mathbf{m}\omega}_i^0$, and $\phi > 0, \theta = 1.618, k = 0$.
while the stopping criterion is not satisfied, **do**
Update $\tilde{\boldsymbol{\beta}}^{k+1}$ by

$$\tilde{\boldsymbol{\beta}}_1^{k+1} = \text{Shrink}\Big(\tilde{\boldsymbol{\beta}}_{1j}^k - \frac{\phi}{\eta_1}\mathbf{X}_{1j}^T(\mathbf{X}_1\tilde{\boldsymbol{\beta}}_1^k + \sum_{g=2}^{G} \tilde{\mathbf{m}\omega}_g^k + \tilde{\mathbf{z}}^k - \mathbf{y} + \frac{\tilde{\boldsymbol{\gamma}}_1^k}{\phi}), \frac{n\upsilon\lambda}{\eta_1}\Big)_{j=1,\ldots,p_1}$$

$$\tilde{\boldsymbol{\beta}}_g^{k+1} = \text{Shrink}\Big(\tilde{\boldsymbol{\beta}}_{gj}^k - \frac{\phi}{\eta_g}\mathbf{X}_{gj}^T(\mathbf{X}_g\tilde{\boldsymbol{\beta}}_g^k - \tilde{\mathbf{m}\omega}_g^k + \frac{\tilde{\boldsymbol{\gamma}}_g^k}{\phi}), \frac{n\upsilon\lambda}{\eta_g}\Big)_{j=1,\ldots,p_g}, \quad g = 2, \ldots, G.$$

Compute $\tilde{\mathbf{m}\omega}^{k+\frac{1}{2}}, \tilde{\mathbf{z}}^{k+1}$ and $\tilde{\mathbf{m}\omega}^{k+1}$ by (4.36).
Update $\tilde{\boldsymbol{\gamma}}^{k+1}$ by (4.37).
end while denote the solution as $\hat{\boldsymbol{\beta}}^{\ell_1}, \hat{\mathbf{z}}^{\ell_1}, \hat{\mathbf{m}\omega}^{\ell_1}$
Require: $\hat{\boldsymbol{\beta}}^0 = \hat{\boldsymbol{\beta}}^{\ell_1}, \hat{\mathbf{z}}^0 = \hat{\mathbf{z}}^{\ell_1}, \hat{\mathbf{m}\omega}^0 = \hat{\mathbf{m}\omega}^{\ell_1}$ and $\phi > 0, \theta = 1.618, k = 0$. Compute $\alpha_j = \lambda^{-1} p_\lambda(|\hat{\beta}_j^0|)$ for $j = 1, \ldots, p$.
while the stopping criterion is not satisfied, **do**
Update $\hat{\boldsymbol{\beta}}^{k+1}$ by (4.38).
Compute $\hat{\mathbf{m}\omega}^{k+\frac{1}{2}}, \hat{\mathbf{z}}^{k+1}$ and $\hat{\mathbf{m}\omega}^{k+1}$ by (4.36).
Update $\hat{\boldsymbol{\gamma}}^{k+1}$ by (4.37).
end while

Table 4.10 Performance of proposed algorithms for PQR when $p = 1000$ and $n = 30000$

$n = 30000$, $p = 1000$	τ	$\|\hat{\boldsymbol{\beta}} - \boldsymbol{\beta}\|_1$	P1 (%)	P2 (%)	Size
FS-QRADMM-CD (Lasso)	0.3	0.031 (0.0004)	100	100	5.06 (0.003)
	0.5	0.020 (0.0004)	100	0.4	4.04 (0.003)
	0.7	0.029 (0.0004)	100	100	5.06 (0.003)
FS-QRADMM-prox (Lasso)	0.3	0.029 (0.0003)	100	100	5.05 (0.003)
	0.5	0.020 (0.0003)	100	0.5	4.03 (0.002)
	0.7	0.029 (0.0003)	100	100	5.05 (0.003)
FS-QRADMM-CD (TS-SCAD)	0.3	0.012 (0.0005)	100	100	5.00 (0.00)
	0.5	0.004 (0.0002)	100	0.2	4.00 (0.00)
	0.7	0.013 (0.0004)	100	100	5.00 (0.00)
FS-QRADMM-prox (TS-SCAD)	0.3	0.011 (0.0003)	100	100	5.00 (0.00)
	0.5	0.004 (0.0002)	100	0	4.00 (0.00)
	0.7	0.012 (0.0003)	100	100	5.00 (0.00)

$$
\begin{aligned}
&\min \sum_{j=1}^{p} u_j \\
\text{s.t.}\ &-\beta_j \le u_j \text{ for all } 1 \le j \le p, \\
&+\beta_j \le u_j \text{ for all } 1 \le j \le p, \\
&-\hat{\boldsymbol{\Sigma}}_k^T \boldsymbol{\beta} + \hat{\mu}_k \le \lambda \text{ for all } 1 \le k \le p, \\
&-\hat{\boldsymbol{\Sigma}}_k^T \boldsymbol{\beta} + \hat{\mu}_k \le \lambda \text{ for all } 1 \le k \le p,
\end{aligned}
\tag{4.114}
$$

where $(\hat{\boldsymbol{\Sigma}}_1, \ldots, \hat{\boldsymbol{\Sigma}}_p)^T = \hat{\boldsymbol{\Sigma}}$, $(\hat{\mu}_1, \ldots, \hat{\mu}_p) = \hat{\boldsymbol{\mu}}$ and $(\beta_1, \ldots, \beta_p) = \boldsymbol{\beta}$. Similar to the implementation of the Dantzig selector in high-dimensional linear regression [69], (4.114) can be solved by the primal-dual interior-point method. The details of the primal-dual interior-point method can be found in the book of [70].

ADMM

The ADMM has a wide range of applications in many statistical and machine learning problems. Boyd et al. [40] gives a systematic review of ADMM. Recently, it has drawn remarkable attention thanks to its simplicity in implementation and scalability to large-scale data. ADMM is well suited to solving a wide range of statistical and machine learning models. For example, a collection of regularization models can be formulated as the following optimization problem,

$$\min_{\boldsymbol{\beta}} L(\mathbf{m}X\boldsymbol{\beta} - \mathbf{y}) + g(\boldsymbol{\beta}), \tag{4.115}$$

where $\boldsymbol{\beta} \in \mathbb{R}^p$ is the learning parameter, $L(\cdot)$ is a convex loss function, not necessarily smooth, and $g(\cdot)$ is a regularization function. We can reformulate (4.115) into a two-block optimization problem that ADMM can handle as follows,

$$\min L(\mathbf{m}X\boldsymbol{\beta} - \mathbf{y}) + g(\mathbf{z}) \quad \text{s.t. } \boldsymbol{\beta} - \mathbf{z} = 0, \tag{4.116}$$

and ADMM can be directly applied to solve (4.116). In this section, we introduce the application of two-block ADMM in solving the problem (4.3) without asset splitting.

We first reformulate (4.3) as

$$\min \|\boldsymbol{\beta}\|_1 \quad \text{s.t.} \quad \hat{\boldsymbol{\Sigma}}\boldsymbol{\beta} - \hat{\boldsymbol{\mu}} = \mathbf{y}, \text{ and } \{\mathbf{y} : \|\mathbf{y}\|_\infty \leq \lambda\}. \tag{4.117}$$

Then we apply the following procedure to update the variable.

$$\boldsymbol{\beta}^{t+1} = \operatorname*{argmin}_{\boldsymbol{\beta}\in\mathbb{R}^p} \{\|\boldsymbol{\beta}\|_1 + \rho(\boldsymbol{\beta} - \boldsymbol{\beta}^t)^T\mathbf{v}^t + \frac{\tau}{2}\|\boldsymbol{\beta} - \boldsymbol{\beta}^t\|^2\}, \tag{4.118}$$

$$\mathbf{y}^{t+1} = \operatorname*{argmin}_{\{\mathbf{y}:\|\mathbf{y}\|_\infty\leq\lambda\}} \frac{\rho}{2}\|\mathbf{y} - (\hat{\boldsymbol{\Sigma}}\boldsymbol{\beta}^{t+1} - \hat{\boldsymbol{\mu}}) + \frac{\boldsymbol{\gamma}^t}{\rho}\|^2, \tag{4.119}$$

$$\boldsymbol{\gamma}^{t+1} = \boldsymbol{\gamma}^t - \rho(\hat{\boldsymbol{\Sigma}}\boldsymbol{\beta}^{t+1} - \boldsymbol{\mu} - \mathbf{y}^{t+1}), \tag{4.120}$$

where $\mathbf{v}^t = \hat{\boldsymbol{\Sigma}}^T(\hat{\boldsymbol{\Sigma}}\boldsymbol{\beta}^t - \hat{\boldsymbol{\mu}} - \mathbf{y}^t - \frac{\boldsymbol{\gamma}^t}{\rho})$, $\tau > \rho\|\hat{\boldsymbol{\Sigma}}^T\hat{\boldsymbol{\Sigma}}\|_2$ and $\|\cdot\|_2$ is the spectral norm of a matrix.

Equation (4.118) is equivalent to

$$\boldsymbol{\beta}^{t+1} = \operatorname*{argmin}_{\boldsymbol{\beta}\in\mathbb{R}^p} \{\|\boldsymbol{\beta}\|_1 + \frac{\rho\tau}{4}\|\boldsymbol{\beta} - (\boldsymbol{\beta}^t - \frac{2}{\tau}\mathbf{v}^t)\|^2\}. \tag{4.121}$$

Let $\mathrm{S}(x, t) = \mathrm{sign}(x)(|x| - t)I(|x| > t)$ be the soft thresholding function, where $\mathrm{sign}(\cdot)$ is the sign function. The closed-form solution of (4.121) is

$$\boldsymbol{\beta}^{t+1} = \mathrm{S}(\boldsymbol{\beta}^t - \frac{2}{\tau}\mathbf{v}^t, \frac{2}{\rho\tau}). \tag{4.122}$$

The solution of (4.119) is

$$\mathbf{y}^{t+1} = \min(\max(\hat{\boldsymbol{\Sigma}}\boldsymbol{\beta}^{t+1} - \hat{\boldsymbol{\mu}} - \frac{\boldsymbol{\gamma}^t}{\rho}, -\lambda), \lambda). \tag{4.123}$$

The procedure of ADMM for solving (4.3) can be summarized in Algorithm 7.

Algorithm 7 ADMM algorithm for LPO

Require: Initialized $\boldsymbol{\beta}_0, \mathbf{y}_0, \boldsymbol{\gamma}_0, \tau, \rho$.

while the stopping criterion is not satisfied, **do**

Compute $\boldsymbol{\beta}^{t+1}$ by (4.122).

Compute $\mathbf{y}^{t+1}$ by (4.123).

Update $\boldsymbol{\gamma}^{t+1}$ by (4.120)

end while

References

1. Markowitz, H.: Portfolio selection. J. Financ. **7**(1), 77–91 (1952)
2. Chopra, V.K., Ziemba, W.T.: The effect of errors in means, variances, and covariances on optimal portfolio choice. In: Handbook of the fundamentals of financial decision making: Part I, pp. 365–373. World Scientific (2013)
3. Michaud, R.O.: The markowitz optimization enigma: Is 'optimized' optimal? Financ. Anal. J. **45**(1), 31–42 (1989)
4. Fan, J., Liao, Y., Shi, X.: Risks of large portfolios. J. Econom. **186**(2), 367–387 (2015)
5. Lam, C., Feng, P.: A nonparametric eigenvalue-regularized integrated covariance matrix estimator for asset return data. J. Econom. **206**(1), 226–257 (2018)
6. Fan, J., Wang, W., Zhong, Y.: Robust covariance estimation for approximate factor models. J. Econom. **208**(1), 5–22 (2019)
7. Hafner, C.M., Linton, O.B., Tang, H.: Estimation of a multiplicative correlation structure in the large dimensional case. J. Econom. **217**(2), 431–470 (2020)
8. A t-Sahalia, Y., Mykland, P.A., Zhang, L.: Ultra high frequency volatility estimation with dependent microstructure noise. J. Econom. **160**(1), 160–175 (2011)
9. Frahm, G., Memmel, C.: Dominating estimators for minimum-variance portfolios. J. Econom. **159**(2), 289–302 (2010)
10. Engle, R.F., Ledoit, O., Wolf, M.: Large dynamic covariance matrices. J. Busi. Econ. Stat. **37**(2), 363–375 (2019)
11. Boudt, K., Laurent, S., Lunde, A., Quaedvlieg, R., Sauri, O.: Positive semidefinite integrated covariance estimation, factorizations and asynchronicity. J. Econom. **196**(2), 347–367 (2017)
12. Cai, T.T., Hu, J., Li, Y., Zheng, X.: High-dimensional minimum variance portfolio estimation based on high-frequency data. J. Econom. **214**(2), 482–494 (2020)
13. Ding, Y., Li, Y., Zheng, X.: High dimensional minimum variance portfolio estimation under statistical factor models. J. Econom. **222**(1, Part B), 502–515 (2021)
14. So, M.K., Chan, T.W., Chu, A.M.: Efficient estimation of high-dimensional dynamic covariance by risk factor mapping: applications for financial risk management. J. Econom. Press (2020)

15. Jagannathan, R., Ma, T.: Risk reduction in large portfolios: Why imposing the wrong constraints helps. J. Financ. **58**(4), 1651–1683 (2003)
16. DeMiguel, V., Garlappi, L., Nogales, F.J., Uppal, R.: A generalized approach to portfolio optimization: Improving performance by constraining portfolio norms. Manage. Sci. **55**(5), 798–812 (2009)
17. Fan, J., Zhang, J., Yu, K.: Vast portfolio selection with gross-exposure constraints. J. Am. Stat. Assoc. **107**(498), 592–606 (2012)
18. Pun, C.S., Wong, H.Y.: A linear programming model for selection of sparse high-dimensional multiperiod portfolios. Eur. J. Oper. Res. **273**(2), 754–771 (2019)
19. Koenker, R., Bassett, G.: Regression quantiles. Econometrica **46**(1), 33–50 (1978)
20. Wang, L., He, X.: Analysis of global and local optima of regularized quantile regression in high dimensions: a subgradient approach. Econom. Theory (2022)
21. Gimenes, N., Guerre, E.: Quantile regression methods for first-price auctions. J. Econom. **226**(2), 224–247 (2022)
22. Cai, Z., Chen, H., Liao, X.: A new robust inference for predictive quantile regression. J. Econom. (2022). In press
23. D'Haultfœuille, X., Maurel, A., Zhang, Y.: Extremal quantile regressions for selection models and the black-white wage gap. J. Econom. **203**(1), 129–142 (2018)
24. Altunbaş, Y., Thornton, J.: The impact of financial development on income inequality: a quantile regression approach. Econ. Lett. **175**, 51–56 (2019)
25. Giessing, A., He, X.: On the predictive risk in misspecified quantile regression. J. Econom. **213**(1), 235–260 (2019)
26. Gu, J., Volgushev, S.: Panel data quantile regression with grouped fixed effects. J. Econom. **213**(1), 68–91 (2019)
27. Firpo, S., Galvao, A.F., Pinto, C., Poirier, A., Sanroman, G.: GMM quantile regression. J. Econom. (2022). In press
28. He, X., Pan, X., Tan, K.M., Zhou, W.X.: Smoothed quantile regression with large-scale inference. J. Econom. (2022). In press
29. Narisetty, N., Koenker, R.: Censored quantile regression survival models with a cure proportion. J. Econom. **226**(1), 192–203 (2022)
30. Li, Y., Zhu, J.: L1-norm quantile regression. J. Comput. Graph. Stat. **17**(1), 163–185 (2008)
31. Wu, Y., Liu, Y.: Variable selection in quantile regression. Stat. Sin. **19**(2), 801–817 (2009)
32. Belloni, A., Chernozhukov, V., et al.: $\ell 1$-penalized quantile regression in high-dimensional sparse models. Ann. Stat. **39**(1), 82–130 (2011)
33. Wang, L., Wu, Y., Li, R.: Quantile regression for analyzing heterogeneity in ultra-high dimension. J. Am. Stat. Assoc. **107**(497), 214–222 (2012)
34. Fan, J., Li, R.: Variable selection via nonconcave penalized likelihood and its oracle properties. J. Am. Stat. Assoc. **96**(456), 1348–1360 (2001)
35. Zhang, C.H.: Nearly unbiased variable selection under minimax concave penalty. The Annals of Statistics, pp. 894–942 (2010)
36. Sherwood, B., Maidman, A.: rqPen: Penalized Quantile Regression (2017). https://CRAN.R-project.org/package=rqPen. R package version 2.0
37. Koenker, R., Mizera, I.: Convex optimization, shape constraints, compound decisions, and empirical bayes rules. J. Am. Stat. Assoc. **109**(506), 674–685 (2014)
38. Peng, B., Wang, L.: An iterative coordinate descent algorithm for high-dimensional nonconvex penalized quantile regression. J. Comput. Graph. Stat. **24**(3), 676–694 (2015)
39. Gu, Y., Fan, J., Kong, L., Ma, S., Zou, H.: ADMM for high-dimensional sparse penalized quantile regression. Technometrics **60**(3), 319–331 (2018)
40. Boyd, S., Parikh, N., Chu, E., Peleato, B., Eckstein, J.: Distributed optimization and statistical learning via the alternating direction method of multipliers. Found. Trends®in Mach. Learn. **3**(1), 1–122 (2011)
41. Yu, L., Lin, N., Wang, L.: A parallel algorithm for large-scale nonconvex penalized quantile regression. J. Comput. Graph. Stat. **26**(4), 935–939 (2017)

42. Fan, Y., Lin, N., Yin, X.: Penalized quantile regression for distributed big data using the slack variable representation. J. Comput. Graph. Stat. **30**(3), 557–565 (2021)
43. Sun, D., Toh, K.C., Yang, L.: A convergent 3-block semiproximal alternating direction method of multipliers for conic programming with 4-type constraints. SIAM J. Optim. **25**(2), 882–915 (2015)
44. Zou, H., Li, R.: One-step sparse estimates in nonconcave penalized likelihood models. Ann. Stat. **36**(4), 1509–1533 (2008)
45. Wang, L., Kim, Y., Li, R.: Calibrating non-convex penalized regression in ultra-high dimension. Ann. Stat. **41**(5), 2505–2536 (2013)
46. Fan, J., Xue, L., Zou, H.: Strong oracle optimality of folded concave penalized estimation. Ann. Stat. **42**(3), 819–849 (2014)
47. Cai, T., Liu, W.: A direct estimation approach to sparse linear discriminant analysis. J. Am. Stat. Assoc. **106**(496), 1566–1577 (2011)
48. Fan, J., Li, R.: Variable selection via nonconcave penalized likelihood and its oracle properties. J. Am. Stat. Assoc. **96**(456), 1348–1360 (2001)
49. Mai, Q., Zou, H., Yuan, M.: A direct approach to sparse discriminant analysis in ultra-high dimensions. Biometrika **99**(1), 29–42 (2012)
50. Kolar, M., Liu, H.: Optimal feature selection in high-dimensional discriminant analysis. IEEE Trans. Inf. Theory **61**(2), 1063–1083 (2014)
51. Zhang, Z., Wang, S., Bian, W.: Sign consistency for the linear programming discriminant rule. Pattern Recogn. **100**, 107083 (2020)
52. Zhao, P., Yu, B.: On model selection consistency of Lasso. J. Mach. Learn. Res. **7**(Nov), 2541–2563 (2006)
53. Fan, J., Li, R., Zhang, C.H., Zou, H.: Statistical Foundations of Data Science. Chapman and Hall/CRC (2020)
54. Fortin, M., Glowinski, R.: Augmented Lagrangian methods: applications to the numerical solution of boundary-value problems, vol. 15. Elsevier (2000)
55. Fazel, M., Pong, T.K., Sun, D., Tseng, P.: Hankel matrix rank minimization with applications to system identification and realization. SIAM J. Matrix Anal. Appl. **34**(3), 946–977 (2013)
56. Chen, C., He, B., Ye, Y., Yuan, X.: The direct extension of ADMM for multi-block convex minimization problems is not necessarily convergent. Math. Program. **155**(1), 57–79 (2016)
57. Koenker, R.: Quantile regression: 40 years on. Ann. Rev. Econom. **9**, 155–176 (2017)
58. Lee, E.R., Noh, H., Park, B.U.: Model selection via Bayesian information criterion for quantile regression models. J. Am. Stat. Assoc. **109**(505), 216–229 (2014)
59. Sharpe, W.F.: Mutual fund performance. J. Bus. **39**(1), 119–138 (1966). http://www.jstor.org/stable/2351741
60. Sharpe, W.F.: The sharpe ratio. J. Portf. Manag. **21**(1), 49–58 (1994)
61. Friedman, J., Hastie, T., Höfling, H., Tibshirani, R.: Pathwise coordinate optimization. Ann. Appl. Stat. **1**(2), 302–332 (2007)
62. Friedman, J., Hastie, T., Tibshirani, R.: Regularization paths for generalized linear models via coordinate descent. J. Stat. Softw. **33**(1), 1–22 (2010)
63. Yi, C., Huang, J.: Semismooth newton coordinate descent algorithm for elastic-net penalized huber loss regression and quantile regression. J. Comput. Graph. Stat. **26**(3), 547–557 (2017)
64. Tan, K.M., Wang, L., Zhou, W.X.: High-dimensional quantile regression: Convolution smoothing and concave regularization. J. R. Stat. Soc.: Ser. B (Statistical Methodology) **84**(1), 205–233 (2022)
65. Wang, H.: Forward regression for ultra-high dimensional variable screening. J. Am. Stat. Assoc. **104**(488), 1512–1524 (2009)
66. Han, D., Sun, D., Zhang, L.: Linear rate convergence of the alternating direction method of multipliers for convex composite programming. Math. Oper. Res. **43**(2), 622–637 (2018)
67. Poliquin, R., Rockafellar, R.: A calculus of epi-derivatives applicable to optimization. Can. J. Math. **45**(4), 879–896 (1993)
68. Han, D., Sun, D., Zhang, L.: Linear rate convergence of the alternating direction method of multipliers for convex composite programming. Math. Oper. Res. **43**(2), 622–637 (2018)

69. Candes, E., Tao, T.: The dantzig selector: statistical estimation when p is much larger than n. Ann. Stat. **35**(6), 2313–2351 (2007)
70. Boyd, S., Boyd, S.P., Vandenberghe, L.: Convex Optimization. Cambridge University Press (2004)

Part III
Financial Risk Management

In the third part, we focus on the financial risk management. *Specifically, we first study the nested simulation method for portfolio risk measurement. We establish a budget allocation algorithm to guarantee that the nested simulation risk estimator could be applied efficiently in practice. Furthermore, we propose a unified confidence interval for nested simulation so that both the mean squared error of the point estimator and confidence interval width attain the optimal convergence rate.*

Next, we concentrate on the deep probabilistic forecasting for market risks, including deep sequential learning of conditional heavy-tailed distributions and ensemble multi-quantile regression with deep learning. Initially, we introduce a generative machine learning approach tailored for modeling heavy-tailed distributions commonly observed in financial markets. Subsequently, this approach is integrated with prevalent neural network architectures for sequential learning, such as Long Short-Term Memory (LSTM), to predict the conditional distribution of market returns and, consequently, forecast associated risk measures. Finally, we propose a novel neural ensemble approach for multi-quantile regression, designed for the general-purpose forecasting of conditional distributions. The approach seeks a balance between the distribution structure and the flexibility when predicting the conditional distributions. Through visualization, we observe that our approach is capable of learning conditional distributions with a diverse range of density shapes, including Gaussian, approximately Gaussian, sharp peak, asymmetry, heavy tails, finite bound, and multi-modality.

Chapter 5
Bootstrap-Based Budget Allocation for Nested Simulation

5.1 Introduction

Nested simulation refers to the problem of estimating a functional of a conditional expectation that cannot be evaluated analytically but requires simulation. In particular, the quantity of interest can be written as

$$\alpha = \rho(\mathbb{E}[Y|X]), \tag{5.1}$$

where X is a random vector in R^d with $d \geq 1$, Y is a random variable in $\mathbb{R}$, and the functional ρ maps a random variable to a real number, e.g., the cumulative distribution function or quantiles.

The estimation of α as depicted in Eq. (5.1) holds significant relevance in operations research, particularly in contexts such as risk assessment for financial portfolios. This application involves Y representing the (discounted) loss of a portfolio upon maturity, typically over one year, while $\mathbb{E}[Y|X]$ signifies its mark-to-market loss given a scenario of risk factors X up to a specified horizon, say, one week before maturity. In such scenarios, risk managers seek to compute various risk metrics associated with the mark-to-market loss $\mathbb{E}[Y|X]$, including its value-at-risk (VaR). VaR, defined as the quantile of the random loss $\mathbb{E}[Y|X]$, stands as a widely used risk measure within the financial industry.

The expression in Eq. (5.1) is quite versatile and encompasses a wide range of applications in nested simulation, accommodating various forms of the functional ρ. For example, in risk measurement, VaR with a confidence level $\beta \in (0, 1)$ is expressed as:

$$\rho(\mathbb{E}[Y|X]) = \inf\{x : F(x) \geq \beta\}, \tag{5.2}$$

where F denotes the cumulative distribution function of $\mathbb{E}[Y|X]$, and another widely used risk measure, the so-called conditional value-at-risk (CVaR) with a confidence

Z. Zhang et al., *Big Data in Economics and Management*, Statistics and Big Data 1,
https://doi.org/10.1007/978-981-95-3125-7_5

level β, is given by

$$\rho(\mathbb{E}[Y|X]) = \inf_{u\in\mathbb{R}} \left\{u + (1-\beta)^{-1}\mathbb{E}(\mathbb{E}[Y|X] - u)^+\right\}. \tag{5.3}$$

In addition to VaR and CVaR, other forms of ρ are of interest in the literature where it can be written as $\rho(\mathbb{E}[Y|X]) = \mathbb{E}\left[g(\mathbb{E}[Y|X])\right]$ for an appropriate function $g(\cdot)$; see, e.g., [1, 2]. In particular, when $g(\cdot)$ is a quadratic function (i.e., $g(z) = (z - z_0)^2$ for a given threshold value z_0), $\rho(\mathbb{E}[Y|X])$ is referred to as squared tracking error in financial engineering. In risk measurement context, when $g(\cdot)$ is a hockey-stick function (i.e., $g(z) = (z - z_0)^+$) or an indicator function (i.e., $g(z) = \mathbb{1}_{\{z\geq z_0\}}$), $\rho(\mathbb{E}[Y|X])$ is referred to as expected excess loss or probability of large losses beyond z_0, respectively.

The classical method for estimating (5.1) is nested simulation. While it has been well recognized that the allocation of simulation budget to outer and inner levels is at the heart of the nested simulation approach, how to decide an allocation rule has not been fully developed yet. Ideally, a good allocation rule should take into account two factors, including both characteristics of the distribution of (X, Y) and the form of ρ. In the existing literature within the nested simulation framework, most works have focused on the development of allocation rules by exploring specific structures of the forms of ρ, e.g., for probability of large losses as in [3, 4], for CVaR as in [5, 6], and for variance of the conditional expectation as in [7]. A notable exception is [8] where several forms of ρ are considered, including probability of large losses, VaR, and CVaR. Under the criterion of minimizing the asymptotic mean squared error (MSE) with a given computational budget Γ, [8] showed that the asymptotically optimal outer- and inner-sample sizes are represented as $n = c^{-1}\Gamma^{2/3} + o(\Gamma^{2/3})$ and $m = c\Gamma^{1/3} + o(\Gamma^{1/3})$, respectively, as $\Gamma \to \infty$. However, the parameter c usually involves unknown quantities such as the density function of $\mathbb{E}[Y|X]$ and its derivatives, and empirical evidence suggests that different c's may lead to substantially different MSEs. As such, a critical issue of how to choose c and thus an effective allocation rule, remains unresolved. Another issue that has not been addressed yet is in regard to the allocation rule when the goal is to construct confidence intervals (CIs), instead of minimizing the MSE of the estimator. Due to the existence of asymptotic bias, it is not clear how to construct asymptotically valid CIs for nested simulation estimators in [8].

This chapter aims to address the aforementioned issues on budget allocation for nested simulation. More specifically, we propose an automatic and sample-driven budget allocation rule to determine the outer- and inner-sample sizes under a unified framework that allows for different forms of ρ. The proposed method leverages asymptotic results in the literature and employs bootstrap sampling and least-squares regression to estimate the unknown parameter c, leading to effective outer- and inner-sample sizes.

The rest of this chapter is organized as follows. We introduce the backgrounds of the nested simulation approach in Sect. 5.2, and propose a sample-driven budget allocation rule in Sect. 5.3.

5.2 Backgrounds

We provide an overview of the nested simulation method in [8]. For ease of notation, let $L(X) = \mathbb{E}[Y|X]$. When estimating the quantity described in (5.1), the nested simulation method operates across two levels, as depicted in Fig. 5.1 for visual reference.

1. *Outer-level scenarios*. Generate n independent and identically distributed (i.i.d.) scenarios of X, denoted by $\{X_i, i = 1, ..., n\}$.
2. *Inner-level samples*. For each X_i, generate m i.i.d. samples of Y, denoted by $\{Y_{ij}, j = 1, ..., m\}$.

   Then $L(X_i)$ can be approximated by

$$L_m(X_i) = \frac{1}{m}\sum_{j=1}^{m} Y_{ij}.$$

As a result, the quantity of interest, i.e., $\alpha = \rho(L(X))$, can be estimated using $\{L_m(X_1), \ldots, L_m(X_n)\}$. For instance, when $\rho(\cdot) = \mathbb{E}g(\cdot)$ for a given function $g(\cdot)$, a nested estimator of α is given by

$$\alpha_{m,n} = \frac{1}{n}\sum_{i=1}^{n} g(L_m(X_i)).$$

When $\rho(L(X))$ is the quantile (or VaR) of $L(X)$ at a confidence level of β, the estimator $\alpha_{m,n}$ can be set as the sample quantile of $\{L_m(X_1), ..., L_m(X_n)\}$, denoted by $L_{m,\lceil \beta n \rceil}$, i.e.,

$$\alpha_{m,n} = L_{m,\lceil \beta n \rceil}, \tag{5.4}$$

where the operator $\lceil x \rceil$ denotes the smallest integer that is not less than x. When $\rho(L(X))$ is the CVaR of $L(X)$ with a confidence level β, the estimator is given by

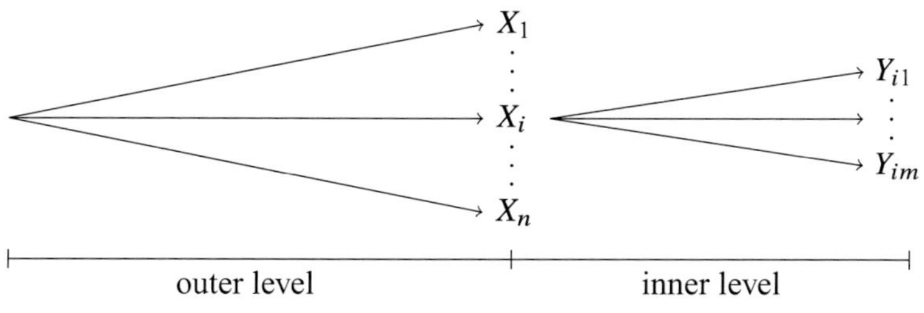

Fig. 5.1 Illustration of the nested simulation method

$$\alpha_{m,n} = \inf_{u \in \mathbb{R}} \left\{ u + \frac{1}{1-\beta} \frac{1}{n} \sum_{i=1}^{n} [L_m(X_i) - u]^+ \right\}. \tag{5.5}$$

Like in [9–11], we consider the following forms of the functional ρ in this chapter. These accommodate commonly used functionals in nested simulation.

($\mathcal{A}$.1) $\rho(\cdot) = \mathbb{E}g(\cdot)$ where $g(\cdot)$ is a smooth function with bounded third-order derivatives,
($\mathcal{A}$.2) $\rho(\cdot) = \mathbb{E}g(\cdot)$ where $g(\cdot)$ is a hockey-stick function,
($\mathcal{A}$.3) $\rho(\cdot) = \mathbb{E}g(\cdot)$ where $g(\cdot)$ is an indicator function,
($\mathcal{A}$.4) $\rho(\cdot)$ is the VaR at a confidence level of β as in (5.2),
($\mathcal{A}$.5) $\rho(\cdot)$ is the CVaR at a confidence level of β as in (5.3).

References [8, 9] provided the following results about asymptotic bias and variance of the nested simulation estimator.

Lemma 5.1 (Lemma 1 in [9]) *Suppose that* $\mathbb{E}[g''(L(X))Y^2] < \infty$ *when* ρ *takes the form as in (*$\mathcal{A}$*.1), and Assumption 5.1 holds when* $\rho(\cdot)$ *takes the form as in (*$\mathcal{A}$*.2)–(*$\mathcal{A}$*.5). Then,*

$$\mathbb{E}\alpha_{m,n} - \alpha = A/m + o(1/m) \quad and \quad \varepsilon(\alpha_{m,n}) = B/n + o(1/n), \tag{5.6}$$

where A and B are constants that depend on the forms of ρ *and the distribution of* (X, Y).

The asymptotic result outlined in (5.6) serves as the foundation for budget allocation strategies aimed at minimizing asymptotic MSE. More specifically, the total simulation budget is given by $\Gamma = n \cdot m$. Then, minimizing the asymptotic MSE leads to the following optimal outer-level sample size n^* and inner-level sample size m^*.

$$n^* = \left(B/(2A^2)\right)^{1/3} \Gamma^{2/3} + o(\Gamma^{2/3}), \text{ and } m^* = \left(2A^2/B\right)^{1/3} \Gamma^{1/3} + o(\Gamma^{1/3}), \tag{5.7}$$

where the operator $o(\cdot)$ denotes a small-order term.

When the constants A and B are known, a practical allocation strategy involves setting the outer- and inner-level sample sizes as outlined in (5.7), disregarding the small-order terms. However, A and B are typically unknown and involve parameters such as the density of $L(X)$ and its derivatives, which can be more challenging to estimate compared to α, the quantity of interest itself. Furthermore, empirical observations from numerical experiments reveal that substituting different values of $2A^2/B$ in (5.7) can lead to significantly divergent MSEs. In the next section, we introduce a sample-driven budget allocation method based on bootstrap proposed in [9].

5.3 A Sample-Driven Budget Allocation Method

This section discusses a budget allocation rule for nested simulation, as proposed by [9]. As discussed above, the constants A and B are typically unknown. A straightforward idea is to estimate these two constants. If one can obtain their estimators, denoted by $\hat{A}$ and $\hat{B}$, then the outer- and inner-sample sizes can be set as follows.

$$\hat{n} = \left\lceil \left(\hat{B}/(2\hat{A}^2)\right)^{1/3} \Gamma^{2/3} \right\rceil, \quad and \quad \hat{m} = \left\lceil \left(2\hat{A}^2/\hat{B}\right)^{1/3} \Gamma^{1/3} \right\rceil. \tag{5.8}$$

Zhang et al. [9] proposed a "bootstrap+regression" method to obtain estimators $\hat{A}$ and $\hat{B}$. Let us introduce a bootstrap procedure based on the initial samples of (X, Y), denoted by $\{(X_i, Y_{ij}), i = 1, ..., n_0, j = 1, ..., m_0\}$, where n_0 and m_0 are relatively small, e.g., $n_0 = m_0 = 100$. For convenience in notation, define $\mathbf{X}_{n_0} \triangleq \{X_1, ..., X_{n_0}\}$ and $\mathbf{Y}_{im_0} \triangleq \{Y_{i1}, ..., Y_{im_0}\}$. A bootstrap sample from $\mathbf{X}_{n_0}$ with size n' for some integer $n' \leq n_0$ refers to a sample obtained by independently and randomly drawing with replacement n' times from $\mathbf{X}_{n_0}$, denoted by $\mathbf{X}^*_{n'} = \{X^*_1, ..., X^*_{n'}\}$. Conditional on a bootstrap observation X^*_i, suppose $X^*_i = X_{i^*}$ for some index $i^* \in \{1, \ldots, n_0\}$, and let $\mathbf{Y}^*_{im_0} \triangleq (Y^*_{i1}, \ldots, Y^*_{im_0})$ denote the observations of Y given X_{i^*}, i.e., $(Y^*_{i1}, \ldots, Y^*_{im_0}) = (Y_{i^*1}, \ldots, Y_{i^*m_0})$. Then, we can similarly obtain a bootstrap sample with size m' from $\mathbf{Y}^*_{im_0}$, denoted by $\mathbf{Y}^{**}_{im'} \triangleq \{Y^{**}_{i1}, ..., Y^{**}_{im'}\}$, where $m' \leq m_0$.

After conducting the described bootstrap sampling procedure, we acquire bootstrap samples $\mathbf{X}^*_{n'}$ and $\mathbf{Y}^{**}_{im'}$ for $i = 1, \ldots, n'$, based on the initial sample. Using these bootstrap samples, a bootstrap estimator for $L(X^*_i), i = 1, \ldots, n'$ is proposed as follows,

$$L^*_{m'}(X^*_i) = \frac{1}{m'} \sum_{j=1}^{m'} Y^{**}_{ij},$$

leading to a bootstrap estimator of α, denoted by $\alpha^b_{m',n'}$. For instance, when $\alpha = \mathbb{E}[g(L(X))]$ for some function $g(\cdot)$, we set

$$\alpha^b_{m',n'} = \frac{1}{n'} \sum_{i=1}^{n'} g\left(L^*_{m'}(X^*_i)\right).$$

When α is the VaR or CVaR of $L(X)$, the bootstrap estimator $\alpha^b_{m',n'}$ can be constructed similarly to that of (5.4) and (5.5), respectively.

Zhang et al. [9] also introduced an alternative way of constructing bootstrap estimators based only on the bootstrap sample of $\mathbf{X}_{n_0}$, while keeping the observations of Y unchanged. In other words, for $i = 1, \ldots, n'$, $L(X^*_i)$ is estimated by

$$L_{m_0}(X_i^*) = \frac{1}{m_0}\sum_{j=1}^{m_0} Y_{ij}^*,$$

leading to another bootstrap estimator of α, denoted by $\alpha_{m_0,n'}^{b'}$. For instance, when $\alpha = \mathbb{E}[g(L(X))]$,

$$\alpha_{m_0,n'}^{b'} = \frac{1}{n'}\sum_{i=1}^{n'} g\left(L_{m_0}(X_i^*)\right).$$

The primary distinction between $\alpha_{m_0,n'}^{b'}$ and $\alpha_{m',n'}^{b}$ lies in the fact that the latter incorporates bootstrap sampling at both the outer and inner levels, whereas the former solely involves bootstrap sampling at the outer level.

It should be pointed out that the bootstrap samples are obtained based on the same initial samples and do not require generating additional samples of (X, Y). Essentially, the bootstrap procedure is based on the reuse of the initial samples. It can be implemented very fast, and the implementation time is often negligible compared to the simulation time required for generating the samples of (X, Y).

Intuitively, it is reasonable to expect that the bootstrap bias and variance of $\alpha_{m',n'}^{b}$ and $\alpha_{m_0,n'}^{b'}$ may have the same asymptotic forms as in (5.6), when m', m_0, n' and n_0 are sufficiently large, although it is not trivial from a theoretical point of view. As one of our contributions, we show in what follows that it is indeed the case. Denote by $\mathbb{E}_*$ and ε_* the expectation and variance taken under the bootstrap probability measure $\mathbb{P}_*(\cdot) = \mathbb{P}(\cdot|\mathbf{X}_{n_0}, \mathbf{Y}_{im_0}, i = 1, ..., n_0)$ given the initial sample of (X, Y). Asymptotic results on bootstrap bias and variance are summarized in the following theorem, whose proof is provided in the Online Appendix.

Theorem 5.1 (Theorem 1 in [9]) *Suppose that $\mathbb{E}Y^4 < \infty$ and $\mathbb{E}|g''(L(X))Y^2|^2 < \infty$ when ρ takes the form of ($\mathcal{A}$.1), and Assumption 5.1 in the Appendix holds for ($\mathcal{A}$.2)–($\mathcal{A}$.5). Moreover, assume that $n_0 = O((m')^{4/5+\delta})$ and $n_0 = O((m')^{5/3+\delta})$ for some $\delta > 0$ for ($\mathcal{A}$.2) and ($\mathcal{A}$.3)–($\mathcal{A}$.5), respectively. Then, as $m', m_0, n', n_0 \to \infty$,*

$$\mathbb{E}_*\alpha_{m',n_0}^{b} - \alpha_{m',n_0} = A/m' + o_p(1/m'), \text{ and } \varepsilon_*\left(\alpha_{m_0,n'}^{b'}\right) = B/n' + o_p(1/n'), \tag{5.9}$$

where A and B are the same constants as in Lemma 5.1, and the notation $o_p(\cdot)$ denotes a small-order term in probability, i.e., $a_n = o_p(b_n)$ if $a_n/b_n \to 0$ in probability.

The bootstrap mean of α_{m',n_0}^{b} and the bootstrap variance of $\alpha_{m_0,n'}^{b'}$ in Theorem 5.1 can be estimated based on repeating the bootstrap procedure multiple times independently. More specifically, for each given pair of (m', n_0), we repeat independently the above bootstrap procedure I times, leading to I versions of α_{m',n_0}^{b}, denoted by $\left\{\alpha_{m',n_0}^{b}(1), \ldots, \alpha_{m',n_0}^{b}(I)\right\}$. Then the bootstrap mean of α_{m',n_0}^{b} can be approximated by

$$\bar{\alpha}^b_{m',n_0} = \frac{1}{I}\sum_{q=1}^{I}\alpha^b_{m',n_0}(q).$$

In a similar manner, for each pair of (m_0, n'), we can obtain I versions of $\alpha^{b'}_{m_0,n'}$, denoted by $\{\alpha^{b'}_{m_0,n'}(1), \ldots, \alpha^{b'}_{m_0,n'}(I)\}$. Then the bootstrap variance estimator is given by

$$s^2_{m_0,n'} = \frac{1}{I}\sum_{q=1}^{I}\left[\alpha^{b'}_{m_0,n'}(q) - \frac{1}{I}\sum_{j=1}^{I}\alpha^{b'}_{m_0,n'}(j)\right]^2. \tag{5.10}$$

With the estimates of the bootstrap mean and variance, results in Theorem 5.1 pave the way for estimating the unknown constants A and B. We describe how to do so as follows. Choose a set of K_1 integer values $m_1 < m_2 < \cdots < m_{K_1} = m_0$, and estimate the bootstrap mean of $\alpha^b_{m_k,n_0}$ for each $k \in \{1, \ldots, K_1\}$, which is shown to be asymptotically linear in $1/m_k$ based on the first half of (5.9). Then, a least-squares regression fitting between the estimated bootstrap mean and $1/m_k$ leads to an estimate of the constant A. Similarly, choose a set of K_2 integer values $n_1 < n_2 < \cdots < n_{K_2} = n_0$, least-squares regression fitting between the estimated bootstrap variance of $\alpha^{b'}_{m_0,n_k}$ and $1/n_k$ leads to an estimate of the constant B. By (5.8), estimates of A and B lead straightforwardly to an allocation rule for outer- and inner-level sample sizes. The resulting algorithm is summarized as follows.

Algorithm 1 Nested estimator based on a sample-driven allocation rule

---

1. Inputs: $\Gamma, m_0, n_0, 1 \le m_1 < \cdots < m_{K_1} = m_0$ and $1 \le n_1 < \cdots < n_{K_2} = n_0$. Generate initial samples $\mathbf{X}_{n_0}$ and $\mathbf{Y}_{im_0}$, $i = 1, ..., n_0$.
2. Compute the bootstrap estimators

$$\left\{\bar{\alpha}^b_{m_1,n_0}, \ldots, \bar{\alpha}^b_{m_{K_1},n_0}\right\}, \quad \text{and} \quad \left\{s^2_{m_0,n_1}, \ldots, s^2_{m_0,n_{K_2}}\right\}.$$

   Regress the vectors $\left(\bar{\alpha}^b_{m_1,n_0}, ..., \bar{\alpha}^b_{m_{K_1},n_0}\right)^\top$ and $\left(s^2_{m_0,n_1}, ..., s^2_{m_0,n_{K_2}}\right)^\top$ on $(1/m_1, \ldots, 1/m_{K_1})^\top$ and $(1/n_1, \ldots, 1/n_{K_2})^\top$, respectively, and set $\hat{A}$ and $\hat{B}$ to be slopes of the regression, respectively.
3. Set outer- and inner-level sample sizes $\hat{n}$ and $\hat{m}$ by (5.8), generate the remaining outer- and inner-level samples, and compute the nested estimator.

During implementation, the initial sample sizes n_0 and m_0 are usually set to be relatively small, and thus it is often the case that $n_0 < \hat{n}$ and $m_0 < \hat{m}$. In practical cases, n_0 and m_0 can be set in such a manner that the initial samples consume a small percentage, e.g., 5 or 10%, of the total simulation budget Γ. With small n_0 and m_0, implementation of the bootstrap procedure is usually fast, and the corresponding computational time is often negligible compared to the overall simulation time of the samples of (X, Y). In a nutshell, the method in Algorithm 1 uses a small initial sample to guide the setting of the outer- and inner-level sample sizes. It is therefore referred to as a sample-driven allocation rule.

It should be emphasized that the above algorithm is applicable to settings with different forms of ρ and different distributions of (X, Y), which is appealing in practical applications.

5.4 Appendix

5.4.1 Regularity Conditions

Notation. Denote the inner-level sample by $Y_{ij} = L(X_i) + Z_j(X_i)$, where $Z_j(X_i)$ is a random error term with mean zero for a given X_i. Then $L_m(X_i) = L(X_i) + \bar{Z}_m(X_i)$, where $\bar{Z}_m(X) \triangleq \frac{1}{m}\sum_{j=1}^{m} Z_j(X)$ converges to 0 almost surely by the strong law of large numbers. Similarly, denote $L_m^*(X) = L(X) + \bar{Z}_m^*(X)$, where $\bar{Z}_m^*(X) = \frac{1}{m}\sum_{j=1}^{m} Z_j^*(X)$, and $\{Z_j^*(X), j = 1, ..., m\}$ are the bootstrap samples from $\{Z_j(X), j = 1, ..., m\}$. Throughout the analysis, the bootstrap probability measure is given by $\mathbb{P}_*(\cdot) = \mathbb{P}\left(\cdot|X_i, Z_j(X_i), i = 1, ..., n, j = 1, ..., m\right)$, and $\mathbb{E}_*$ denotes the expectation taken under this measure.

The central limit theorem suggests that $\widetilde{Z}_m(X) \triangleq \sqrt{m}\bar{Z}_m(X)$ has a non-trivial limiting distribution as $m \to \infty$. Similarly define $\widetilde{Z}_m^*(X) \triangleq \sqrt{m}\bar{Z}_m^*(X)$. It has been well known that $\widetilde{Z}_m^*(X)$ and $\widetilde{Z}_m(X)$ have the same asymptotic distribution; see, e.g., [12]. To facilitate analysis, we make the following assumption that imposes smoothness and boundedness on the joint densities of $(L(X), \widetilde{Z}_m(X))$ and $(L(X), \widetilde{Z}_m^*(X))$. Similar assumptions have been made in the literature on asymptotic analysis of nested simulation, and interested readers may refer to [8] for further discussion on similar regularity conditions.

Assumption 5.1 *Let $f_m(\cdot, \cdot)$ and $\tilde{f}_m(\cdot, \cdot)$ denote the joint densities of $(L(X), \widetilde{Z}_m(X))$ and $(L(X), \widetilde{Z}_m^*(X))$, respectively. Suppose that $f_m(y, z)$ and $\tilde{f}_m(y, z)$ are thrice differentiable in y for each m and all (y, z), and for $m \geq 1$, there exist nonnegative functions $p_{k,m}(\cdot)$ and $\tilde{p}_{k,m}(\cdot)$, $k = 0, 1, 2, 3$, such that $f_m(y, z) \leq p_{0,m}(z)$ and $\tilde{f}_m(y, z) \leq \tilde{p}_{0,m}(z)$, and for $k = 1, 2, 3$,*

$$\left|\frac{\partial^k}{\partial y^k} f_m(y, z)\right| \leq p_{k,m}(z), \text{ and } \left|\frac{\partial^k}{\partial y^k} \tilde{f}_m(y, z)\right| \leq \tilde{p}_{k,m}(z)$$

for all (y, z). Moreover, for $k = 0, 1, 2, 3$, and $0 \leq r \leq 10$,

$$\sup_m \int_{\mathbb{R}} |z|^r p_{k,m}(z) dz < \infty.$$

References

1. Broadie, M., Du, Y., Moallemi, C.C.: Risk estimation via regression. Oper. Res. **63**(5), 1077–1097 (2015)
2. Hong, L.J., Juneja, S., Liu, G.: Kernel smoothing for nested estimation with application to portfolio risk measurement. Oper. Res. **65**(3), 657–673 (2017)
3. Lee, S.H., Glynn, P.W.: Computing the distribution function of a conditional expectation via monte carlo: Discrete conditioning spaces. ACM Trans. Model. Comput. Simul. **13**(3), 238–258 (2003)
4. Broadie, M., Du, Y., Moallemi, C.C.: Efficient risk estimation via nested sequential simulation. Manage. Sci. **57**(6), 1172–1194 (2011)
5. Lan, H., Nelson, B.L., Staum, J.: A confidence interval procedure for expected shortfall risk measurement via two-level simulation. Oper. Res. **58**(5), 1481–1490 (2010)
6. Liu, M., Nelson, B.L., Staum, J.: An efficient simulation procedure for point estimation of expected shortfall. In: Proceedings of the 2010 Winter Simulation Conference, pp. 2821–2831. IEEE (2010)
7. Sun, Y., Apley, D.W., Staum, J.: Efficient nested simulation for estimating the variance of a conditional expectation. Oper. Res. **59**(4), 998–1007 (2011)
8. Gordy, M.B., Juneja, S.: Nested simulation in portfolio risk measurement. Manage. Sci. **56**(10), 1833–1848 (2010)
9. Zhang, K., Liu, G., Wang, S.: Bootstrap-based budget allocation for nested simulation. Oper. Res. **70**(2), 1128–1142 (2022)
10. Cheng, H.F., Liu, X., Zhang, K.: Constructing confidence intervals for nested simulation. Nav. Res. Logist. **69**, 1138–1149 (2022)
11. Wang, W., Yanyuan, W., Xiaowei, Z.: Smooth nested simulation: bridging cubic and square root convergence rates in high dimensions. Manag. Sci. **To appear** (2023)
12. Shao, J., Tu, D.: The Jackknife and Bootstrap. Springer Science and Business Media (2012)

Chapter 6
Constructing Confidence Intervals for Nested Simulation

6.1 Introduction

In the last chapter, we introduced the budget allocation rule for the nested simulation point estimator. In this chapter, we shift to focus on the construction of the confidence interval of nested simulation, which is proposed in [1].

The remaining chapter is structured as follows: Sect. 6.2 describes the formulations of nested simulation. In Sect. 6.3, we show CLTs for the five forms of ρ and provide a UCI framework, then we propose a detailed algorithm for constructing UCIs using the sample-driven budget allocation rule.

6.2 Formulations

There are some extra notations in this chapter. We denote L_β by the β-quantile of $L(X)$; $L_{\lceil \beta n \rceil}$ is the sample β-quantile of $\{L(X_1), ..., L(X_n)\}$; $(L_m)_\beta$ is the β-quantile of $L_m(X)$; and $L_{m,\lceil \beta n \rceil}$ is the sample β-quantile of $\{L_m(X_1), ..., L_m(X_n)\}$. Nested simulation proceeds in two levels, as shown in Fig. 5.1.

We also examine five specifications of ρ in Sect. 5.2 of Chap. 5, i.e., $(\mathcal{A}.1)$–$(\mathcal{A}.5)$. In this chapter, we specify these five estimators. That is, the corresponding expressions of α and their nested simulation estimators based on $\{L_m(X_1), ..., L_m(X_n)\}$ can be listed as follows.

$$\alpha^{(1)} = \mathbb{E}[g(L(X))], \qquad \alpha^{(1)}_{m,n} = \frac{1}{n}\sum_{i=1}^{n} g(L_m(X_i)),$$

$$\alpha^{(2)} = \mathbb{E}[(L(X) - x_0)^+], \qquad \alpha^{(2)}_{m,n} = \frac{1}{n}\sum_{i=1}^{n} (L_m(X_i) - x_0)^+,$$

$$\alpha^{(3)} = \mathbb{E}\left[\mathbb{1}_{\{L(X)\geq x_0\}}\right], \qquad \alpha^{(3)}_{m,n} = \frac{1}{n}\sum_{i=1}^{n} \mathbb{1}_{\{L_m(X_i)\geq x_0\}},$$

$$\alpha^{(4)} = \inf\{x : F(x) \geq \beta\}, \qquad \alpha^{(4)}_{m,n} = L_{m,\lceil \beta n \rceil},$$

Z. Zhang et al., *Big Data in Economics and Management*, Statistics and Big Data 1,
https://doi.org/10.1007/978-981-95-3125-7_6

$$\alpha^{(5)} = L_\beta + (1-\beta)^{-1}\mathbb{E}\left[L(X) - L_\beta\right]^+, \qquad \alpha^{(5)}_{m,n} = L_{m,\lceil \beta n\rceil} + \frac{1}{1-\beta}\frac{1}{n}\sum_{i=1}^{n}\left[L_m(X_i) - L_{m,\lceil \beta n\rceil}\right]^+,$$

where the forms of $\alpha^{(5)}$ and $\alpha^{(5)}_{m,n}$ hold because (5.3) attains its infimum if $u = L_\beta$.

Cheng et al. [1] provided the following lemma, which summarizes the convergence rates of the biases and variances of the nested estimators for the five cases.

Lemma 6.1 (Lemma 1 in [1]) *For any fixed $\lambda \in (0, 1)$, let L_λ be the λ-quantile of $L(X)$ and L'_λ denote the derivative of L_λ with regard to λ. Suppose that $\mathbb{E}[g''(L(X))Y^2] < \infty$ when ρ takes the form as in ($\mathcal{A}$.1), and Assumption A.1 holds when ρ takes the form as in ($\mathcal{A}$.2)–($\mathcal{A}$.5).*

- *When ρ takes the form as in ($\mathcal{A}$.1)–($\mathcal{A}$.4), as $m, n \to \infty$ and $m/n \to 0$, the following conclusion holds.*
- *When ρ takes the form as in ($\mathcal{A}$.5), assume that L'_λ is continuous over $(0, 1)$ and the order of the Riemann sum $\sum_{i=1}^{n-1} L'_{i/n}$ is no more than $O_n(n^{1+\delta})$ for some fixed $\delta \in (0, 1)$. Moreover, assume that $\theta(x) \in IR$ and is of BCV (the definitions of IR and BCV are presented in Online Appendix A.1. In this case, as $m, n \to \infty$ and $m/n^{1-\delta} < \infty$, the following conclusion holds.*

$$\mathbb{E}\left[\alpha^{(k)}_{m,n}\right] - \alpha^{(k)} = A_k/m + o_m(1/m) \text{ and } \mathrm{Var}\left[\alpha^{(k)}_{m,n}\right] = B_k/n + o_n(1/n),$$

where $k \in \{1, 2, 3, 4, 5\}$, and the functional forms of A_k and B_k are listed in Online Appendix A.1.

It could be beneficial to note that the findings concerning case ($\mathcal{A}$.5), namely CVaR, as detailed in Lemma 6.1, rely on the practically attainable estimator $\alpha^{(5)}_{m,n}$. In contrast, the results presented by [2] are based on a simplified estimator. Interested readers are encouraged to consult [1] for further insights.

Similar to (5.7), we have

$$m_k^* = \left(2A_k^2/B_k\right)^{1/3}\Gamma^{1/3}, \; n_k^* = \left(B_k/(2A_k^2)\right)^{1/3}\Gamma^{2/3}. \tag{6.1}$$

6.3 Confidence Intervals

In this section, we introduce the construction of CI for nested simulation. First, we derive the CLTs for all $\{\alpha^{(k)}_{m,n}\}_{k=1}^5$ with the asymptotically optimal convergence rate of $\Gamma^{-1/3}$.

From the derivation of (5.7) or (6.1), to minimize the MSE, we have to balance the bias and variance of the nested simulation estimator. Therefore, the CLT of the nested simulation may have the following expression.

$$r(\Gamma)^{-1/2}\left(\alpha^{(k)}_{m,n} - \alpha^{(k)}\right) \Rightarrow \text{bias term} + \text{noise term}.$$

Based on this intuition, [1] provide the following CLT result.

6.3.1 Central Limit Theorems

The following CLT result achieves the optimal convergence rate of $\Gamma^{-1/3}$ for nested simulation.

Theorem 6.1 (Theorem 1 in [1]) *Suppose that $\sqrt{n}/m \to C_k$, as $n, m \to \infty$ with a positive constant C_k for $k = 1, ..., 5$. Consider the following conditions:*

1. *ρ takes the form in ($\mathcal{A}$.1) and $\mathbb{E}\left[Y^8\right]$, $\mathbb{E}\left[g'(L(X))^4\right]$, $\mathbb{E}\left[g''(L(X))^4\right]$, and* $\mathrm{Var}\left[g(L(X))\right]$ *are finite;*
2. *ρ takes the form in ($\mathcal{A}$.2), $\mathbb{E}\left[Y^2\right] < \infty$, and $\sup_{m\geq 1}\mathbb{E}\left[|L_m(X)|^{2+\delta}\right] < \infty$ for some $\delta > 0$;*
3. *ρ takes the form in ($\mathcal{A}$.3), Assumption A.1 holds, and $\mathbb{E}\left[Y^2\right] < \infty$;*
4. *ρ takes the form in ($\mathcal{A}$.4), Assumption A.1 holds, $\mathbb{E}\left[Y^2\right] < \infty$, and $L(X)$ has a positive and continuous density in a neighborhood of L_β;*
5. *ρ takes the form in ($\mathcal{A}$.5), Assumption A.1 holds, $\mathbb{E}[Y^4] < \infty$, the optimal value of u in (5.3) is in a closed and bounded interval, $\theta(x) \in IR$, $\theta(x)$ is of BCV, and $L(X)$, $L_m(X)$ have positive and continuous densities.*

If one of the above-mentioned conditions holds, and $m, n \to \infty$,

$$\sqrt{n}\left(\alpha_{m,n}^{(k)} - \alpha^{(k)}\right) \Rightarrow C_k A_k + \mathcal{N}\left(0, B_k\right). \tag{6.2}$$

The findings outlined in Theorem 6.1 align with those in Lemma 6.1 in an intuitive manner. Specifically, the convergence rates of the bias and variance as elucidated in Lemma 6.1 lead to the optimal budget allocation, where n is proportional to the square of m. This allocation coincides with the condition $\sqrt{n}/m \to C_k$, as $n, m \to \infty$. Consequently, it follows from Lemma 6.1 that the bias and variance of $\sqrt{n}\left(\alpha_{m,n}^{(k)} - \alpha^{(k)}\right)$ converge to $C_k A_k$ and B_k, respectively.

Corollary 6.1 (Corollary 1 in [1]) *We obtain the following results for the five forms of ρ, under the same conditions as those considered in Theorem 6.1 except $\sqrt{n}/m \to C_k$:*

(1). If $\sqrt{n}/m \to 0$, then $\sqrt{n}\left(\alpha_{m,n}^{(k)} - \alpha^{(k)}\right) \Rightarrow \mathcal{N}\left(0, B_k\right)$;
(2). If $\sqrt{n}/m \to \infty$, then $m\left(\alpha_{m,n}^{(k)} - \alpha^{(k)}\right) \Rightarrow A_k$, i.e., $\sqrt{n}\left(\alpha_{m,n}^{(k)} - \alpha^{(k)}\right) \Rightarrow \infty$.

6.3.2 Unified Confidence Interval Framework

Theorem 6.1 provides a framework for constructing CIs. However, a challenge arises in estimating the parameters A_k and B_k in CIs, as well as selecting the appropriate value for C_k. To address this, [1] propose a nuanced solution by aiming to minimize the MSE of the point estimation. Given that the budget allocation rule (n_k^*, m_k^*)

outlined in (6.1) minimizes the MSE, we set $C_k = \sqrt{n_k^*/m_k^*} = \sqrt{B_k/2}/A_k$ as our selection criterion. Surprisingly, the CLTs in Theorem 6.1 reduce to

$$\sqrt{n}\left(\alpha_{m,n}^{(k)} - \alpha^{(k)}\right) \Rightarrow \sqrt{B_k/2} + \mathcal{N}(0, B_k). \tag{6.3}$$

The selection of C_k ensures that the point estimators have the minimum asymptotic MSE, while also guaranteeing the validity of the CLT (6.3). The implementation logic can be summarized as follows:

- Estimation of (A_k, B_k) and Budget Allocation: Initially, we apply the sample-driven budget allocation rule detailed in Sect. 5.3 to obtain estimators for (A_k, B_k), denoted by $(\hat{A}_k, \hat{B}_k)$, along with the budget allocation. Samples are generated based on these estimators, and point estimators with asymptotically optimal MSEs are constructed. This involves selecting $C_k = \sqrt{\hat{B}_k/2/\hat{A}_k}$ due to the budget allocation being determined by $(\hat{A}_k, \hat{B}_k)$.
- Estimation of B_k for Constructing CIs: The total samples are utilized to estimate B_k in (6.3) to construct the CIs. The estimator is denoted by $\hat{B}_k'$. Theoretically, as the sample sizes n_k and m_k approach infinity, $(\hat{A}_k, \hat{B}_k)$ converge to (A_k, B_k) according to Theorem 5.1, and $\hat{B}_k'$ converges to B_k according to Proposition 1 presented by [3].

In particular, an advantage of CLT (6.3) is that it is free of the bias constant A_k. Note that B_k can be typically estimated by sample variance, but the estimation of A_k is considerably more challenging. Broadie et al. [4] stated that "bias in particular is notoriously difficult to estimate." Furthermore, sample variance is a more robust and efficient estimate than that using the sample-driven budget allocation rule for B_k. Therefore, CI construction based on (6.3) is easier than that based on (6.2).

To construct the CIs, we must estimate B_k $(k = 1, ..., 5)$. Specifically, in cases $(\mathcal{A}.1)$–$(\mathcal{A}.3)$, the estimators of $\{B_k\}_{k=1}^3$ are established as

$$\xi_{m,n}^{2,(k)} = \frac{1}{n}\sum_{i=1}^{n} g^2(L_m(X_i)) - \left[\frac{1}{n}\sum_{i=1}^{n} g(L_m(X_i))\right]^2.$$

In case $(\mathcal{A}.4)$, the estimation of B_4 by the sample variance is challenging, and thus, we estimate B_4 through the sample-driven budget allocation rule. Because the allocation problem has been solved, the estimation uses the entire budget Γ instead of the initial budget Γ_0, thereby obtaining a more accurate estimate of B_4. The estimator of B_4 can be expressed as

$$\xi_{m,n}^{2,(4)} = n \cdot s_{m,n}^{2,(4)},$$

where $s_{m,n}^{2,(4)}$ is defined as

$$s^{2,(4)}_{m_0,n'} = \frac{1}{I}\sum_{q=1}^{I}\left[\alpha^{(4),b}_{m_0,n'}(q) - \frac{1}{I}\sum_{j=1}^{I}\alpha^{(4),b}_{m_0,n'}(j)\right]^2. \tag{6.4}$$

In case $(\mathcal{A}.5)$,

$$\xi^{2,(5)}_{m,n} = \frac{1}{(1-\beta)^2}\left\{\frac{1}{n}\sum_{i=1}^{n}\left[(L_m(X_i) - L_{m,\lceil \beta n\rceil})^+\right]^2 - \left[\frac{1}{n}\sum_{i=1}^{n}(L_m(X_i) - L_{m,\lceil \beta n\rceil})^+\right]^2\right\}.$$

[3] showed the consistency of these estimators. Therefore, (6.3) is used to obtain the asymptotically valid CIs of $100(1-\gamma)\%$ level, i.e.,

$$\left(\alpha_{m,n} + z_{\gamma/2}\xi^{(k)}_{m,n}/\sqrt{n},\ \alpha_{m,n} - z_{\gamma/2}\xi^{(k)}_{m,n}/\sqrt{n}\right) - (\sqrt{2}/2)\xi^{(k)}_{m,n}/\sqrt{n}, \tag{6.5}$$

where $z_{\gamma/2}$ is the $\gamma/2$ quantile of the standard normal distribution.

In summary, CIs (6.5) demonstrate that the convergence rate of the CI width is of order $n^{-1/2}$, i.e., $\Gamma^{-1/3}$. An important conclusion from (6.5) is that $\alpha_{m,n} - (\sqrt{2}/2)\xi^{(k)}_{m,n}/\sqrt{n}$ is a point estimator with bias correction which can reduce the convergence rate of the bias. However, the dominant term of MSE and convergence rate of MSE between $\alpha_{m,n}$ and $\alpha_{m,n} - (\sqrt{2}/2)\xi^{(k)}_{m,n}/\sqrt{n}$ are identical. In other words, both $\alpha_{m,n}$ and $\alpha_{m,n} - (\sqrt{2}/2)\xi^{(k)}_{m,n}/\sqrt{n}$ in (6.5) obtain the asymptotically optimal MSE. Therefore, the CI, which is constructed using the point estimator and CI width with optimal convergence rates, is termed the UCI. Observe that

$$\left(\alpha_{m,n} + z_{\gamma/2}\xi^{(k)}_{m,n}/\sqrt{n},\ \alpha_{m,n} - z_{\gamma/2}\xi^{(k)}_{m,n}/\sqrt{n}\right), \tag{6.6}$$

the first part in (6.5), represents the CIs constructed by [3] under the allocation assumption $\sqrt{n}/m \to 0$. The original CIs (6.6) become CIs with bias terms by rebalancing the sample sizes, i.e., $\sqrt{n}/m \to C_k$, and are improved by CIs (6.5) in the sense of convergence rate of the CI width.

The implementation of the UCI is expressed as Algorithm 1. Certain notations from Sect. 5.3 is reused. Notably, the UCI procedure is a general method to construct CIs for nested simulation only if the optimal budget allocation for point estimation could be obtained, and does not rely on a specific budget allocation method. The sample-driven budget allocation rule proposed by [3], as a robust and efficient method, is used in this chapter.

Algorithm 1 Procedures of the UCI

Step 1: *Use sample-driven rule to allocate outer- and inner-level sample sizes n and m;* Initialize sample sizes n_0 and m_0, generate samples $\mathbf{X}_{n_0}$ and $\{\mathbf{Y}_{1m_0}, \mathbf{Y}_{2m_0}, ..., \mathbf{Y}_{n_0m_0}\}$;

for $k = 1, ..., 5$ **do**

1. Compute the bootstrap estimator of bias constant A_k;
   Choose integers $m_1 < \cdots < m_{K_1} = m_0$;
   for $k_1 = 1, 2, ..., K_1$ **do**
   (a) **for** $i = 1, ..., n_0$ **do**
   Resample from $\mathbf{Y}_{im_0}$ and obtain $\mathbf{Y}^*_{im_{k_1}}$;
   Average $\mathbf{Y}^*_{im_{k_1}}$ to obtain loss estimator $L^*_{m_{k_1}}(X_i)$;
   (b) Use $\left\{L^*_{m_{k_1}}(X_i)\right\}_{i=1}^{n_0}$ to construct an estimator of α: $\alpha^{(k),b}_{m_{k_1},n_0}$;
   Repeat (a) and (b) I times to obtain I estimators of α, denoted by $\left\{\alpha^{(k),b}_{m_{k_1},n_0}(p)\right\}_{p=1}^{I}$;
   Take the average and get $\bar{\alpha}^{(k),b}_{m_{k_1},n_0} = \frac{1}{I}\sum_{p=1}^{I}\alpha^{(k),b}_{m_{k_1},n_0}(p)$;
   Regress the vector $\left(\bar{\alpha}^{(k),b}_{m_1,n_0}, ..., \bar{\alpha}^{(k),b}_{m_{K_1},n_0}\right)^\top$ on $\left[(\underbrace{1, 1, ..., 1}_{K_1})^\top, \left(1/m_1, ..., 1/m_{K_1}\right)^\top\right]$;
   Set the slope of the regression as an estimate of A_k;
2. Compute the bootstrap estimator of the variance constant B_k;
   Choose integers $n_1 < \cdots < n_{K_2} = n_0$;
   for $k_2 = 1, 2, ..., K_2$ **do**
   Resample from $\mathbf{X}_{n_0}$ and obtain $\mathbf{X}^*_{n_{k_2}}$;
   (c) **for** $i = 1, ..., n_{k_2}$ **do**
   Average the inner-level samples corresponding to X^*_i in $\mathbf{X}^*_{n_{k_2}}$;
   Use the above average to get the loss estimator $L_{m_0}(X^*_i)$;
   (d) Use $\{L_{m_0}(X^*_i)\}_{i=1}^{n_{k_2}}$ to construct estimator of α: $\alpha^{(k),b}_{m_0,n_{k_2}}$;
   Repeat (c) and (d) I times to obtain I estimators of α, denoted by $\left\{\alpha^{(k),b}_{m_0,n_{k_2}}(p)\right\}_{p=1}^{I}$;
   The bootstrap variance is $s^{2,(k)}_{m_0,n_{k_2}} = \frac{1}{I}\sum_{p=1}^{I}\left[\alpha^{(k),b}_{m_0,n_{k_2}}(p) - \frac{1}{I}\sum_{q=1}^{I}\alpha^{(k),b}_{m_0,n_{k_2}}(q)\right]^2$;
   Regress the vector $\left(s^{2,(k)}_{m_0,n_1}, s^{2,(k)}_{m_0,n_2}, ..., s^{2,(k)}_{m_0,n_{K_2}}\right)^\top$ on $\left(1/n_1, 1/n_2, ..., 1/n_{K_2}\right)^\top$;
   Set the slope of the regression as an estimate of B_k;
3. Take m and n by (6.1), then generate the remaining $\Gamma - n_0 * m_0$ samples;

Step 2: *Use the above m and n to construct the UCI;*

for $k = 1, ..., 5$ **do**

Compute a point estimator $\alpha^{(k)}_{m,n}$, and a consistent estimator $\xi^{2,(k)}_{m,n}$ for B_k;

Construct an asymptotically valid CI based on (6.5).

References

1. Cheng, H.F., Liu, X., Zhang, K.: Constructing confidence intervals for nested simulation. Nav. Res. Logist. **69**, 1138–1149 (2022)
2. Gordy, M.B., Juneja, S.: Nested simulation in portfolio risk measurement. Manag. Sci. **56**(10), 1833–1848 (2010)

3. Zhang, K., Liu, G., Wang, S.: Bootstrap-based budget allocation for nested simulation. Oper. Res. **70**(2), 1128–1142 (2022)
4. Broadie, M., Du, Y., Moallemi, C.C.: Efficient risk estimation via nested sequential simulation. Manag. Sci. **57**(6), 1172–1194 (2011)

Chapter 7
Deep Probabilistic Forecasting for Market Risks

7.1 Background of Market Risk Forecasting

In financial risk management, numerous institutional decisions hinge upon accurate forecasts of the conditional distributions of asset returns, particularly their left tails. These crucial decisions include the determination of bank capital requirements [1] and the stipulation of collateral requirements in lending and clearing operations [2]. What primarily concerns risk managers are not the routine daily price fluctuations, but rather the occurrence of unexpected and significantly large downfalls. There is a pervasive worry that such downturns could potentially trigger systemic spirals, leading to the destabilization of the financial system. Extensive research has been conducted to ascertain what constitutes an effective measure of tail risk [3, 4]. Furthermore, there is a widespread interest in developing methodologies to forecast extreme events in the far-left tail across various risk classes and asset types. However, it is important to note that significant deviations in asset prices are infrequent occurrences. By definition, an event breaching the 1%-quantile is expected to occur approximately twice a year. Therefore, it is imperative to first understand the underlying causes of such events before delving into the methodologies for forecasting them using historical data.

The primary catalysts for significant price drawdowns are often extreme events that have a widespread impact on the market. Notable examples include the stock market crash of 1987, the collapse of the dot-com bubble in 2000, the default of Lehman Brothers, and the subsequent financial crisis of 2008. However, idiosyncratic shocks, such as short selling [5], fire sales [6], and flash crashes [7], can also cause comparable damages. These events can lead to rapid market destabilization, either through positive feedback loops between diminishing market liquidity and funding liquidity [8], or as a result of complex interactions among market participants [9]. Moreover, research indicates that even in the absence of severe external shocks, accumulations of ordinary-sized shocks can still precipitate large price swings, as evidenced by the well-documented phenomenon of volatility clustering [10]. Glasserman and Wu

Z. Zhang et al., *Big Data in Economics and Management*, Statistics and Big Data 1,
https://doi.org/10.1007/978-981-95-3125-7_7

demonstrated that the self-exciting nature of volatility can transform conditionally light-tailed innovations into unconditionally heavy-tailed ones [11].

Empirical evidence suggests that the memory of asset returns is not confined solely to lower moments such as volatility. Rather, it is plausible that markets retain memories of more extreme episodes, albeit less frequent, which are reflected in the dynamics of return skewness, return kurtosis, and other higher moments. This hypothesis prompts an investigation into whether real data exhibit varying degrees of serial dependence at different quantiles in the conditional distribution of asset returns. If that is the case, risk managers would possess such insights during their forecasting activities. In essence, the ability to capture the underlying drivers of tail events, beyond those influencing volatility clustering, could substantially improve the prediction of near-extreme price movements.

To achieve this, a dynamic model is required that can exhibit varying lengths of memory across different probability levels in the conditional distribution. Incorporating a nonparametric component into the model is essential to allow the data to speak freely, distinguishing this approach from the parametric methodologies such as the GARCH family and those based on Extreme Value Theory. Consequently, this study adopts a semiparametric machine learning approach. The developed model leverages the flexibility of a recurrent neural network (RNN) to capture nonlinear dynamics and memory effects of varying lengths. Additionally, we address the typical black-box limitation of end-to-end neural networks by integrating a parametrization of the quantile function into the output layer, thereby rendering the learning process statistically interpretable. By framing the model training as a quantile regression problem, our preliminary studies demonstrate that the serial dependencies of conditional quantiles at low probability levels indeed contain risk factors distinct from those driving the second moments [12].

7.2 Background of Uncertainty Quantification in Machine Learning

Machine learning algorithms have experienced significant advancements in recent years, largely attributed to the successes of deep neural networks. Advanced deep learning models have achieved state-of-the-art performance across a variety of tasks, as measured by prediction accuracy [13]. However, deep learning has faced criticism for its opacity in making predictions and its tendency toward overconfidence in these predictions [14, 15]. Overconfidence can arise from an incorrect estimation of uncertainty associated with predictions. To address this issue, it is essential to provide a reliable measure of the uncertainty level for each prediction made by the model. For instance, in a regression task, this could involve providing a standard deviation or a prediction interval alongside the predicted value, indicating the range within which the true label is expected to fall with a certain probability. This would enhance the interpretability and reliability of the predictions made by deep learning models.

Uncertainty quantification occupies a critical role in machine learning, particularly in real-world applications of deep learning where decision-making relies heavily on model predictions. An inaccurate estimation of uncertainty can lead to highly risky decisions. Effective uncertainty estimation enables us to gauge the reliability of the model's predictions. For instance, in disease detection [16], the model can defer decision-making to human experts when it assesses the uncertainty to be high. Similarly, an autopilot system may transfer control to a human operator when the predicted situation is deemed risky. Furthermore, robust uncertainty quantification is advantageous in identifying and addressing out-of-distribution samples [17], which should be characterized by high uncertainty. Another illustrative example is in financial risk management, where decision-makers may need to adjust the margin rate or take hedging actions if they perceive high market uncertainty. This uncertainty is often defined by distributional characteristics of market variables, such as the fluctuation of the S&P 500 index in a future period. Quantitatively, these characteristics include volatility, kurtosis, Value-at-Risk, expected shortfall, and others, all of which are crucial and widely utilized risk measures in financial markets [18–20].

Numerous studies have highlighted that deep learning models may exhibit poor performance in uncertainty estimation [15, 17, 21, 22]. This issue is partly attributable to the absence of a definitive ground truth for uncertainty, posing significant challenges in addressing this critical problem. In classification tasks, where the softmax function is commonly employed, it is often straightforward to interpret the prediction score of a class as a measure of uncertainty. However, this practice has been criticized in the literature, as seen in works such as [21, 23]. While uncertainty quantification in classification tasks warrants considerable attention, this study primarily focuses on the regression problem, where uncertainty estimation presents its own set of challenges. In regression, aleatoric uncertainty quantification can be conceptualized as the task of estimating certain distributional characteristics of the conditional distribution $\mathbb{P}(\mathbf{y}|\mathbf{X} = x)$, where $\mathbf{X}$ represents the feature vector and $\mathbf{y}$ denotes the continuous label.

In recent years, a diverse array of methods has been developed for uncertainty quantification in neural networks. A common approach involves assuming a Gaussian distribution for $\mathbb{P}(\mathbf{y}|\mathbf{X} = x)$ and attempting to estimate its standard deviation as a function of x using a neural network trained with maximum likelihood and specific training schemes, as demonstrated in works such as [17, 24–26]. This straightforward strategy has proven to be highly effective when combined with robust training and ensemble techniques. Furthermore, Bayesian methods offer a natural way to obtain the density function of $\mathbb{P}(\mathbf{y}|\mathbf{X} = x)$, as seen in [21, 22, 27]. However, these methods can be challenging to implement and computationally intensive [17], and they require accurate specification of the prior distribution, which is often assumed for convenience. Conformal prediction [28, 29] represents another popular framework, which involves separating a calibration set from the training set and quantifying uncertainty using a conformity score. Several variants [30–32] have been proposed to enhance the flexibility of conformal prediction and adapt it to heteroscedasticity. Recent approaches have sought to estimate the conditional distribution by designing novel loss functions between the predicted $\mathbb{P}(\mathbf{y}|\mathbf{X} = x)$ and the observed data, as in [33, 34]. Additionally, ensembles over multiple Gaussians [35] or prediction intervals [36] have been proposed to improve uncertainty estimation, employing parallel ensemble techniques rather than the boosting-like approach presented in this study. In summary, there exists a wide variety of methods for uncertainty quantification in machine learning, each with its own advantages and limitations.

7.3 Deep Sequential Learning of Conditional Heavy-Tailed Distributions

In a discrete-time setting, forecasting the conditional quantiles of asset returns involves determining the time-t conditional α-quantile q_t of a scalar $\mathbf{y}_t$, where $\mathbb{P}(\mathbf{y}_t \leq q_t|\mathcal{I}_{t-1}) = \alpha$, given the probability $\alpha \in (0, 1)$ and the information set $\mathcal{I}_{t-1}$ up to time $t-1$. There are three primary approaches to addressing this problem: fully parametric, nonparametric, and semiparametric. The classification of approach depends on the method used to model the conditional distribution of $\mathbf{y}_t$: $F(y, \theta_t|\mathcal{I}_{t-1})$, where θ_t represents the unknown parameter of F, which governs the dynamics of the conditional distribution.

Fully parametric models, which are inherently interpretable, presuppose specific functional forms for the conditional distribution F and delineate parametric dynamics governing the temporal evolution of F. Prominent among these models for modeling discrete-time asset return dynamics is the GARCH family. Nonetheless, GARCH models are limited by the assumption that the dynamics of tail behavior are solely driven by the factors influencing clustered conditional volatility, with both conditional kurtosis and skewness remaining constant over time. Within the parametric framework, a notable endeavor to enhance the GARCH model involves introducing time-varying higher conditional moments. This initiative originated with the work of Hansen [37] and was subsequently extended by [38–40]. These advancements permit the parameters controlling the tail heaviness in $F(y, \theta_t|\mathcal{I}_{t-1})$ to exhibit autoregressive structures, thereby becoming time-varying. However, the autoregressive structures for different moments are governed by a single stochastic term, namely, the innovation term.

Nonparametric models, in contrast, do not presuppose specific functional forms for the conditional distribution. Instead, these models employ kernel methods for estimating the conditional distribution, a notable example being the Historical Simulation (HS) method, particularly used for Value-at-Risk (VaR) forecasts. HS is primarily based on the rectangular kernel. In practice, however, the filtered Historical Simulation (FHS) approach is more commonly utilized to account for data that are not independent and identically distributed (i.i.d.) [41]. The FHS methodology incorporates a location-scale model, such as GARCH, to pre-filter the data, enhancing its adaptability to real-world financial time series. Nonetheless, a significant limitation of the FHS approach is its sensitivity to the length of historical data used for forecasting, exemplified by the decision of whether to include events such as the 1987 stock market crash in the analysis.

Semiparametric models represent a compromise between the flexibility of nonparametric approaches and the structured nature of parametric models. These models typically concentrate on specific quantile levels of interest, while leaving the remainder of the conditional distribution unspecified. Among the semiparametric methodologies, two approaches stand out: the Conditional Extreme Value Theory (CEVT) and Dynamic Quantile Regression (DQR). The CEVT approach, as outlined by [42], characterizes the tail distribution according to one of three potential forms for the limiting distribution of extreme order statistics. A critical concern is the adequacy of extreme distributions as proxies for near-extreme quantiles at relatively small probability levels, such as 1% or 0.1%, rather than exceedingly rare events with probabilities on the order of 10^{-10}. To address this, [42] suggests fitting a GARCH model to the data initially, followed by the application of extreme value distributions to the standardized residuals, which are presumed to be i.i.d., mirroring the philosophy of the FHS approach. Conversely, the DQR approach places less emphasis on parameterizing the tail distribution and instead focuses on the dynamics of specific quantiles. A prominent example is the Conditionally Autoregressive Value-at-Risk (CAViaR) model proposed by [43]. However, a common challenge faced by DQR models is their susceptibility to the issue of quantile crossing.

7.3.1 Summary of Contributions

In this section, we adopt a semiparametric approach to develop a dynamic model for asset returns, acknowledging the need for a balance between the interpretability provided by parametrization and the flexibility required by data-driven considerations. Specifically, our contributions are threefold.

(i) A novel parametric construction of conditional quantile function: In our framework, we propose a novel construction of the conditional quantile function for asset returns. This construction is distinct in its ability to accommodate a broad spectrum of tail heaviness while separately addressing the behaviors of the left and right tails. Being fully parametric, it circumvents the challenging decision of selecting an appropriate conditional density. Moreover, this parametric quantile function is both parsimonious and sufficiently flexible to capture asymmetric heavy-tailed distributions.

(ii) A machine learning approach to estimate parameter dynamics: The nonparametric component of our methodology is manifested in the specification of the dynamics of the quantile function's parameters. We employ a sequential neural network, specifically a Long Short-Term Memory (LSTM) model, to learn these dynamics from historical data. Compared to traditional GARCH models, this machine learning approach offers several advantages. It is capable of capturing a) complex, potentially nonlinear dynamics that are unspecified in the dataset, and b) memory effects that may vary in length across different probability levels. Furthermore,

this approach provides the flexibility to incorporate expanded information sets. The training of the neural network can also be efficiently formulated within a quantile regression framework.

(iii) Capturing high-order memories beyond volatility persistence: Through a combination of simulation experiments and comprehensive empirical analysis on real data, we demonstrate that our model effectively captures long-term dependencies in the dynamics of the conditional distribution of asset returns. Our focus extends to the dynamics of higher moments and quantiles of the left tail. Specifically, for conditional skewness and kurtosis, our model identifies the presence of long-term memory that is distinct from the persistence commonly associated with volatility. Similarly, the left-tail quantiles, such as the 0.01-quantile, exhibit prolonged memories that deviate from the dynamics of volatility clustering. An ARMA model can numerically account for these additional serial dependencies. These findings suggest the existence of additional risk factors influencing the dynamics of higher moments or tail-side quantiles.

Please refer to [44] for more details on this topic.

7.3.2 Related Works

Let P_t denote the asset price at time t, and let $r_t = \log(P_t) - \log(P_{t-1})$ represent the asset return over the period from $t-1$ to t. The existing models under consideration assume that r_t follows the process described by [45]:

$$r_t = \mu_t + \sigma_t \varepsilon_t, \quad \varepsilon_t \sim \text{i.i.d. } F(\cdot), \tag{7.1}$$

where μ_t and σ_t represent the conditional mean and conditional volatility, respectively, and F is the cumulative distribution function of ε_t. A comprehensive model should specify the distribution F and delineate how μ_t and σ_t are dependent on the historical information set. For instance, the GARCH(1, 1) model [46, 47] assumes a standard normal distribution for ε_t and dictates the evolution of σ_t according to

$$\sigma_t^2 = \beta_0 + \beta_1(\sigma_{t-1}\varepsilon_{t-1})^2 + \beta_2\sigma_{t-1}^2. \tag{7.2}$$

Extensions such as EGARCH [48] and GJR-GARCH [49] introduce modifications to the aforementioned equation. Alternatives can also be applied to the choice of distribution F, for instance, the heavy-tailed t-distribution or the (generalized) skewed-t distribution as proposed by [37]. However, the degrees of freedom representing the tail heaviness and the degree of asymmetry, remain constant over time in these models. The rationale for employing asymmetric heavy-tailed distributions is rooted in the empirical evidence of heavy tails and asymmetry in financial returns, as summarized in [50]. While the heavy-tail phenomenon is widely acknowledged, the evidence for asymmetry is not as statistically robust. Nevertheless, the presence of moderate asymmetry is non-negligible. For instance, [51, 52] have documented empirical observations regarding the return skewness of single stocks and stock indices, indicating that stock index returns are typically negatively skewed, whereas single stock returns are positively skewed. These studies also propose theoretical explanations for these phenomena.

The conditional mean μ_t was usually modeled using a linear autoregressive approach, denoted as $\mu_t = \gamma_0 + \gamma_1 r_{t-1}$. Throughout the study, we will refer to GARCH-type models with t-distribution innovations and a linear autoregressive conditional mean as AR-GARCH-t, AR-EGARCH-t, and so forth. Unless explicitly specified, we assume that the orders of the GARCH and ARCH terms are both set to 1 for all GARCH-type models, analogous to the GARCH(1, 1) model described previously.

Beyond the GARCH family, both the FHS and CEVT methods, which initially filter the original return series using a GARCH model, can also be represented within the framework of Eq. (7.1). The distinctions between these methods lie in the characterization of the error term ε. In an FHS model, the empirical quantiles of ε are estimated nonparametrically using the samples of ε_t, which are the residuals obtained after the GARCH filtering process. On the other hand, the CEVT approach models ε by employing a nonparametric kernel distribution for the central part of the distribution and an extreme value distribution, such as the generalized Pareto distribution, for the tails.

Equation (7.1) provides a crucial insight into the relationship between different probability levels α and the corresponding α-quantiles of r_t, denoted as $q_t(\alpha)$. This quantile is linearly dependent on the α-quantile of ε:

$$q_t(\alpha) = \mu_t + \sigma_t F^{-1}(\alpha). \tag{7.3}$$

By fixing α and examining the temporal behavior of $q_t(\alpha)$, it is observed that the series $q_t(\alpha)$ is proportional to the volatility series σ_t (plus μ_t), resulting in similar quantile dynamics across different values of α. This limitation is present in the GARCH family as well as in models such as FHS and CEVT, potentially constraining the model's ability to accurately represent real data. Addressing this issue, several studies, including [37–40], have employed skewed heavy-tailed innovation distributions and allowed for time-varying skewness and kurtosis. The dynamics of skewness and kurtosis are modeled similarly to linear autoregression, akin to volatility. However, challenges arise in selecting an appropriate probability density function and managing the complexity of its mathematical formulation.

In the CAViaR model, Engle and Manganelli [43] critique the GARCH framework for its assumption that negative extremes follow the same process as volatility. They argue that different quantiles may exhibit distinct dynamics. Consequently, they propose modeling the dynamics of each quantile separately for different values of α, rather than specifying the full conditional distribution. The estimation of CAViaR is conducted through quantile regression with the loss function $L_\alpha(r, q) = (\alpha - I(r < q))(r - q)$. However, a potential limitation of CAViaR is its susceptibility to the common issue of quantile regression, known as quantile crossing. This phenomenon occurs when $q_t(\alpha_1) > q_t(\alpha_2)$ despite $\alpha_1 < \alpha_2$, leading to inconsistencies in the estimated quantiles.

Stochastic volatility (SV) models constitute another significant category of models pertinent to forecasting the conditional distribution of asset returns. Various studies have compared GARCH-type models with SV models [53–56]. SV models are particularly applicable in scenarios where volatility is influenced by additional risk factors. In a continuous-time framework, when driven by Brownian motion, SV models exhibit Markovian properties, which fundamentally distinguish them from the GARCH family and our proposed model. Such Markovian characteristics may limit their suitability for capturing serial dependence in volatility. Models that align more closely with our proposed approach and the focus of this section are those that incorporate long-memory volatility, driven by, for example, fractional Brownian motion or the Hawkes process, and are preferably formulated in discrete time.

7.3.3 Long Short-Term Memory

Since the AI program AlphaGo triumphed over a top human player in the game of Go, deep learning [13] has gained increasing prominence across various fields beyond computer science and engineering. In the financial industry, leading investment banks and hedge funds have begun to incorporate AI expertise into their newly established AI research groups. The aim is to leverage machine learning techniques across various business units, including trading execution, risk management, and portfolio management. For economists and management scientists, machine learning, encompassing deep learning, offers a novel set of econometric or statistical methods [57] that potentially hold significant advantages over traditional models due to their data-driven nature and fewer restrictions on model assumptions. Consequently, there is potential for machine learning to provide fresh insights into longstanding research problems in finance.

Deep learning models, characterized by their numerous parameters, are adept at capturing complex nonlinear relationships inherent in extensive datasets, albeit at the cost of some interpretability. The prevalent assumption of linearity in many econometric models may not accurately represent the complexities of the real world, particularly when handling large volumes of data. As such, there is a willingness to trade off a degree of interpretability for enhanced model capability and performance. Recent advancements in optimization algorithms and computational hardware have facilitated the efficient fitting of deep learning models. Moreover, contrary to intuition, the abundance of parameters does not invariably lead to overfitting, thanks to the implementation of effective training techniques. Deep learning has led to significant breakthroughs in various domains, including computer vision, machine translation, and bioinformatics. However, its successful applications in finance are comparatively less documented.

The basic architecture in deep learning is the deep neural network, wherein nonlinearity is introduced through the composition of selected nonlinear activation functions. Within the domain of time series modeling, Long Short-Term Memory (LSTM) networks, a type of sequential neural network, have gained popularity for their ability to capture both long-term and short-term dependencies or complex dynamics in sequences. LSTMs have achieved remarkable successes in various applications, including speech recognition, machine translation, and protein structure prediction. Consequently, they are a natural choice for modeling the dynamics of the conditional distribution of financial asset return series. Mathematically, an LSTM is a complex nonlinear function that maps a sequence of input vectors $x_1, \ldots, x_n$ to a corresponding sequence of output vectors $y_1, \ldots, y_n$ (or to a single output vector y), via hidden state vectors $h_1, \ldots, h_n$. Examples of applications include the translation of a sentence from Chinese to English and the classification of a music clip by genre.

The Formulation of LSTM

Before describing the full mathematics of it, we first introduce the simple recurrent neural network (RNN), which is the understructure of LSTM and has the formulation:

$$h_j = \sigma_h(W_h x_j + U_h h_{j-1} + b_h), \tag{7.4}$$
$$y_j = \sigma_y(W_y h_j + b_y), \tag{7.5}$$

for $j = 1, \ldots, n$. W_h, U_h, b_h, W_y, b_y are the parameters that need to be learned from the data, and σ_h and σ_y are nonlinear activation functions. It models a nonlinear functional relationship between $x_1, \ldots, x_n$ and $y_1, \ldots, y_n$. One can stack this structure multiple times to get a multi-layered or deep RNN, i.e., obtain layer k's hidden state vectors $h_1^k, \ldots, h_n^k$ using $h_1^{k-1}, \ldots, h_n^{k-1}$ via $h_j^k = \sigma_h(W_h^k h_j^{k-1} + U_h^k h_{j-1}^k + b_h^k)$.

LSTM extends this understructure and has the formulation:

$$f_j = \sigma_g(W_f x_j + U_f h_{j-1} + b_f), \tag{7.6}$$
$$i_j = \sigma_g(W_i x_j + U_i h_{j-1} + b_i), \tag{7.7}$$
$$o_j = \sigma_g(W_o x_j + U_o h_{j-1} + b_o), \tag{7.8}$$
$$g_j = \sigma_h(W_g x_j + U_g h_{j-1} + b_g), \tag{7.9}$$
$$c_j = f_j * c_{j-1} + i_j * g_j, \tag{7.10}$$
$$h_j = o_j * \sigma_h(c_j), \tag{7.11}$$

where $*$ represents element-wise multiplication of two vectors. At last, the output y_j is a chosen nonlinear function of h_j:

$$y_j = \sigma_h(W_y h_j + b_y). \tag{7.12}$$

All parameters W, U, b need to be learned, while σ_g and σ_h are nonlinear activation functions, chosen as the S-shaped logistic function and the tanh function, respectively. In scenarios where only one output vector y is required, one may utilize either the average of all hidden state vectors $\frac{1}{n}\sum h_j$ or simply the last one h_n, as exemplified by $y = \sigma_h(W_y h_n + b_y)$ in our Eq. (7.19). The logistic function and tanh function are the most widely employed activation functions in deep neural network models, serving as nonlinear transformations that map the support of their arguments to a bounded domain. The multiple compositions of these activation functions enable the approximation of complex nonlinear relationships between input vectors $x_1, \ldots, x_n$ and output vectors $y_1, \ldots, y_n$ or y.

The Long Memory

To demonstrate the capacity of LSTM networks to represent flexible serial dependence structures, particularly their ability to capture long-term memory, we investigate the derivative of the hidden state h_j with respect to a preceding input x_{j-D} as the lag D increases. For simplicity of analysis, we assume that all vectors and matrices in the network are reduced to scalars. In the case of a simple RNN:

$$\frac{\partial h_j}{\partial x_{j-D}} = \frac{\partial h_j}{\partial h_{j-1}} \frac{\partial h_{j-1}}{\partial x_{j-D}} = \cdots = \frac{\partial h_{j-D}}{\partial x_{j-D}} \prod_{k=1}^{D} \frac{\partial h_{j-k+1}}{\partial h_{j-k}} \tag{7.13}$$

$$= (1 - h_{j-D}^2) W_h U_h^D \prod_{k=1}^{D} (1 - h_{j-k+1}^2). \tag{7.14}$$

The usual situation is $|U_h| < 1$. Now $|\partial h_j / \partial x_{j-D}| \leq |W_h||U_h|^D$ is exponentially decaying with respect to D, suggesting a short memory of the network only. If $|U_h| > 1$, the derivative may explode, which is another known issue of the simple RNN.

For LSTM, to make things simple, we replace $h_j = o_j * \sigma_h(c_j)$ by $h_j = c_j$, and set f_j and i_j to be constant across different j (this can be done by setting W_f, U_f, W_i, and U_i to 0). Now $h_j = f_j h_{j-1} + i_j g_j$, and

$$\frac{\partial h_j}{\partial x_{j-D}} = \frac{\partial h_j}{\partial h_{j-1}} \frac{\partial h_{j-1}}{\partial x_{j-D}} = \cdots = \frac{\partial h_{j-D}}{\partial x_{j-D}} \prod_{k=1}^{D} \frac{\partial h_{j-k+1}}{\partial h_{j-k}} \tag{7.15}$$

$$= \frac{\partial h_{j-D}}{\partial x_{j-D}} \prod_{k=1}^{D} \left(f_{j-k+1} + i_{j-k+1}(1 - g_{j-k+1}^2) U_g \right). \tag{7.16}$$

We can interpret $f_{j-k+1}, i_{j-k+1} \in (0, 1)$ as the weights of 1 and $(1 - g_{j-k+1}^2) U_g$, respectively. The term within the multiplicative product represents a rebalancing between 1 and $(1 - g_{j-k+1}^2) U_g$, regardless of whether $|U_g| < 1$ or $|U_g| > 1$. If f_{j-k+1}

is close to 1 and i_{j-k+1} is close to 0, the phenomena of vanishing or exploding derivatives are mitigated as D increases. This characteristic suggests the presence of long-term memory in the LSTM network. Furthermore, allowing f_{j-k+1} and i_{j-k+1} to vary enables a more flexible memory structure within the LSTM framework.

7.3.4 The New Methodology

In this subsection, we begin by presenting the proposed parametric quantile function. Subsequently, we employ this function to model the conditional distribution $\mathbb{P}(r_t|\mathcal{I}_{t-1})$ of financial return series, illustrating how it depends on the past information set. Finally, we conclude the discussion of the proposed methodology by detailing the model fitting process, which is accomplished through a quantile regression formulation.

A Novel Heavy-Tailed Quantile Function

Three common approaches exist for comprehensively representing a continuous distribution: the probability density function (PDF), cumulative distribution function (CDF), and quantile function. In the context of modeling financial data, considerable emphasis is placed on selecting an appropriate parametric PDF that aligns with the empirical characteristics of financial returns, such as heavy tails. To our knowledge, there is no existing literature that achieves this through a suitable CDF or quantile function. In this study, we introduce a parsimonious parametric quantile function designed to accommodate varying degrees of tail heaviness with intuitive parameters.

Our approach is inspired by the Q-Q plot, a widely used method for assessing whether a set of observations conforms to a normal distribution. The underlying principle is straightforward: the α-quantile of a normal distribution $\mathcal{N}(\mu, \sigma^2)$ is given by $\mu + \sigma Z_\alpha$, where Z_α represents the α-quantile of the standard normal distribution. As α varies within the interval $(0, 1)$, the corresponding Q-Q plot forms a straight line. An inverted S-shaped Q-Q plot, on the other hand, suggests that the distribution in question exhibits heavy tails. For instance, Fig. 7.1a illustrates the Q-Q plot of a t-distribution with 2° of freedom against $\mathcal{N}(0, 1)$, indicative of a heavy-tailed distribution.

We develop a parsimonious parametric quantile function as a function of Z_α, allowing for a controllable-shape Q-Q plot when compared against the standard normal distribution. Specifically, the upper and lower tails of the inverted S-shaped Q-Q plot are governed by two distinct parameters, respectively. Our proposed heavy-tailed quantile function (HTQF) is expressed as follows:

$$Q(\alpha|\mu, \sigma, u, v) = \mu + \sigma Z_\alpha \left(\frac{e^{uZ_\alpha}}{A} + \frac{e^{-vZ_\alpha}}{A} + 1 \right), \tag{7.17}$$

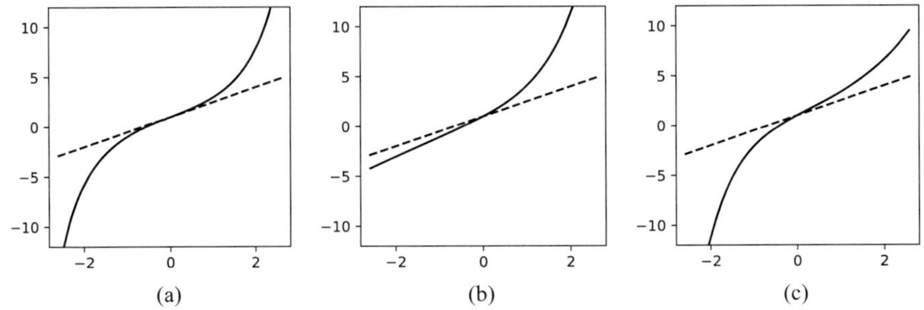

Fig. 7.1 Q-Q plots against $\mathcal{N}(0, 1)$: **a** $t(2)$; **b** HTQF with $u = 1.0$ and $v = 0.1$; **c** HTQF with $u = 0.6$ and $v = 1.2$. For all three distributions, $\mu = 1$ and $\sigma = 1.5$. For HTQF, $A = 4$

where μ and σ represent the location and scale parameters, respectively, while A denotes a relatively large positive constant. The parameter $u \geq 0$ governs the upper tail of the inverted S-shape, corresponding to the right tail of the distribution, whereas $v \geq 0$ regulates the lower tail, corresponding to the left tail of the distribution. The magnitude of u or v is directly proportional to the heaviness of the respective tail. In the case where $u = v = 0$, the HTQF simplifies to the quantile function of a normal distribution.

To elucidate this, consider that in Eq. (7.17), Z_α is initially multiplied by a simpler factor $f_u(Z_\alpha) = e^{uZ_\alpha}/A + 1$, then scaled by σ and shifted by μ (for simplicity, one may set $\mu = 0$ and $\sigma = 1$). The factor f_u is a monotonically increasing and convex function of Z_α, and it satisfies $f_u \to 1$ as $Z_\alpha \to -\infty$. Consequently, $Z_\alpha f_u(Z_\alpha)$ will only exhibit the upper tail of the inverted S. A similar analysis applies to $Z_\alpha f_v(Z_\alpha) = Z_\alpha(e^{-vZ_\alpha}/A + 1)$ as well. Therefore, $Z_\alpha(f_u(Z_\alpha) + f_v(Z_\alpha))/2$ displays the entire inverted S-shaped Q-Q plot (with $2A$ replaced by A in Eq. (7.17)). The role of A is to ensure that $f_u(0)$ and $f_v(0)$ are close to 1, and to guarantee that the HTQF is monotonically increasing with Z_α. Figure 7.1b and c illustrate the Q-Q plots of HTQF with various values of u and v against $\mathcal{N}(0, 1)$. These plots demonstrate different degrees of tail heaviness, with the tails flexibly adjusting based on u and v. Moreover, an HTQF with fixed parameters represents the quantile function of a unique probability distribution, as its inverse function exists and constitutes a CDF. For the proof, please refer to Appendix A.

The LSTM-HTQF Model

Unlike GARCH-type models, we do not make specific assumptions regarding the probability density function of the distribution $\mathbb{P}(r_t|\mathcal{I}_{t-1})$. Instead, we propose that its quantile function follows the heavy-tailed quantile function (HTQF), denoted by $Q(\alpha|\mu_t, \sigma_t, u_t, v_t)$, where μ_t and σ_t are time-varying parameters representing the location and scale, respectively, while u_t and v_t control the shapes of the left and right tails of the corresponding distribution.

We posit that the time-t parameters μ_t, σ_t, u_t, and v_t are functions of the historical return series. To model this dependency, we select a fixed-length subsequence from the past returns $r_{t-1}, r_{t-2}, \ldots$ to construct a feature vector sequence. We then employ the LSTM network to learn the mapping between the feature vectors and the four HTQF parameters. LSTM networks [58] are a popular and effective choice for sequential learning and are particularly suited for our methodology (see [59] for a comprehensive review of LSTM). Specifically, we choose a fixed-length L, and construct a feature vector sequence of length L from $r_{t-1}, \ldots, r_{t-L}$:

$$x_1^t, \ldots, x_L^t = \begin{bmatrix} r_{t-L} \\ (r_{t-L} - \bar{r}_t)^2 \\ (r_{t-L} - \bar{r}_t)^3 \\ (r_{t-L} - \bar{r}_t)^4 \end{bmatrix}, \ldots, \begin{bmatrix} r_{t-1} \\ (r_{t-1} - \bar{r}_t)^2 \\ (r_{t-1} - \bar{r}_t)^3 \\ (r_{t-1} - \bar{r}_t)^4 \end{bmatrix}, \tag{7.18}$$

where $\bar{r}_t = \frac{1}{L}\sum_{i=1}^{L} r_{t-i}$. The rationale behind this construction is to extract information encapsulated in the raw quantities associated with the first four central moments (mean, variance, skewness, and kurtosis) of the past L samples. We posit that these high-order moments provide direct insights into the future conditional distribution, particularly concerning the heaviness of the future left and right tails. This construction is designed to facilitate the extraction of relevant information by the neural network, as opposed to relying solely on the first two moments. Subsequently, we model the four parameters of HTQF, namely μ_t, σ_t, u_t, and v_t, as the outputs of an LSTM when fed with the input feature vectors $x_1^t, \ldots, x_L^t$:

$$[\mu_t, \sigma_t, u_t, v_t]^\top = \tanh(W^o h_t + b^o), \qquad h_t = \mathrm{LSTM}_\Theta(x_1^t, \ldots, x_L^t). \tag{7.19}$$

Θ is the LSTM parameters, and h_t is the last hidden state vector of LSTM. W^o and b^o are the output layer parameters.

Finally, for model training, we select K fixed probability levels $0 < \alpha_1 < \alpha_2 < \cdots < \alpha_K < 1$ and minimize the average quantile regression loss between r_t and its conditional quantiles $Q(\alpha_k|\mu_t, \sigma_t, u_t, v_t)$ across all k and t, akin to traditional quantile regression:

$$\min_{\Theta, W^o, b^o} \frac{1}{K}\frac{1}{T-L}\sum_{k=1}^{K}\sum_{t=L+1}^{T} L_{\alpha_k}\left(r_t, Q(\alpha_k|\mu_t, \sigma_t, u_t, v_t)\right). \tag{7.20}$$

$L_\alpha(r, q) = (\alpha - I(r < q))(r - q)$ is the quantile regression loss. Combining Eqs. (7.17), (7.18), (7.19), and (7.20) completes the formulation of our proposed LSTM-HTQF model and its fitting process. After fitting, for subsequent out-of-sample series $\{r_{t'}\}_{t'>T}$, the time-varying HTQF parameters $\mu_{t'}, \sigma_{t'}, u_{t'}, v_{t'}$ can be directly calculated from historical returns using the learned model parameters $\hat{\Theta}$, $\hat{W}^o$, $\hat{b}^o$. Consequently, the full conditional distribution at time t' can be estimated, along with the conditional quantiles or any moments of interest. This enables an analysis of the dynamics of the conditional distribution learned from the model.

Furthermore, our study also emphasizes quantile or Value-at-Risk (VaR) forecasting. To assess performance on the out-of-sample set, one can obtain the forecasted quantile sequence and apply common VaR backtesting procedures as recommended by regulatory authorities, or evaluate the forecasts using a loss function.

Remarks

One of the advantages of the LSTM-HTQF model is that the proposed HTQF is more intuitive to understand and flexible enough to model asymmetric heavy tails. It has a simpler mathematical form compared to the distributions in Extreme Value Theory and the probability densities in GARCH-type models, such as the generalized skewed t-distribution in [37, 40]. Additionally, the LSTM in our model is data-driven and can learn nonlinear dependence and long memory from the past information set of the data, while the linear autoregressive FHS, CEVT, CAViaR, and GARCH family may not. Furthermore, from another perspective, compared to traditional quantile regression, our model overcomes the issue of quantile crossing naturally since the HTQF is monotonically increasing with α.

The feature vector sequence $x_1^t, x_2^t, \ldots, x_L^t$ can be designed to incorporate any information that is pertinent to the conditional distribution of r_t or that enhances prediction accuracy, such as trading volume, related assets, or fundamental indicators. To maintain consistency with the GARCH family and other related models, and to ensure fair comparisons, we construct $x_1^t, x_2^t, \ldots, x_L^t$ solely from past returns $r_{t-1}, r_{t-2}, \ldots$. However, in practical applications of our method, additional information can be integrated into the feature vector sequence.

Our method has broad applicability in quantile prediction and time series modeling across various non-financial fields. It is particularly well suited for time series data that exhibit time-varying asymmetrical tail behavior and nonlinear serial dependence in the conditional distribution, such as hydrologic data, internet traffic data, or electricity price and demand. Additionally, the standard normal distribution in the Q-Q plot can be replaced with other baseline distributions, allowing the HTQF to have a controllable-shaped Q-Q plot against the chosen distribution, such as exponential or lognormal distributions. The selection of the baseline distribution should be based on specific domain knowledge.

7.3.5 Simulation Studies

The objective of the simulation experiment is to ascertain whether our method is capable of capturing the serial dependence of a conditional distribution, particularly when its higher moments demonstrate significant time-varying effects. Analogous to a GARCH specification, we generate a simulated time series in discrete time according to

$$r_t = \mu_t + \sigma_t z_t, \quad (7.21)$$
$$\mu_t = 0.052 + 0.172 r_{t-1}, \quad (7.22)$$
$$\sigma_t^2 = 0.293 + 0.161(\sigma_{t-1} z_{t-1})^2 + 0.575\sigma_{t-1}^2, \quad (7.23)$$

but with a skew-t distributed innovation where λ and η are parameters controlling the skewness and kurtosis (degrees of freedom):

$$z \sim \text{skew-}t(\lambda, \eta).$$

Its density is given by:

$$f(z|\lambda,\eta) = c\left(1 + \frac{2(g-\rho^2)\left(z+\rho/\sqrt{g-\rho^2}\right)^2}{(\eta+1)\left(1+\lambda\,\text{sign}\left(z+\rho/\sqrt{g-\rho^2}\right)\right)^2}\right)^{-\frac{\eta+1}{2}}, \text{ where} \quad (7.24)$$
$$c = \left(\frac{2(g-\rho^2)}{\eta+1}\right)^{1/2} B\left(\frac{\eta}{2},\frac{1}{2}\right)^{-1}, \quad (7.25)$$
$$\rho = 2\lambda\left(\frac{\eta+1}{2}\right)^{1/2} B\left(\frac{\eta-1}{2},1\right) B\left(\frac{\eta}{2},\frac{1}{2}\right)^{-1}, \quad (7.26)$$
$$g = (1+3\lambda^2)\frac{\eta+1}{2} B\left(\frac{\eta}{2},\frac{1}{2}\right)^{-1} B\left(\frac{\eta-2}{2},\frac{3}{2}\right), \quad (7.27)$$

and $B(x,y) := \int_0^1 t^{x-1}(1-t)^{y-1}dt$ is the Beta function.

To impose parametric dynamics on the skewness parameter λ_t and the kurtosis parameter η_t so that their evolutions are known as a prior, we modify the setup in [40] as follows:

$$\lambda_t = -1 + 2/(1+\exp(-\tilde{\lambda}_t)), \text{ where} \quad (7.28)$$
$$\tilde{\lambda}_t = -0.038 + 0.076 z_{t-1}^3 + 0.463\tilde{\lambda}_{t-1}; \quad (7.29)$$
$$\eta_t = 2 + 2\exp(3 - \tilde{\eta}_t), \text{ where} \quad (7.30)$$
$$\tilde{\eta}_t = 0.136 + 0.057 z_{t-1}^4 + 0.717\tilde{\eta}_{t-1}. \quad (7.31)$$

We generate a total of 30,000 data points, with the last one-tenth designated as the out-of-sample data set. Another one-tenth is extracted to form the validation set, which is utilized to halt the fitting process when the loss on this set begins to increase, thereby preventing overfitting. For the LSTM configuration, we set $L = 25$ and $H = 8$, while for the HTQF, we fix $A = 4$. We choose $K = 21$ probability levels for model fitting, with the set of α defined as $[\alpha_1, \ldots, \alpha_{21}] = [0.01, 0.05, 0.1, \ldots, 0.9, 0.95, 0.99]$.

After fitting, we forecast the conditional distribution for each day in the out-of-sample set by predicting the four HTQF parameters and compare them with the true parameters that generated the time series: μ_t, σ_t, λ_t, η_t. Denoting the forecasted HTQF parameters as $\hat{\mu}_t$, $\hat{\sigma}_t$, $\hat{u}_t$, $\hat{v}_t$, we employ $\hat{u}_t - \hat{v}_t$ as a proxy for skewness and location parameters: $\{\mu_t\}$ v.s. $\{\hat{\mu}_t\}$ scale parameters: $\{\sigma_t\}$ v.s. $\{\hat{\sigma}_t\}$

skewness parameters: $\{\lambda_t\}$ v.s. $\{\hat{u}_t - \hat{v}_t\}$ kurtosis parameters: $\{\eta_t\}$ v.s. $\{\hat{u}_t + \hat{v}_t\}$

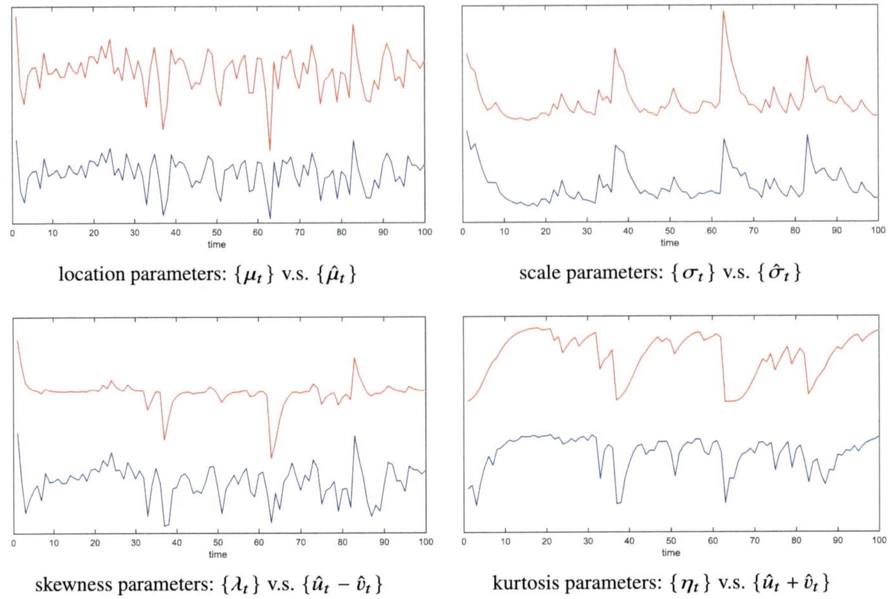

Fig. 7.2 Comparisons between the true parameters (red lines) of the simulated time series and the forecasted HTQF parameters (blue lines) by our model on the out-of-sample set. The linear correlation coefficients between the four pairs of parameters are 0.9780, 0.9104, 0.8014, and $-$0.7867, respectively. Linear transformations are made before plotting

$\hat{u}_t + \hat{v}_t$ as a proxy for kurtosis. These proxies are chosen for their intuitiveness and because the true skewness or kurtosis may not be well defined for certain values of λ_t and η_t, which are themselves proxies. In Fig. 7.2, we present the pairs of true and forecasted parameters for location, scale, skewness, and kurtosis, where the red lines represent the true parameters and the blue lines represent the forecasted ones. Linear transformations are applied prior to plotting to ensure that the parameters are within similar ranges. The high linear correlation between the forecasted and true parameters indicates that our model has successfully captured the temporal dynamics of the conditional distribution of r_t. The linear correlation coefficients between the four pairs of parameters are 0.9780, 0.9104, 0.8014, and -0.7867, respectively. The negative sign in the last coefficient arises because a heavier tail corresponds to larger values of $\hat{u}_t$ or $\hat{v}_t$, but smaller values of η_t.

7.3.6   *Empirical Studies*

We collect the time series of NASDAQ 100 index's daily returns of the maximum possible length. The start date and end date are 1985-10-01 and 2018-07-02, respectively, with totally 8257 data points. We employ a rolling-window forecast setting, wherein 250 days are used as the out-of-sample set for forecasting. Subsequently, these 250 days are incorporated into the in-sample set, and the model is re-fitted. The in-sample set consistently begins from the start of the time series. The start date of the entire out-of-sample set is 1996-08-26, and the length is 5500. All returns are calculated using $r_t = \ln(P_t/P_{t-1})$, where P_t is the price at time t. The time series is normalized to have zero sample mean and unit sample variance.

Upon fitting the model, a quarter of the in-sample set is extracted to form the validation set. This machine learning technique is employed for selecting hyperparameters and for halting the optimization iterations when the loss on the validation set starts to increase, thereby preventing overfitting. Our model has two hyper-parameters: the length L of the past series $r_{t-1}, \ldots, r_{t-L}$, which influences the time-t HTQF parameters $\mu_t, \sigma_t, u_t, v_t$, and the hidden state dimension H of the LSTM. In subsequent empirical studies, we set $L = 100$ and $H = 16$ without change. The parameter A in the HTQF is fixed at 4. For model fitting, we select $K = 21$ probability levels for the set of α: $[\alpha_1, \ldots, \alpha_{21}] = [0.01, 0.05, 0.1, \ldots, 0.9, 0.95, 0.99]$, to use in the optimization in Eq. (7.20).

To expedite the training of the neural network, we employ the commonly used mini-batch technique, which obviates performing gradient descent over the entire dataset. Initially, we randomly partition the in-sample set into five equal parts, referred to as batches. Subsequently, for every five iterations, we apply gradient descent to each of the five batches in sequence to update the neural network parameters. This approach is viable because the loss function to be minimized in Eq. (7.20) is a summation of the loss for each data point, or for each time t, with shared parameters. In each iteration, only a fraction of all t is utilized to compute the gradient and update the model parameters. For further details on the mini-batch technique, please refer to [60]. We employ the Adam optimization algorithm [61] with the recommended default settings provided by TensorFlow [62]. Typically, the training requires a few hours on a desktop PC equipped with two 3.30GHz quad-core CPUs.

Given that our utilization of LSTM does not presuppose a specific parametric form for the dynamics of the conditional distribution, this study aims to explore the dynamics that our model has learned from the daily return data. These dynamics include the behaviors of both moments and quantiles of the conditional distribution.

Moment Dynamics

In the out-of-sample set $\{r_{t'}\}$, our model forecasts the HTQF parameters $\hat{\mu}_{t'}, \hat{\sigma}_{t'}, \hat{u}_{t'}, \hat{v}_{t'}$ for each t', which fully specify the conditional distribution of $r_{t'}$ via its quantile function $Q(\alpha|\hat{\mu}_{t'}, \hat{\sigma}_{t'}, \hat{u}_{t'}, \hat{v}_{t'})$. Consequently, we can randomly sample from this distribution and compute the sample moments to estimate the true moments of the distribution. We then analyze the temporal behaviors of these moments to discern the dynamics our model has extracted from the data. The sampling process is straightforward, as it merely involves drawing a number z from the standard normal distribution and then inputting it into the HTQF to obtain $\hat{\mu}_{t'} + \hat{\sigma}_{t'} z \left(e^{\hat{u}_{t'} z}/A + e^{-\hat{v}_{t'} z}/A + 1\right)$. This sample follows the distribution defined by the

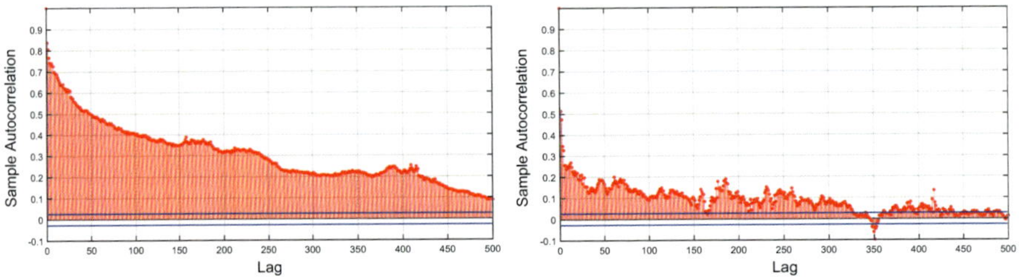

Fig. 7.3 Autocorrelation functions of the volatility and skewness sequences forecasted by our model on the NASDAQ 100 out-of-sample set. The skewness is regressed to the mean and volatility first, and the residual sequence is used instead

quantile function $Q(\alpha|\hat{\mu}_{t'}, \hat{\sigma}_{t'}, \hat{u}_{t'}, \hat{v}_{t'})$. Taking the equity index NASDAQ 100 as an example, for each t', after sampling 100,000 times, we calculate the sample mean, volatility, skewness, and kurtosis, respectively. These sample statistics serve as estimations of the conditional mean, volatility, skewness, and kurtosis of $r_{t'}$. Consequently, we obtain four sequences of moment estimations for the out-of-sample set.

Subsequently, we assess the length of dependence or memory within these moment sequences using the autocorrelation function. To evaluate the volatility dynamics captured by our model, we present in Fig. 7.3 the autocorrelation function of the volatility sequence forecasted by our model. It is evident that our model produces highly persistent volatility, as indicated by the slowly decaying autocorrelation function. In the second subplot of Fig. 7.3, we investigate the memory length of the skewness sequence. After linearly regressing skewness against mean and volatility to eliminate correlations and obtaining the residuals, we plot the autocorrelation function of this residual sequence. This subplot reveals that the skewness also possesses long memory, remaining significant even at a 300-day lag, despite having a relatively smaller magnitude compared to the memory of volatility.

Based on these observations, we can infer that the dynamics of skewness are distinct from those of volatility. This is implicitly evidenced by the memory length of the skewness sequence after the removal of mean and volatility effects. It suggests that existing models may overlook additional risk drivers, and new stochastic terms might be required to adequately model the skewness process. Due to the potential non-existence of kurtosis at certain time points, we are unable to demonstrate its memory length using the autocorrelation function. However, in the following, we will explore the memory characteristics of the left-tail quantiles.

Quantile Dynamics

Our LSTM-HTQF model also predicts conditional quantiles for all probability levels. We investigate the length of dependence or memory of the forecasted quantile sequence through the autocorrelation function, using the NASDAQ 100 index as

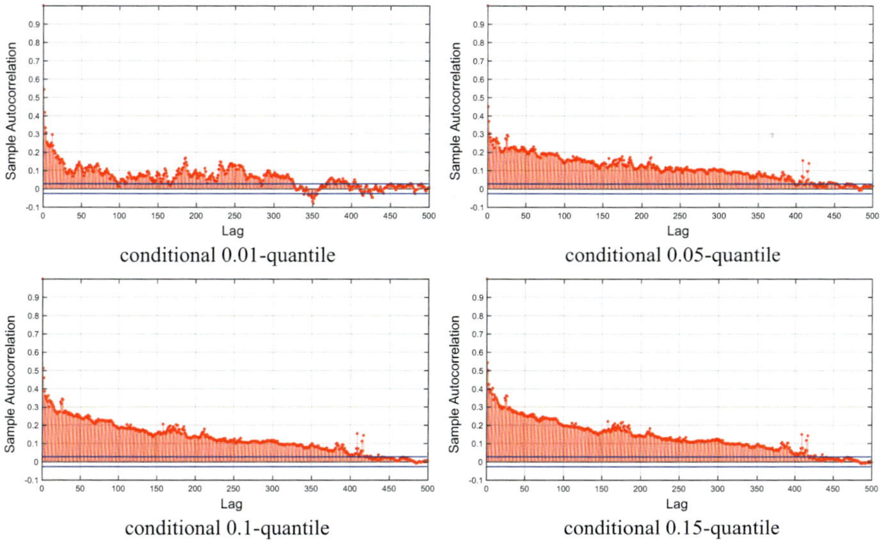

Fig. 7.4 Autocorrelation functions of the forecasted quantile sequences (regressed to mean and volatility) on the out-of-sample set of NASDAQ 100

an illustrative example. It is important to note that the quantiles are likely to be significantly correlated with the mean and volatility. Therefore, we perform a linear regression of the forecasted quantiles against the forecasted mean and volatility and examine the autocorrelation function of the residual sequence. This analysis helps to uncover the independent dynamics of the quantiles that are distinct from those of the mean and volatility.

We select four left-tail quantiles with probability levels of 0.01, 0.05, 0.1, and 0.15, and plot the autocorrelation functions of their residuals, obtained through regression, in Fig. 7.4. The plots reveal very long dependencies in all four quantile sequences, indicating that the quantiles possess independent dynamics. Furthermore, as the probability level increases from 0.01 to 0.15, the magnitude of the dependence progressively intensifies. This observation suggests that different quantiles may exhibit distinct dynamics depending on their probability levels.

To ascertain the likely dynamics of the regressed quantile sequences, we fit an ARMA model to each sequence. Subsequently, we examine the autocorrelation function of the ARMA residual sequence to detect any remaining serial dependence. After fitting an ARMA(2, 2) model, the ARMA residuals of the quantile sequences (regressed to mean and volatility) for all probability levels exhibit no serial dependence, as demonstrated by their autocorrelation functions in Fig. 7.5 (taking the 0.01- and 0.05-quantiles as examples). The autocorrelations are consistently within the significance bounds. An ARMA(1, 1) model does not produce ARMA residuals with similar autocorrelation functions, suggesting that the dynamics of the regressed quantile sequences closely resemble those of an ARMA(2, 2) model. The equations

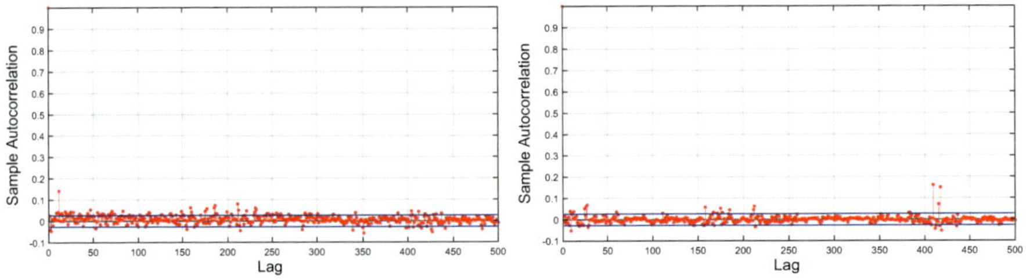

conditional 0.01-quantile conditional 0.05-quantile

Fig. 7.5 Fit an ARMA(2, 2) to the regressed 0.01-and 0.05-quantile sequences of NASDAQ 100 out-of-sample set and plot the autocorrelation functions of the ARMA residuals. The fitted ARMA equation for 0.01-quantile is $y_t = -0.0000 - 0.1502y_{t-1} + 0.7814y_{t-2} + \varepsilon_t + 0.5512\varepsilon_{t-1} - 0.3888\varepsilon_{t-2}$, and for 0.05-quantile is $y_t = -0.0000 + 1.5439y_{t-1} - 0.5457y_{t-2} + \varepsilon_t - 1.2713\varepsilon_{t-1} + 0.2958\varepsilon_{t-2}$

of the fitted ARMA models are provided in the caption of Fig. 7.5. Lastly, the findings discussed in this study apply not only to the NASDAQ 100 but also to many other assets.

7.3.7 Conclusion

In conclusion, we introduced a novel parametric Heavy-Tailed Quantile Function (HTQF) to represent the asymmetric heavy-tailed conditional distribution of financial return series. The dependency of the HTQF's four parameters on the past information set was modeled using a nonparametric, data-driven machine learning approach, specifically the sequential neural network Long Short-Term Memory (LSTM). The training of our LSTM-HTQF model was formulated within a quantile regression framework. After learning from the data, our model successfully captured the dynamics of the conditional distribution. We analyzed the dynamics of higher moments and tail-side quantiles, which all exhibited quite long memories independent of those responsible for volatility clustering. This suggests the existence of additional risk factors driving the dynamics of higher moments or tail-side quantiles.

In future research, the development of more sophisticated models capable of capturing the dynamics of the conditional distribution of financial time series in greater detail will be essential. This may involve enhancing the flexibility of the HTQF or refining the application of LSTM. Additionally, the feature vector sequence inputted into the LSTM could be expanded to include information beyond merely past return history. The overarching goal of these advancements is to achieve a deeper understanding of the conditional distribution dynamics and to generate more accurate Value-at-Risk (VaR) forecasts.

7.4 Ensemble Multi-Quantile Regression with Deep Learning

Another class of methods for uncertainty quantification involves estimating quantiles to construct prediction intervals [32, 63–66], which is the primary focus of this study. Most existing works in this area utilize the quantile regression loss from [67, 68] for model training. These methods typically produce a prediction interval $[q_\tau(x), q_{1-\tau}(x)]$ formed by the τ-quantile and $(1-\tau)$-quantile of $\mathbb{P}(\mathbf{y}|\mathbf{X}=x)$. However, our focus is on situations where multiple intervals $[q_\tau(x), q_{1-\tau}(x)]$ need to be predicted for various values of τ, to quantify uncertainty with different levels of confidence. From another perspective, these quantiles can form an approximation of $\mathbb{P}(\mathbf{y}|\mathbf{X}=x)$. Consequently, our attention shifts toward the prediction of the conditional distribution itself.

In the absence of an assumed distribution structure, this approach is tantamount to presuming an extremely flexible distribution for $\mathbb{P}(\mathbf{y}|\mathbf{X}=x)$, aligning with the nonparametric methodology in statistical analysis. Conversely, the opposite extreme involves adopting a well-established distribution structure, such as a Gaussian distribution, and subsequently predicting its parameters. We contend that both extremes may be detrimental to achieving more accurate uncertainty estimation. On one hand, an overly flexible distribution may lack the necessary constraints to provide meaningful and reliable uncertainty estimates. On the other hand, a rigid distributional assumption, such as Gaussian, may not adequately capture the complexity and variability inherent in the data, leading to overly simplistic and potentially misleading uncertainty estimates.

In the approach where no distribution structure is assumed, the resulting implied density function, which can be derived from the predicted quantiles, may exhibit a chaotic function graph lacking any discernible distribution shape, even with a large training dataset. This phenomenon is illustrated clearly in Fig. 7.8. Some methods have developed neural network representations for $\mathbb{P}(\mathbf{y}|\mathbf{X}=x)$, such as modeling $q_\tau(x)$ as a neural network that takes x and $\tau \in (0, 1)$ as inputs, thereby incorporating significant flexibility in distribution prediction. We contend that excessive flexibility does not necessarily lead to improved performance for several reasons: (i) it can make the training process more challenging; (ii) it may increase the possibility of overfitting; and (iii) it can result in density functions with peculiar shapes or even invalid forms that lack interpretability. Unwarranted flexibility may be detrimental and has been introduced into the modeling of $\mathbb{P}(\mathbf{y}|\mathbf{X}=x)$ in some existing methods. We demonstrate that striking a balance between flexibility and distribution structure leads to substantially better distribution prediction. By proposing a novel Ensemble Multi-Quantiles (EMQ) method, we achieve adaptively flexible distribution prediction. This approach can dynamically find the optimal balance and thus accurately approximate $\mathbb{P}(\mathbf{y}|\mathbf{X}=x)$, as will be shown in the remainder of the paper.

Conversely, the alternative approach of assuming a succinct distribution structure for $\mathbb{P}(\mathbf{y}|\mathbf{X}=x)$ exhibits clear shortcomings. It suffers from a lack of flexibility, and the model's performance is heavily dependent on the suitability of the chosen distribution structure. For convenience, a Gaussian distribution is often selected. However, conditional distributions in real data may exhibit characteristics such as multimodality, asymmetry, and heavy tails, and may even display heterogeneity within a dataset. The use of mixtures of Gaussians in Bayesian approaches or model ensembles does not significantly mitigate these deficiencies. Other distribution structures, such as the t-distribution, skew-t distribution, and Generalized Beta distribution, may have complex mathematical expressions for their density functions. This complexity increases the difficulty of training a model with them using maximum likelihood loss, as the gradient with respect to the distribution parameters can be challenging to obtain. Moreover, the choice among these distribution structures often relies on specific domain knowledge.

Another motivation of this paper is to address the quantile-crossing issue when estimating multiple quantiles of $\mathbb{P}(\mathbf{y}|\mathbf{X} = x)$. Although numerous methods have been proposed in the statistical literature to tackle this problem [69–73], they often do not address the issue of unnecessary flexibility mentioned earlier (peculiar or invalid density function shapes). And they may introduce other challenges, such as computational burden or difficulty, particularly if they are based on constraint optimization. Besides constraint optimization, post-processing or re-sorting represents another imperfect solution to the quantile-crossing issue. We argue that more natural solutions, which do not require additional effort to address this issue, are essential. In this paper, the proposed EMQ method naturally overcomes the quantile-crossing issue through thoughtful design. Our solution is built-in and does not require additional efforts such as constraint optimization or post-processing, as will be demonstrated in the following.

For a complete story of the methodology and more details on the analysis of it, please refer to the research work presented in [74].

7.4.1 The New Methodology

In this section, we describe the proposed Ensemble Multi-Quantiles (EMQ) approach for uncertainty quantification. Throughout the remainder of the chapter, we use $\mathbf{X}$ and $\mathbf{y}$ to denote the random variables representing the features and the continuous label, respectively. Given a feature vector $\mathbf{X} = x$, the objective is to predict the conditional τ_k-quantile of the variable $\mathbf{y}$, denoted as $q_k(x)$, where $k = 1, \ldots, K$ and $0 < \tau_1 < \cdots < \tau_K < 1$. The probability levels $\tau_1, \ldots, \tau_K$ are specified by us and are distributed densely in the interval $(0, 1)$, such as $0.01, 0.02, \ldots, 0.98, 0.99$, forming the set $\Pi = \{\tau_k\}_{k=1}^{K}$. These K conditional quantiles provide an approximation of the conditional distribution $\mathbb{P}(\mathbf{y}|\mathbf{X} = x)$. The learning process is conducted with a dataset $\{(x_i, y_i)\}_{i \in D_{tr}}$. For brevity, we sometimes omit the variable x and simplify the notation $q_k(x)$ to q_k, assuming x is given and remains constant in this context.

The Framework

Our overarching approach involves iteratively updating the quantile predictions in multiple steps in a boosting-like manner. Specifically, we update the quantiles $q_1^t, \ldots, q_K^t$ at each step $t = 0, 1, \ldots, T$ using additive models, and we consider the final-step predictions $q_1^T, \ldots, q_K^T$ as the desired estimates. To elaborate, let us denote $Q^t(x) = [q_1^t(x), \ldots, q_K^t(x)]^\top$. We propose the following ensemble framework:

$$Q^0(x) = G_0\left(F_0(x; \Theta_0)\right), \tag{7.32}$$

$$Q^t(x) = Q^{t-1}(x) + G\left(F_t(x; \Theta_t), Q^{t-1}(x)\right), \quad t = 1, \ldots, T. \tag{7.33}$$

Here, F_0 and F_t represent neural network models with learnable parameters Θ_0 and Θ_t, respectively. G_0 and G are fixed transformations that map the neural network outputs to the quantile predictions, as will be described in subsequent sections.

Iteratively, at every step $t = 0, \ldots, T$, a learning objective is set by us to train the model F_t with $F_0, \ldots, F_{t-1}$ fixed:

$$\min_{\Theta_t} \mathbb{E}_{\mathbb{P}(\mathbf{X},\mathbf{y})}\left[L(\mathbf{y}, Q^t(\mathbf{X}))\right]. \tag{7.34}$$

L is a proper loss function between y and $Q^t(x)$. Given the probability level set $\Pi = \{\tau_k\}_{k=1}^K$, the loss L is chosen as the sum of quantile regression losses:

$$L(y, Q^t(x)) = \sum_{k=1}^{K} L_{\tau_k}(y, q_k^t(x)), \tag{7.35}$$

$$L_\tau(y, q) = (y - q)(\tau - \mathbb{I}_{\{y<q\}}). \tag{7.36}$$

Given the training dataset, the empirical version of the learning objective that we should optimize is

$$\min_{\Theta_t} \frac{1}{|D_{tr}|} \sum_{i \in D_{tr}} L(y_i, Q^t(x_i)) + \lambda_t R(\Theta_t). \tag{7.37}$$

In the proposed framework, $R(\Theta_t)$ represents a regularization term, and λ_t is a hyper-parameter that controls the trade-off between the model's complexity and its fit to the data. This framework is general and requires further elaboration. In the subsequent sections, we meticulously design each component of the framework, with a particular focus on the choices for F_0, G_0, F_t, G, and the handling of the regularization term $\lambda_t R(\Theta_t)$.

Intuitions

The Initial Step $\mathbf{t} = \mathbf{0}$: In traditional boosting methods, weak learners are iteratively combined to form a strong learner. At the initial step $t = 0$, the weak learner might be a constant (as in Gradient Boosted Decision Trees, GBDT) or may not differ significantly from those at subsequent steps $t > 0$ (all are trees). However, our situation is distinct due to two motivations: (i) the need for adaptively flexible distribution that deviates from the Gaussian assumption, and (ii) the observation that assuming a Gaussian distribution can sometimes yield acceptable uncertainty estimation. Therefore, starting with a Gaussian assumption is a natural and reasonable approach. As a result, our aim is to ensure that the initial quantile predictions $Q^0(x)$ are as accurate as possible under the Gaussian assumption. Consequently, F_0 and G_0 are designed to align with this objective. In this context, F_0 cannot be considered a weak learner. This approach facilitates the subsequent learning process at $t > 0$, making it relatively easier.

The Ensemble Steps $\mathbf{t} > \mathbf{0}$: After establishing a Gaussian-distributed $\mathbb{P}(\mathbf{y}|\mathbf{X} = x)$, it is natural to consider imposing consecutive and moderate modifications or adjustments on the distribution. This approach aligns with the motivations of this study. Therefore, for $t > 0$, we configure F_t as a relatively weak learner that aims to incrementally update the quantile predictions $Q^{t-1}(x)$ to improved ones $Q^t(x)$. This update should not be completely unconstrained or nonparametric, as it is essential to partially preserve the previous distribution to maintain adaptive flexibility. Both F_t and G are designed with this consideration in mind. The transformation function G, a pivotal component of the framework, establishes a connection between the weak learner F_t and the differences in quantile predictions $Q^t(x) - Q^{t-1}(x)$ between two consecutive steps. The specifics of this transformation will be discussed later.

The Regularization Term: Traditional machine learning models often incorporate explicit regularization terms into the learning objective to reduce model complexity and mitigate the risk of overfitting. For deep neural networks, regularization can take implicit forms, such as batch normalization, dropout, and early stopping, all of which are widely used strategies. In our framework, both F_0 and F_t are neural networks. Consequently, we employ the following regularization strategies in lieu of the explicit term $\lambda_t R(\Theta_t)$: (i) Restrict the number of layers and the size of hidden layers in F_t for $t > 0$; (ii) Limit the output dimension of F_t for $t \geq 0$, with the assistance of non-learnable transformations G_0 and G to map the output to the quantiles; and (iii) Utilize strategies such as early stopping. Most importantly, to further prevent overfitting and to strike a balance between flexibility and distribution structure, we adopt a regularization strategy that involves adaptively determining the number of ensemble steps T, referred to as the adaptive T strategy.

The Monotonic Property of Quantiles: As discussed before, our approach requires a natural and inherent solution to the quantile-crossing issue. Since $Q^0(x)$ follows a Gaussian distribution, it naturally satisfies the monotonic property. Assuming that $Q^{t-1}(x)$ adheres to the monotonic property, i.e., $q_1^{t-1} < \cdots < q_K^{t-1}$, and provided that we employ a suitable transformation function G whose output represents the difference $Q^t(x) - Q^{t-1}(x)$, we can ensure that $Q^t(x)$ is monotonically increasing:

$q_1^t < \cdots < q_K^t$. Consequently, we can address the quantile-crossing issue naturally without additional efforts. Once again, it is evident that G plays a critical role in our framework. We will design a suitable G that achieves this objective.

In summary, our approach begins with a Gaussian assumption at $t = 0$ and progressively, yet moderately, updates $q_1^t, \ldots, q_K^t$ to improve their accuracy, while ensuring that $q_1^t, \ldots, q_K^t$ are strictly increasing with respect to k. The relatively strong learner F_0 and the weak learners F_t for $t > 0$ will be trained. Appropriate transformations G_0 and G will be designed, and effective regularization strategies, including the adaptive T strategy, will be implemented. The final-step predictions $q_1^T, \ldots, q_K^T$ will be monotonically increasing and provide a good approximation of the conditional distribution $\mathbb{P}(\mathbf{y}|\mathbf{X} = x)$. These aspects will be discussed in detail in the following sections.

The Initial Step

At $t = 0$, Gaussian assumption gives $\mathbb{P}(\mathbf{y}|\mathbf{X} = x) \sim \mathcal{N}(\mu(x), \sigma(x)^2)$. We choose F_0 in Eq. (7.32) as a neural network which maps the feature vector x to the parameters of the Gaussian $\mu(x), \sigma(x)$:

$$[\mu(x), \sigma(x)]^\top = F_0(x; \Theta_0). \tag{7.38}$$

The conditional τ_k-quantile of $\mathbf{y}$ given $\mathbf{X} = x$ is then $q_k^0(x) = \mu(x) + \sigma(x)\Phi^{-1}(\tau_k)$, where $\Phi^{-1}(\tau_k)$ is the τ_k-quantile of standard Gaussian. Thus, G_0 in Eq. (7.32) is

$$\begin{aligned} Q^0(x) &= G_0(\mu(x), \sigma(x)) \\ &= \mu(x) + \sigma(x)[\Phi^{-1}(\tau_1), \ldots, \Phi^{-1}(\tau_K)]^\top. \end{aligned}$$

Given the training set $\{(x_i, y_i)\}_{i\in D_{tr}}$ and considering the concept of implicit regularizations, the learning objective at $t = 0$ is $\min_{\Theta_0} \frac{1}{|D_{tr}|} \sum_{i\in D_{tr}} L(y_i, Q^0(x_i))$, as expressed in Eq. (7.37). After this initial step, $Q^0(x)$ might serve as a coarse approximation of the conditional distribution $\mathbb{P}(\mathbf{y}|\mathbf{X} = x)$. However, it may lack the flexibility to adequately capture characteristics such as multi-modality, asymmetry, and heavy tails. Merely selecting a non-Gaussian assumption based on expertise is insufficient for a robust solution. Therefore, in our approach, we enhance the predictions through ensemble steps, iteratively refining the approximation of the conditional distribution to better represent its underlying complexity.

It is important to note that our framework is adaptable to various application domains, as it does not solely rely on the Gaussian assumption at $t = 0$. Depending on the requirements of specific tasks, alternative choices can be made for the initial distribution assumption. For example, the standard Gaussian could be replaced by the standard exponential or other suitable distributions, allowing for greater flexibility in modeling the initial conditional distribution $\mathbb{P}(\mathbf{y}|\mathbf{X} = x)$.

The Ensemble Steps

Assuming that $Q^{t-1}(x)$, or equivalently $q_1^{t-1}, \ldots, q_K^{t-1}$, obtained at step $t-1$ is known, we now provide the model details for Eq. (7.33) at the ensemble step $t \geq 1$. Given that $q_1^0, \ldots, q_K^0$ are strictly increasing due to their adherence to a Gaussian distribution, we consistently assume that $q_1^{t-1}, \ldots, q_K^{t-1}$ are also strictly increasing. Specifically, when $t \geq 1$, to generate strictly increasing and improved predictions $q_1^t, \ldots, q_K^t$, we add a new quantity to each q_k^{t-1}:

$$q_k^t = q_k^{t-1} + g_k^t(\lambda_k^t), \quad k = 1, \ldots, K, \tag{7.39}$$

where $\lambda_k^t \in (-1, 1)$ is a quantity that we need to determine later. $g_k^t(\cdot)$ is a continuous function we set that is strictly increasing over $[-1, 1]$ and satisfies

$$g_k^t(0) = 0, \tag{7.40}$$

$$g_k^t(1) = r_k^{t-1} := \frac{q_{k+1}^{t-1} - q_k^{t-1}}{2}, \tag{7.41}$$

$$g_k^t(-1) = l_k^{t-1} := \frac{q_{k-1}^{t-1} - q_k^{t-1}}{2}. \tag{7.42}$$

In this study, we choose a piecewise linear $g_k^t(\cdot)$ to satisfy the above conditions:

$$g_k^t(\lambda) = \left(-l_k^{t-1} + \mathbb{I}_{\{\lambda>0\}} \cdot (r_k^{t-1} + l_k^{t-1})\right) \lambda. \tag{7.43}$$

For an intuitive understanding, it can be easily verified that, given $\lambda_k^t \in (-1, 1)$, the left and right bounds of q_k^t are the midpoints of the intervals $(q_{k-1}^{t-1}, q_k^{t-1})$ and $(q_k^{t-1}, q_{k+1}^{t-1})$, respectively. Moreover, $q_k^t = q_k^{t-1}$ when $\lambda_k^t = 0$. This design ensures that q_k^t is strictly increasing with respect to k, as demonstrated by the following equation:

$$q_k^t < q_k^{t-1} + g_k^t(1) = \frac{q_{k+1}^{t-1} + q_k^{t-1}}{2} = q_{k+1}^{t-1} + g_{k+1}^t(-1) < q_{k+1}^t. \tag{7.44}$$

The underlying concept is to shift q_k^{t-1} toward a better prediction q_k^t, but only by a small distance to (i) maintain the monotonic property, and (ii) control the flexibility introduced in step t (further elaborated later). This concept is also illustrated in Fig. 7.6. When $k = 0$ or $k = K + 1$, we naturally set q_0^{t-1} or q_{K+1}^{t-1} to be the left/right boundary of the support set of $\mathbb{P}(\mathbf{y}|\mathbf{X} = x)$ if the boundary is finite. In this case, $g_1^t(\cdot)$ or $g_K^t(\cdot)$ is well defined such that $q_1^{t-1} + g_1^t(-1)$ or $q_K^{t-1} + g_K^t(1)$ is a feasible bound. If the boundary is not finite, then $q_0^{t-1} = -\infty$ or $q_{K+1}^{t-1} = +\infty$, and we can replace q_0^{t-1} by $-B$ or replace q_{K+1}^{t-1} by $+B$ where B is a large positive constant to ensure that $g_1^t(\cdot)$ or $g_K^t(\cdot)$ is well defined.

To complete the formulation for step t, we define λ_k^t as a cubic polynomial function of τ_k followed by a hyperbolic tangent (tanh) transformation to map the values to the

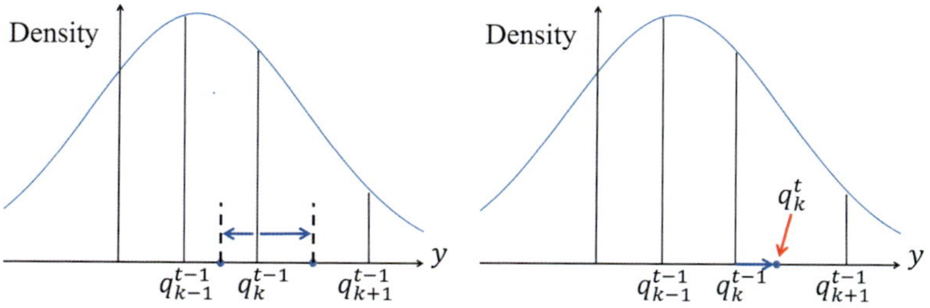

Fig. 7.6 The illustration of the ensemble step $t \geq 1$. q_k^t is selected in the range between the middle point of $(q_{k-1}^{t-1}, q_k^{t-1})$ and the middle point of $(q_k^{t-1}, q_{k+1}^{t-1})$. By this design, $q_1^t, \ldots, q_K^t$ are ensured to be strictly increasing if $q_1^{t-1}, \ldots, q_K^{t-1}$ are strictly increasing

interval $(-1, 1)$, i.e., $\lambda_k^t = \tanh\left(\sum_{l=0}^{3} a_l^t (\tau_k)^l\right)$, where a_l^t is an output of the model $F_t(x; \Theta_t)$:

$$[a_0^t, a_1^t, a_2^t, a_3^t]^\top = F_t(x; \Theta_t). \tag{7.45}$$

With this specification, the function G in Eq. (7.33) is now defined. We refrain from allowing λ_k^t to be entirely unconstrained and directly outputted by $F_t(x; \Theta_t)$, as this would result in a total of K outputs for F_t (e.g., $K = 99$). The use of a cubic polynomial serves as an implicit regularization, imposing restrictions on λ_k^t, $k = 1, \ldots, K$. This setting helps control the flexibility introduced in step t and is crucial for adaptively balancing the distribution structure and the flexibility for $\mathbb{P}(\mathbf{y}|\mathbf{X} = x)$.

From an alternative perspective, our approach yields a smooth density function for $\mathbb{P}(\mathbf{y}|\mathbf{X} = x)$ at each step t. In contrast, allowing λ_k^t to be entirely unconstrained can result in an irregular or chaotic density function that lacks interpretability, a common pitfall of methods that exhibit excessive flexibility, as discussed previously. Returning to our method, given the training dataset $\{(x_i, y_i)\}_{i \in D_{tr}}$, the learning objective at step t is defined as $\min_{\Theta_t} \frac{1}{|D_{tr}|} \sum_{i \in D_{tr}} L(y_i, Q^t(x_i))$, incorporating implicit regularizations. The entire procedure of our approach is complete now.

Adaptive *T* Strategy

From the introduction above, it is evident that the only newly introduced hyper-parameter in the EMQ approach is the number of ensemble steps T. This hyper-parameter is crucial for the performance of the approach, as it serves as a trade-off between the distribution structure and the flexibility. Specifically, a larger number of ensemble steps leads to a more flexible distribution that deviates further from the Gaussian assumption. To address this, we introduce the adaptive T strategy, which determines the value of T adaptively. This strategy enables the EMQ approach to find an optimal balance between the distribution structure and the flexibility, enhancing the overall effectiveness of the method.

The adaptive T strategy operates in a manner analogous to early stopping in deep learning and thus serves as a form of implicit regularization. In this strategy, the training dataset D_{tr} is divided into a new training dataset and a validation set (which may be the same validation set used for early stopping when training F_t, $t \geq 0$). The performance of the predictions $Q^t(x)$ on the validation set is monitored for all $t \geq 0$ and $t \leq T_{max}$. Based on certain criteria, the EMQ approach halts the ensemble steps when the performance of $Q^t(x)$ begins to decline. The value of T_{ada} is set to be the step t at which $Q^t(x)$ exhibits the best performance. Consequently, the EMQ model will ultimately consist of T_{ada} ensemble steps, and $Q^{T_{ada}}(x)$ will be the final quantile predictions. To further mitigate the risk of overfitting, the strategy evaluates $Q^t(x)$ on the validation set using a metric different from the quantile loss L in Eq. (7.35) employed for training F_t, $t \geq 0$. For instance, the metric could be the expected calibration error (ECE), computed by calibrating for every $\tau_k \in \Pi$ and taking the average.

EMQW

In many scenarios, there is a keen interest in tail-side prediction intervals, such as the 90%-interval, 80%-interval, etc. However, predicting quantiles at the tail sides is more challenging. Therefore, we extend the EMQ approach to a new version, referred to as EMQW (Ensemble Multi-Quantiles with Weights), which places greater emphasis on the tail sides than the original EMQ. The primary distinction between EMQW and EMQ is that EMQW employs a weighted quantile loss function in place of Eq. (7.35):

$$L^W(y, Q^t(x)) = \sum_{k=1}^{K} w_{\tau_k} L_{\tau_k}(y, q_k^t(x)), \tag{7.46}$$

$$L_\tau(y, q) = (y - q)(\tau - \mathbb{I}_{\{y<q\}}), \tag{7.47}$$

$$w_\tau = 1/\mathbb{E}_{y\sim\mathcal{N}(0,1)}\left[L_\tau(y, \Phi^{-1}(\tau))\right]. \tag{7.48}$$

Here, Φ denotes the cumulative distribution function of the standard normal distribution $\mathcal{N}(0, 1)$. The goal of this weighting scheme is to normalize the losses for different τ_k to a similar magnitude, as $\mathbb{E}_{y\sim\mathcal{N}(0,1)}\left[L_\tau(y, \Phi^{-1}(\tau))\right]$ can vary significantly for different values of τ. This approach is also suggested by [72, 75]. Except for this modification, EMQW and EMQ share the same design and settings. Our experiments demonstrate that EMQW is a preferable choice when tail-side quantiles are of interest, and one may use it for comparisons with other methods.

Computational Complexity

At step 0, the computational complexity of training F_0 is equivalent to that of training a standard multi-layer perceptron with a two-dimensional output layer. We denote the computational cost of training F_0 as C_0. At step $t > 0$, the network model F_t has a significantly smaller structure compared to F_0. Consequently, training F_t requires considerably less time. If the computational cost of training F_t is C_1 (where $C_1 \ll C_0$), the total computational cost of training EMQ or EMQW will be $C_0 + T_{ada}C_1$, and at most $C_0 + T_{max}C_1$. As will be demonstrated in the experimental section, setting $T_{max} = 40$ is sufficient for most datasets.

Regarding space complexity, since we need to store the predicted quantiles at every step, the additional space required beyond that consumed in training neural networks is $O(|D_{tr}| \cdot |\Pi| \cdot T_{ada})$, and at most $O(|D_{tr}| \cdot |\Pi| \cdot T_{max})$ (where $|\Pi| = 99$ in our experiments). It is noteworthy that training T_{ada} networks requires only the same amount of space as training F_0, because F_0 is the largest model.

7.4.2 *Experimental Studies*

In this section, we implement the EMQ approach and its variant EMQW, and conduct a comparative analysis with both contemporary popular and classical methods across a diverse array of datasets. Our extensive evaluations reveal that EMQW achieves state-of-the-art performance. Additionally, we visualize the learning process and results of EMQW to enhance interpretability and empirically demonstrate that it performs as anticipated. It is shown to effectively capture a range of distribution shapes present in real data. Consequently, we can conclude that the proposed approach possesses the advantages previously outlined.

Datasets and Evaluations

We have selected a diverse set of 20 datasets from the UCI Machine Learning Repository [76]. These datasets are from various domains, facilitating a comprehensive evaluation of regression models. The sample sizes range from 10,000 to 515,345, while the number of features varies between 5 and 465. This collection includes both medium-sized and large-sized datasets, with feature dimensions extending from low to high, thereby ensuring a broad scope for regression analysis.

For evaluation purposes, we adopt a random sampling strategy, allocating 20% of the full dataset as the testing set. To ensure robustness, we perform training and testing N_{te} times with distinct training/testing splits, subsequently averaging the evaluation metrics. We set $N_{te} = 5$ for datasets with a sample size less than 100,000, and $N_{te} = 1$ for the remaining ones to mitigate computational demands. Furthermore, for the two largest datasets, we increase the proportion of the testing set from 20% to 50% to further alleviate computational constraints. Prior to analysis, all features and labels are normalized to achieve a sample mean of 0 and a sample variance of 1.

The performance of the proposed method is measured by two metrics: the calibration error and the sharpness of the prediction intervals constructed by the predicted quantiles. For a comprehensive performance evaluation and analysis, the reader is referred to [74]. The referenced work demonstrates that our method, EMQW, achieves state-of-the-art performance in comparison to existing approaches. Moreover, it conducts an analysis of the learning process and concludes that the proposed method performs in accordance with expectations.

Interpretation of Learning Results

We provide visualizations of the selected results of EMQW to facilitate better interpretation and understanding of its workings. These visualizations aid in comprehending the mechanisms underlying EMQW's superior performance.

Densities Given by EMQW: Consistent with our assertions, the conditional distribution provided by EMQW achieves adaptive flexibility. In Fig. 7.7, we present the densities implied by the conditional quantiles predicted by EMQW to evaluate this claim (noting that the density is the derivative of the inverse of the quantile function, which can be easily obtained numerically). In Fig. 7.7, it is evident that EMQW can learn distributions with various shapes, including but not limited to exactly Gaussian, approximately Gaussian, sharp peak, asymmetry, heavy tails, finite bound, and multi-modality. This diversity can even manifest within a single dataset, highlighting the complexity of real data and the capability of EMQW to capture such variability.

Densities Given by Other Methods: Contrary to the adaptive flexibility achieved by EMQW, we have asserted that many quantile regression-based methods with

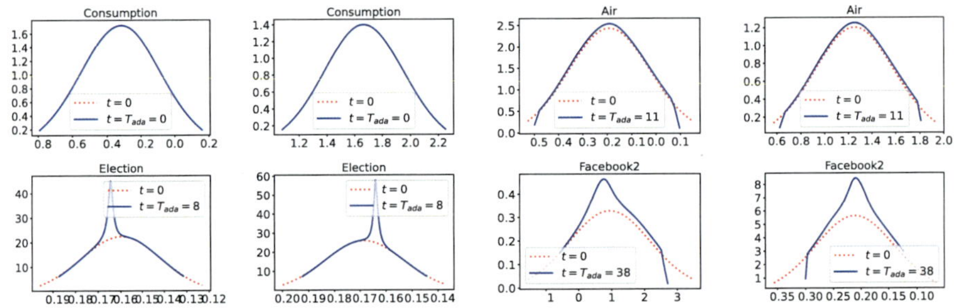

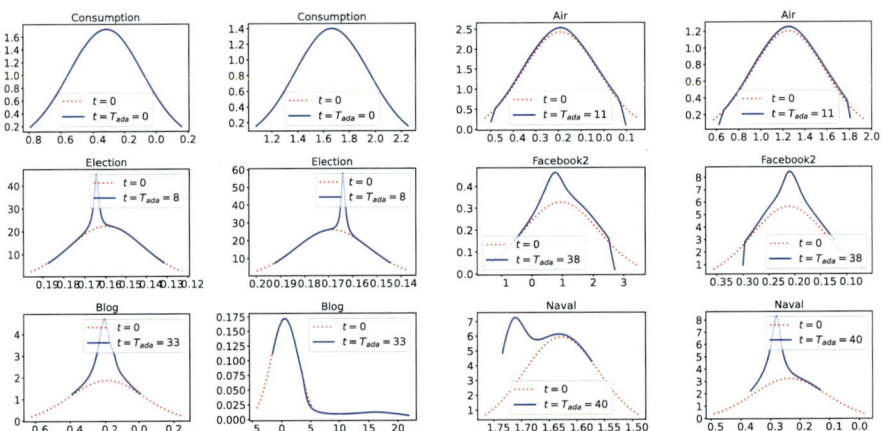

Fig. 7.7 Densities inferred from the 99 conditional quantiles predicted by EMQW. For each dataset (with name on the top), two data points are selected from the testing set, and their densities of $\mathbb{P}(\mathbf{y}|\mathbf{X}=x)$ are plotted. The red lines represent the results at $t=0$, while the blue lines correspond to the results at $t=T_{ada}$. In the first row of subfigures, EMQW produces distributions that are either exactly Gaussian or approximately Gaussian with minor deviations. The second row illustrates densities with various peaks and asymmetric shapes. The third row depicts densities with more diverse characteristics, such as a long right tail and a multi-peak shape. The observed variety within a single dataset highlights the complexity of real-world data and the capability of EMQW

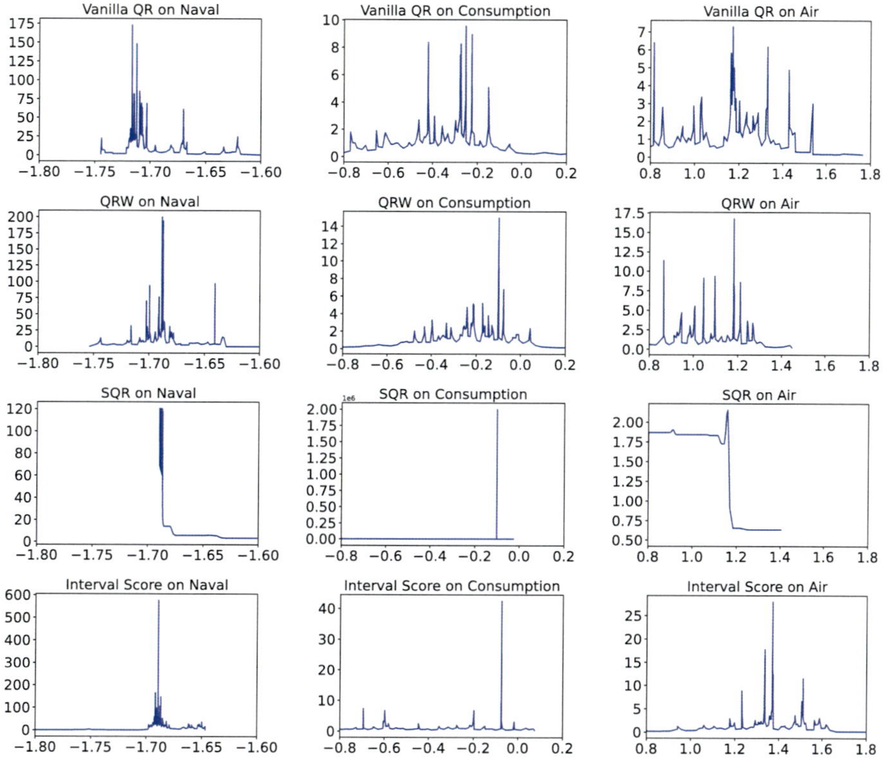

Fig. 7.8 Densities derived from the 99 conditional quantiles predicted by four quantile regression-based methods: Vanilla QR, QRW, SQR, and Interval Score. For each of the three datasets, one data point is selected from the testing set, and its density $\mathbb{P}(\mathbf{y}|\mathbf{X} = x)$, as predicted by the four methods, is plotted. It is observed that all methods result in densities that are rough or even invalid

excessive flexibility tend to yield chaotic or invalid densities. To substantiate this claim, we examine the densities implied by the quantiles predicted by four quantile regression-based methods: the vanilla quantile regression (QR), QR with weights (QRW), simultaneous QR (SQR) [64], and Interval Score [77]. We select one data point from the testing set of each of the three datasets: Naval, Consumption, and Air. The densities provided by the four methods for these data points are illustrated in Fig. 7.8. From the plots, it is apparent that all four methods produce rough or even invalid densities. This phenomenon is not unique to these datasets but is commonly observed across various datasets.

7.4.3 Discussion on Asymptotic Guarantee

It is appealing to know for sure that the conditional quantiles (or conditional densities implied) approximate the true ones asymptotically. Actually, we are looking for a function $f(\cdot) = [f_1(\cdot), \cdots, f_K(\cdot)]^\top$ to minimize: $\min_{f \in \mathcal{F}} \mathbb{E}\left[L(\mathbf{y}, f(\mathbf{X}))\right] = \sum_{k=1}^{K} \mathbb{E}\left[L_{\tau_k}(\mathbf{y}, f_k(\mathbf{X}))\right]$. Obviously, if we consider a sufficiently rich $\mathcal{F}$, the optimal $f_k(x)$ should be the τ_k-quantile of $\mathbb{P}(\mathbf{y}|\mathbf{X} = x)$. In our method, $\mathcal{F}$ consists of those functions defined via a complex ensemble of $T + 1$ neural networks. If we denote the actual τ_k-quantile of $\mathbb{P}(\mathbf{y}|\mathbf{X} = x)$ as $q_k(x)$ and further denote $Q(x) = [q_1(x), \cdots, q_K(x)]^\top$, the consistency means that the estimated $Q^n(\cdot)$ given by our algorithm based on the dataset $\mathcal{D}_n = \{(\mathbf{X}_i, \mathbf{y}_i)\}_{i=1}^n$ and function class $\mathcal{F}_n$ should satisfy: $\lim_{n\to\infty} \mathbb{E}\left[\int_{\mathbb{R}^d} \|Q^n(x) - Q(x)\|^2 \mu(dx)\right] = 0$, for all distributions of $(\mathbf{X}, \mathbf{y})$ with $\mathbb{E}[\mathbf{y}^2] < \infty$. Here μ denotes the distribution of $\mathbf{X}$. This is called the weak universal consistency of Q^n (see [78] for the strong version). We assume $\mathcal{F}_n$ consists of those functions having T_n ensemble steps, and $T_n \to \infty$ as $n \to \infty$.

Multi-quantile regression is not a novel concept in the literature. Works such as [72, 75, 79, 80] have all adopted similar objectives for this type of regression. However, these approaches share some common limitations: (i) they assume a function class $\mathcal{F}$ that is linear, linear spline, or kernel-based; (ii) they incorporate explicit monotonic constraints on quantiles in the optimization process; and (iii) they lack asymptotic analysis for nonparametric estimators. Our method addresses the first two limitations. Regarding the third, while [81, 82] established the consistency of polynomial and neural network estimators of conditional quantile, their analysis was limited to only one quantile level. In our case, we face two challenges: (i) multi-quantile regression with multiple levels; and (ii) the function class we consider is the ensemble of many (small) neural networks, which is a novel and unique design in our approach.

The study of the asymptotic properties of neural networks (NNs) has been a topic of interest in recent literature. Györfi et al. [78] provided consistency results for infinitely wide NNs in ordinary regression problems with least square loss. More recently, [83] demonstrated that wide and deep NNs achieve consistency for classification, and that excessively deep NNs are detrimental to regression. Lin et al. [84] analyzed the consistency of deep convolutional NNs. Regarding our specific problem, to establish the consistency of our estimator, we may need to accomplish the following three steps: (i) Decompose $\int_{\mathbb{R}^d} |Q^n(x) - Q(x)|^2 \mu(dx)$ into the approximation error and estimation error; (ii) Prove that our function class is dense, meaning that our ensemble of NNs can approximate any continuous function with a compact domain; and (iii) Address the bound on the covering numbers of our function class. Completing each step will be non-trivial and challenging, especially the third step, due to the uniqueness of our function class, which is an ensemble of many (small) NNs, as opposed to the infinitely wide NNs commonly used in theoretical works. Given these considerations, we leave the analysis of asymptotic guarantees for future work.

7.4.4 Conclusion

In conclusion, we introduce an ensemble multi-quantiles approach capable of adaptively flexible distribution prediction. This approach is motivated by two key observations: (i) the Gaussian assumption for the conditional distribution lacks flexibility, and (ii) excessive flexibility can impair generalization, as evidenced by the invalid densities produced by many quantile regression methods. Our method employs a boosting framework for multi-quantile prediction, consisting of an initial step and subsequent ensemble steps. Additionally, our approach is driven by the need for an inherent solution to the quantile-crossing issue.

The ensemble approach presented here is both intuitive and interpretable. In the initial step, a relatively strong learner generates a Gaussian distribution, while in the subsequent ensemble steps, T_{ada} weak learners adaptively seek a balance between the distribution structure and flexibility. We validate these characteristics through experimental analysis and compare our approach with recent popular competing methods. In conclusion, the EMQW approach achieves state-of-the-art performance among all evaluated methods, as measured by metrics of calibration and sharpness.

Through visualization, we observe that our approach is capable of learning distributions with a diverse range of density shapes, including Gaussian, approximately Gaussian, sharp peak, asymmetry, heavy tails, finite bound, and multi-modality. This diversity can manifest within a single dataset, underscoring the complexity of real-world data and highlighting the necessity and advantages of such an adaptively flexible ensemble approach. While our experiments focus on tabular data, the proposed approach is generalizable and can be applied to other data formats, such as images and texts.

References

1. Committee, B., et al.: Minimum capital requirements for market risk. Consultative Document (2016)
2. BIS, I.: Margin requirements for non-centrally cleared derivatives (2013)
3. Kou, S., Peng, X., Heyde, C.C.: External risk measures and Basel accords. Math. Oper. Res. **38**(3), 393–417 (2013)
4. Kou, S., Peng, X.: On the measurement of economic tail risk. Oper. Res. **64**(5), 1056–1072 (2016)
5. Geraci, M., Garbaravicius, T., Veredas, D.: Short selling in extreme events. J. Financ. Stab. (2018)
6. Cont, R., Wagalath, L.: Fire sales forensics: measuring endogenous risk. Math. Financ. **26**(4), 835–866 (2016)
7. Kirilenko, A., Kyle, A.S., Samadi, M., Tuzun, T.: The flash crash: high-frequency trading in an electronic market. J. Financ. **72**(3), 967–998 (2017)
8. Brunnermeier, M.K., Pedersen, L.H.: Market liquidity and funding liquidity. Rev. Financ. Stud. **22**(6), 2201–2238 (2008)
9. Easley, D., De Prado, M.L., O'Hara, M.: The microstructure of the flash crash: flow toxicity, liquidity crashes and the probability of informed trading. J. Portf. Manag. **37**(2), 118–128 (2011)

10. Cont, R.: Volatility clustering in financial markets: empirical facts and agent-based models. In: Long Memory in Economics, pp. 289–309. Springer (2007)
11. Glasserman, P., Wu, Q.: Persistence and procyclicality in margin requirements. Manag. Sci. (2018)
12. Yan, X., Zhang, W., Ma, L., Liu, W., Wu, Q.: Parsimonious quantile regression of financial asset tail dynamics via sequential learning. In: Advances in Neural Information Processing Systems, pp. 1582–1592 (2018)
13. LeCun, Y., Bengio, Y., Hinton, G.: Deep learning. Nature **521**(7553), 436–444 (2015)
14. Guo, C., Pleiss, G., Sun, Y., Weinberger, K.Q.: On calibration of modern neural networks. In: International Conference on Machine Learning, pp. 1321–1330. PMLR (2017)
15. Rahaman, R., Thiery, A.: Uncertainty quantification and deep ensembles. Adv. Neural Inf. Process. Syst. **34**, (2021)
16. Begoli, E., Bhattacharya, T., Kusnezov, D.: The need for uncertainty quantification in machine-assisted medical decision making. Nat. Mach. Intell. **1**(1), 20–23 (2019)
17. Lakshminarayanan, B., Pritzel, A., Blundell, C.: Simple and scalable predictive uncertainty estimation using deep ensembles. Adv. Neural Inf. Process. Syst. **30**, (2017)
18. McNeil, A.J., Frey, R., Embrechts, P.: Quantitative Risk Management: Concepts, Techniques and Tools-Revised Edition. Princeton University Press (2015)
19. Duffie, D., Pan, J.: An overview of value at risk. J. Deriv. **4**(3), 7–49 (1997)
20. Acerbi, C., Tasche, D.: On the coherence of expected shortfall. J. Bank. & Financ. **26**(7), 1487–1503 (2002)
21. Gal, Y., Ghahramani, Z.: Dropout as a bayesian approximation: Representing model uncertainty in deep learning. In: International Conference on Machine Learning, pp. 1050–1059. PMLR (2016)
22. Maddox, W.J., Izmailov, P., Garipov, T., Vetrov, D.P., Wilson, A.G.: A simple baseline for bayesian uncertainty in deep learning. Adv. Neural Inf. Process. Syst. **32**, (2019)
23. Sensoy, M., Kaplan, L., Kandemir, M.: Evidential deep learning to quantify classification uncertainty. Adv. Neural Inf. Process. Syst. **31**, (2018)
24. Kuleshov, V., Fenner, N., Ermon, S.: Accurate uncertainties for deep learning using calibrated regression. In: International Conference on Machine Learning, pp. 2796–2804. PMLR (2018)
25. Skafte, N., Jørgensen, M., Hauberg, S.: Reliable training and estimation of variance networks. Adv. Neural Inf. Process. Syst. **32**, (2019)
26. Zhao, S., Ma, T., Ermon, S.: Individual calibration with randomized forecasting. In: International Conference on Machine Learning, pp. 11387–11397. PMLR (2020)
27. Hernández-Lobato, J.M., Adams, R.: Probabilistic backpropagation for scalable learning of bayesian neural networks. In: International Conference on Machine Learning, pp. 1861–1869. PMLR (2015)
28. Shafer, G., Vovk, V.: A tutorial on conformal prediction. J. Mach. Learn. Res. **9**(3), (2008)
29. Balasubramanian, V., Ho, S.S., Vovk, V.: Conformal Prediction for Reliable Machine Learning: Theory, Adaptations and Applications. Newnes (2014)
30. Papadopoulos, H., Vovk, V., Gammerman, A.: Regression conformal prediction with nearest neighbours. J. Artif. Intell. Res. **40**, 815–840 (2011)
31. Lei, J., Robins, J., Wasserman, L.: Distribution-free prediction sets. J. Am. Stat. Assoc. **108**(501), 278–287 (2013)
32. Romano, Y., Patterson, E., Candes, E.: Conformalized quantile regression. Adv. Neural Inf. Process. Syst. **32**, (2019)
33. Cui, P., Hu, W., Zhu, J.: Calibrated reliable regression using maximum mean discrepancy. Adv. Neural Inf. Process. Syst. **33**, 17164–17175 (2020)
34. Zhou, T., Li, Y., Wu, Y., Carlson, D.: Estimating uncertainty intervals from collaborating networks. J. Mach. Learn. Res. **22**(257), 1–47 (2021)
35. Wenzel, F., Snoek, J., Tran, D., Jenatton, R.: Hyperparameter ensembles for robustness and uncertainty quantification. Adv. Neural. Inf. Process. Syst. **33**, 6514–6527 (2020)
36. Pearce, T., Brintrup, A., Zaki, M., Neely, A.: High-quality prediction intervals for deep learning: a distribution-free, ensembled approach. In: International Conference on Machine Learning, pp. 4075–4084. PMLR (2018)

37. Hansen, B.E.: Autoregressive conditional density estimation. Int. Econ. Rev. 705–730 (1994)
38. Rockinger, M., Jondeau, E.: Entropy densities with an application to autoregressive conditional skewness and kurtosis. J. Econ. **106**(1), 119–142 (2002)
39. León, Á., Rubio, G., Serna, G.: Autoregresive conditional volatility, skewness and kurtosis. Q. Rev. Econ. Financ. **45**(4–5), 599–618 (2005)
40. Bali, T.G., Mo, H., Tang, Y.: The role of autoregressive conditional skewness and kurtosis in the estimation of conditional var. J. Bank. & Financ. **32**(2), 269–282 (2008)
41. Barone-Adesi, G., Giannopoulos, K., Vosper, L.: Var without correlations for portfolios of derivative securities. J. Futur. Mark. **19**(5), 583–602 (1999)
42. McNeil, A.J., Frey, R.: Estimation of tail-related risk measures for heteroscedastic financial time series: an extreme value approach. J. Empir. Financ. **7**(3–4), 271–300 (2000)
43. Engle, R.F., Manganelli, S.: Caviar: conditional autoregressive value at risk by regression quantiles. J. Bus. & Econ. Stat. **22**(4), 367–381 (2004)
44. Wu, Q., Yan, X.: Capturing deep tail risk via sequential learning of quantile dynamics. J. Econ. Dyn. Control **109**, 103771 (2019)
45. Engle, R.F., Patton, A.J.: What good is a volatility model? Quant. Financ. **1**, 237–245 (2001)
46. Engle, R.F.: Autoregressive conditional heteroscedasticity with estimates of the variance of united kingdom inflation. Econ. J. Econ. Soc. 987–1007 (1982)
47. Bollerslev, T.: Generalized autoregressive conditional heteroskedasticity. J. Econ. **31**(3), 307–327 (1986)
48. Nelson, D.B.: Conditional heteroskedasticity in asset returns: a new approach. Econ. J. Econ. Soc. 347–370 (1991)
49. Glosten, L.R., Jagannathan, R., Runkle, D.E.: On the relation between the expected value and the volatility of the nominal excess return on stocks. J. Financ. **48**(5), 1779–1801 (1993)
50. Cont, R.: Empirical properties of asset returns: stylized facts and statistical issues. Quant. Financ. **1**(2), 223–236 (2001)
51. Chen, J., Hong, H., Stein, J.C.: Forecasting crashes: trading volume, past returns, and conditional skewness in stock prices. J. Financ. Econ. **61**(3), 345–381 (2001)
52. Albuquerque, R.: Skewness in stock returns: reconciling the evidence on firm versus aggregate returns. Rev. Financ. Stud. **25**(5), 1630–1673 (2012)
53. Taylor, S.J.: Modeling stochastic volatility: a review and comparative study. Math. Financ. **4**(2), 183–204 (1994)
54. Fleming, J., Kirby, C.: A closer look at the relation between GARCH and stochastic autoregressive volatility. J. Financ. Econ. **1**(3), 365–419 (2003)
55. Carnero, M.A., Peña, D., Ruiz, E.: Persistence and kurtosis in GARCH and stochastic volatility models. J. Financ. Econ. **2**(2), 319–342 (2004)
56. Franses, P.H., Van Der Leij, M., Paap, R.: A simple test for GARCH against a stochastic volatility model. J. Financ. Econ. **6**(3), 291–306 (2007)
57. Mullainathan, S., Spiess, J.: Machine learning: an applied econometric approach. J. Econ. Perspect. **31**(2), 87–106 (2017)
58. Hochreiter, S., Schmidhuber, J.: Long short-term memory. Neural Comput. **9**(8), 1735–1780 (1997)
59. Lipton, Z.C., Berkowitz, J., Elkan, C.: A critical review of recurrent neural networks for sequence learning (2015). arXiv:1506.00019
60. Keskar, N.S., Mudigere, D., Nocedal, J., Smelyanskiy, M., Tang, P.T.P.: On large-batch training for deep learning: generalization gap and sharp minima. In: International Conference on Learning Representations (2016)
61. Kingma, D.P., Ba, J.: Adam: a method for stochastic optimization. In: International Conference on Learning Representations (2015)
62. Abadi, M., Barham, P., Chen, J., Chen, Z., Davis, A., Dean, J., Devin, M., Ghemawat, S., Irving, G., Isard, M., et al.: Tensorflow: a system for large-scale machine learning. In: 12th USENIX Symposium on Operating Systems Design and Implementation (OSDI 16), pp. 265–283 (2016)
63. Song, H., Diethe, T., Kull, M., Flach, P.: Distribution calibration for regression. In: International Conference on Machine Learning, pp. 5897–5906. PMLR (2019)

64. Tagasovska, N., Lopez-Paz, D.: Single-model uncertainties for deep learning. Adv. Neural Inf. Process. Syst. **32**, (2019)
65. Chung, Y., Neiswanger, W., Char, I., Schneider, J.: Beyond pinball loss: quantile methods for calibrated uncertainty quantification. Adv. Neural Inf. Process. Syst. **34**, (2021)
66. Feldman, S., Bates, S., Romano, Y.: Improving conditional coverage via orthogonal quantile regression. Adv. Neural Inf. Process. Syst. **34**, (2021)
67. Koenker, R., Bassett, G.: Regression quantiles. Econometrica **46**(1), 33–50 (1978)
68. Koenker, R., Hallock, K.F.: Quantile regression. J. Econ. Perspect. **15**(4), 143–156 (2001)
69. He, X.: Quantile curves without crossing. Am. Stat. **51**(2), 186–192 (1997)
70. Takeuchi, I., Le, Q.V., Sears, T.D., Smola, A.J.: Nonparametric quantile estimation. J. Mach. Learn. Res. **7**(Jul), 1231–1264 (2006)
71. Liu, Y., Wu, Y.: Stepwise multiple quantile regression estimation using non-crossing constraints. Stat. Interface **2**(3), 299–310 (2009)
72. Bondell, H.D., Reich, B.J., Wang, H.: Noncrossing quantile regression curve estimation. Biometrika **97**(4), 825–838 (2010)
73. Chernozhukov, V., Fernández-Val, I., Galichon, A.: Quantile and probability curves without crossing. Econometrica **78**(3), 1093–1125 (2010)
74. Yan, X., Su, Y., Ma, W.: Ensemble multi-quantiles: Adaptively flexible distribution prediction for uncertainty quantification. IEEE Trans. Pattern Anal. Mach. Intell. (2023)
75. Liu, Y., Wu, Y.: Simultaneous multiple non-crossing quantile regression estimation using kernel constraints. J. Nonparametric Stat. **23**(2), 415–437 (2011)
76. Dua, D., Graff, C.: Machine learning repository (2017). http://archive.ics.uci.edu/ml
77. Gneiting, T., Raftery, A.E.: Strictly proper scoring rules, prediction, and estimation. J. Am. Stat. Assoc. **102**(477), 359–378 (2007)
78. Györfi, L., Köhler, M., Krzyżak, A., Walk, H.: A Distribution-Free Theory of Nonparametric Regression, vol. 1. Springer (2002)
79. Zou, H., Yuan, M.: Composite quantile regression and the oracle model selection theory1. Ann. Stat. **36**(3), 1108–1126 (2008)
80. Moon, S.J., Jeon, J.J., Lee, J.S.H., Kim, Y.: Learning multiple quantiles with neural networks. J. Comput. Graph. Stat. **30**(4), 1238–1248 (2021)
81. Chaudhuri, P.: Nonparametric estimates of regression quantiles and their local bahadur representation. Ann. Stat. **19**(2), 760–777 (1991)
82. White, H.: Nonparametric estimation of conditional quantiles using neural networks. In: Computing Science and Statistics: Statistics of Many Parameters: Curves, Images, Spatial Models, pp. 190–199. Springer (1992)
83. Radhakrishnan, A., Belkin, M., Uhler, C.: Wide and deep neural networks achieve consistency for classification. Proc. Natl. Acad. Sci. **120**(14), e2208779120 (2023)
84. Lin, S.B., Wang, K., Wang, Y., Zhou, D.X.: Universal consistency of deep convolutional neural networks. IEEE Trans. Inf. Theory **68**(7), 4610–4617 (2022)

Appendix
The Proof of the Existence of HTQF's Unique Probability Distribution

The proof idea is to show that HTQF is continuously differentiable, is strictly monotonically increasing over $(0, 1)$, and approaches $-\infty/+\infty$ as α tends to $0/1$. So the inverse function of HTQF exists and is a cumulative distribution function.

The HTQF has the specification:

$$Q(\alpha|\mu, \sigma, u, v) = \mu + \sigma Z_\alpha \left(\frac{e^{uZ_\alpha}}{A} + \frac{e^{-vZ_\alpha}}{A} + 1 \right) = \mu + \sigma g(Z_\alpha), \tag{A.1}$$

where $g(x) = x(\frac{e^{ux}}{A} + \frac{e^{-vx}}{A} + 1)$. Z_α is the quantile function of the standard normal distribution. We only need to prove that $g(x)$ is continuously differentiable, is strictly monotonically increasing over $(-\infty, +\infty)$, and approaches $-\infty/+\infty$ as x tends to $-\infty/+\infty$. Obviously $g(x)$ is continuously differentiable and $\lim_{x\to-\infty/+\infty} g(x) = -\infty/+\infty$. To prove the monotonicity, we calculate the derivative of $g(x)$:

$$g'(x) = \left(\frac{e^{ux}}{A} + \frac{e^{-vx}}{A} + 1 \right) + x \left(u\frac{e^{ux}}{A} - v\frac{e^{-vx}}{A} \right) \tag{A.2}$$

$$= \frac{e^{ux}}{A}(1 + ux) + \frac{e^{-vx}}{A}(1 - vx) + 1 \tag{A.3}$$

$$= \frac{1}{A}h(ux) + \frac{1}{A}h(-vx) + 1. \quad (h(x) = e^x(1 + x)) \tag{A.4}$$

Next we prove $h(x) \geq -1, \forall x$. This is equivalent to $1 + x \geq -e^{-x}$, or $1 + x + e^{-x} \geq 0$, $\forall x$. A simple monotonic analysis on the function $1 + x + e^{-x}$ can reveal that

Z. Zhang et al., *Big Data in Economics and Management*, Statistics and Big Data 1,
https://doi.org/10.1007/978-981-95-3125-7

its global minimum is reached at $x = 0$, hence $1 + x + e^{-x} \geq 2 \geq 0$. Therefore, $h(x) \geq -1$ and

$$g'(x) \geq -\frac{1}{A} - \frac{1}{A} + 1. \tag{A.5}$$

If we choose $A \geq 3$, then $g'(x) \geq -\frac{1}{3} - \frac{1}{3} + 1 = \frac{1}{3} > 0$ holds for all x. Consequently, $g(x)$ is strictly monotonically increasing, and our proof is completed.

FSC
www.fsc.org
MIX
Papier | Fördert
gute Waldnutzung
FSC® C083411

Zeitfracht Medien GmbH
Ferdinand-Jühlke-Straße 7
99095 Erfurt, Deutschland
produktsicherheit@kolibri360.de